THE SPIRITUAL TECHNOLOGY OF ANCIENT EGYPT

"Ed Malkowski's *The Spritual Technology of Ancient Egypt* is a meticulously researched master work that carefully documents the historical and philosophical underpinnings of modern science, philosophy and religion as they should rightfully be viewed, in the context of the Mystery Tradition of Ancient Egypt."

LAIRD SCRANTON, AUTHOR OF *SCIENCE OF THE DOGON*

"Edward F. Malkowski brilliantly examines how the sacred science and spiritual technology of Ancient Egypt parallel today's most cutting edge theories about quantum physics and human consciousness. This book proves that by looking back to the past, we can indeed see into our future."

MARIE D. JONES, AUTHOR OF *PSIENCE: HOW NEW DISCOVERIES IN QUANTUM PHYSICS AND NEW SCIENCE MAY EXPLAIN THE EXISTENCE OF PARANORMAL PHENOMENA*

THE SPIRITUAL TECHNOLOGY OF ANCIENT EGYPT

Sacred Science and the Mystery of Consciousness

EDWARD F. MALKOWSKI

A Historical and Philosophical Approach in Tribute to the Life and Work of René A. Schwaller de Lubicz

Inner Traditions
Rochester, Vermont

Inner Traditions
One Park Street
Rochester, Vermont 05767
www.InnerTraditions.com

Library of Congress Cataloging-in-Publication Data
Malkowski, Edward F.
The spiritual technology of ancient Egypt : sacred science and the mystery of consciousness / Edward F. Malkowski.
p. cm.
Includes bibliographical references and index.
ISBN-13: 978-1-59477-186-6 (pbk.)
ISBN-10: 1-59477-186-3 (pbk.)
1. Physics—Ancient Egypt. 2. Quantum theory. 3. Technology—Ancient Egypt. 4. Ancient Egypt—Religious life and customs. I. Title.
QC9. E3M35 2007
530.0932—dc22

2007024469

Printed and bound in the United States by Lake Book Manufacturing

10 9 8 7 6 5 4 3 2

Text design by Virginia Scott Bowman and layout by Priscilla Baker
This book was typeset in Garamond, with Post Medieval used as a display typeface

To send correspondence to the author of this book, mail a first-class letter to the author c/o Inner Traditions • Bear & Company, One Park Street, Rochester, VT 05767, and we will forward the communication.

To my sons and daughter: Joe, Thomas, Veronica, and Shawn.
Without the uncountable hours of discussion and debate,
this work would have not been possible.

Contents

Foreword

At first glance, the title on this book seemed to be either a contradiction in terms or New Age idealism with no grounding in reality. Surprisingly, though, after reading Edward Malkowski's *The Spiritual Technology of Ancient Egypt,* I began to realize that there are inherent concepts within the title itself that, when viewed in the context of the book and with contemplation of personal experiences and historical progress related to technology, cause me to question why the appearance of "spiritual" and "technology" in the same sentence prompted my own reaction. In wrangling with the answer to this question, I felt it appropriate to begin with a discussion of what the concept of "spiritual technology" might realistically mean to our relationships with technologies of the past, with each other and, ultimately, how it could influence technologies of the future.

There is a tendency by lovers of history to look into the distant past and become charmed by great civilizations and the legacy we have inherited from them. We are inspired by their architecture, while idealized tales of struggle and heroism command our admiration and respect. Many see a utopian lifestyle in which great people lived more closely with the earth and benefited from a more natural relationship with the environment. Sometimes we yearn for great masters to emerge in our midst like a Mozart or a Leonardo so we can claim membership in a golden age or renaissance.

The manner in which industry has developed has given some people reason to express unreserved distrust and resentment toward the machine's existence. This could be mainly because of the general impression of automation where the influence of human spirit has either taken a back seat to the machine or may not even be on the bus! Nostalgia for a time when life was simpler and cleaner, without the machine, invariably accompanies such dark thoughts.

Without a doubt, it is healthy to envision and strive for a better world. Conversely, it is harmful to be consumed with regrets while reliving the past. Sometimes, though, as a point of reference, it is useful to review one's own past to gain true appreciation for currently owned circumstances while holding hope for a positive future.

In 1961, when I was a mere fifteen years old, I entered the cavernous bays of Mather & Platt engineering company in Manchester, England, with great trepidation. There, concepts of manufacturing engineering were instilled through schooling and practical application. The machines were complex and dangerous. The environment was unfettered by strict adherence to occupational safety and health, anti-harassment policies, or other employment laws—the violation of which, or even the prospect of violation, has cost contemporary employers billions of dollars over the years.

After six months in the apprentice training school and six years toward becoming a journeyman, the concept of "spiritual technology" did not enter even the most remote regions of my mind. As I tended a vertical lathe, considered a man-killer to inattentive and careless operators, my mind would occasionally drift into Dickensian contemplations regarding industry and the technological products, which through the sweat and pain of myriad workers had spilled out of factory doors for over a century.

Born of the industrial revolution, the dissonant shriek of the morning steam whistle, while in many ways a symbol of technology and progress, was a fiendish invention for civilization's race toward wealth. It impressed on workers the fear of being late and distasteful accompanying consequences: a reduction in pay, a sarcastic supervisor looking at his watch as the tardy worker rushed to his workstation, and jeering co-workers asking how many pints he downed the night before. Like the last order bell being rung at a local pub or an obnoxious alarm clock, there was nothing spiritually uplifting about the plant's demonic banshee. A system of communication had been created that clashed with natural harmony and jarred the soul.

Compared to my ancestors, however, I was extremely lucky to enjoy these particular working conditions. The old hands told stories of when they started in the trade, and of how conditions were less favorable back then. Eighty years before that, the country's factory youth were not so well treated. The abuses in the mines and factories of children who were only ten or twelve years of age were horrifying and real.

From a historical humanistic viewpoint, the desire to provide reparation for abuses toward native people apart from our own (now being discussed by apologists on both sides of the Atlantic) might be tempered somewhat with consideration of the abuses of children in seventeenth- and eighteenth-century England. Those who would consider such issues might learn from a short story by John Hartley in a compendium of stories about the English county Yorkshire. *Yorkshire Pudding* was published in 1876 and within the reader finds a heartrending story entitled "Frozen to Death."

"Frozen to Death" is about a family whose husband and father has been stricken by illness for many years. There are three children in the family, one a baby. Siblings Tom and Susy work at the mill to put food on the table and, absent a welfare system, produce the only income for this poor family.

Departing for work one cold and snowy morning, Tom and Susy are late for work after Susy's foot becomes injured. After the whistle blows, the gates are closed and the timekeeper does not open them to admit the children.

"My orders iz az nubdy mun come in after a quarter past." The book continued, "Poor Tom! There had still lingered some little faith in the goodness of human nature in his breast, but as he turned away, the last spark died out."

Following the rules of the factory owners, the timekeeper did not consider that his actions would bring to him a rebuke from management for leaving two children out in the freezing cold. As John Hartley poignantly harkens:

> . . . He [Tom] laid his head against the cold stone wall, and the snow still fell, so softly, so very gently, that he dozed away and dreamed of sunny lands where all was bright and warm; and in a short time the passer-by could not have told that a brother and sister lay quietly slumbering there, wrapped in the shroud of snow.
>
> The hum of wheels has ceased; the crowd of laborers hurry out to their morning's meal; a few short minutes and the discordant whistles again shriek out the call to work. Tom and Susy, where are they? The gates will soon be closed again!
>
> Well, let them close! Other gates have opened for those little suffering ones. The gates of pearl have swung upon their golden hinges; no harsh voice of unkind taskmaster greets them on their entrance, but that glorious welcome . . .

> But those pearly gates are not forever open. The time may come when those shall stand before them unto whom the words "Inasmuch as he did it not to one the least of these, ye did it not to me" shall sound the death-knell of all hopes throughout an inconceivable eternity.

Why was factory work so sought after, despite the suffering found in pursuing a job there? Because it was the best, sometimes the only, work available and was infinitely better than scratching a living out of working as a serf in the fields and, absent a common welfare system, allowed better living standards than those of our ancestors. Nonetheless, conditions were not perfect, and needed the introduction of compassion into the workforce at all levels.

From these comfortless origins of modernization, industry has progressed with lurch and sputter prompted by internal and external impositions of a social consciousness that was most famously heralded by Charles Dickens in the nineteenth century. The triumph of spirit over adversity has long been a successful formula for writers and the media, which in many ways has benefited society as a whole. It has given hope where it did not previously exist and has stimulated debate and an overall improvement of circumstances. The machine has played a dominant role in improving people's lives, so much so that as Isaac Asimov wrote in his foreword to *Thinking by Machine* in 1956, "If the machine were to vanish tomorrow, most of the earth's population would starve and the remnant be reduced to wretchedness and ruin."

Remarkably, memories of the negative aspects of industrialization still fuel voices that revile manufacturing while receiving both facility and pleasure from its products—the availability of which have become, in many ways, expected in economically correct form and function. It is, however, an unfortunate perception that modern factories are soulless, thunderous behemoths, belching pollutants into the atmosphere with careless abandon, with little concern for drudgery and danger, like our view of old factories.

While today's young protégés of progressive education tend to avoid such toil and threaten to starve manufacturing machines of the intelligence they will need in the future, those who do find their way into a modern manufacturing plant face a different environment from that of their forefathers. Recognition of the individuality of the worker and the unique creative spirit that lies within each person has prompted remarkable changes over the past century. Having been witness to this evolution over the past forty-six years, it bears noting that philosophies relating to humanistic issues that might be considered close to the spirit of man were evident and espoused well before I entered the trade as a teenager and have blossomed in modern factories under a practice of Kaizen (Japanese for "continuous improvement").

In 1912, the Jones & Lamson Machine Company published a book entitled *Hartness Flat Turret Lathe Manual,* which is prefaced by a philosophical discussion about the workers in a machine shop and their relationship, more toward each other than with the machine.

> Since the machine is only an implement, it cannot be considered as a thing entirely apart from the man. In fact, the man is the greater part. The personal welfare of the operator must be considered. This is something more than the man's relation to the machine. It includes an equally important phase—his relation to other men and to his environment in general.
>
> To disregard the personal interests of the operator would be to miss the principal element in the consideration of the use of the machine.
>
> The mechanical problems are getting more and more complex, but still the greatest problem for each man is how to understand his fellow creatures. No attempt will be made to give the last word on this subject, in fact little more than an inkling can be offered. But this inkling may establish a line of thought that will accomplish great things for each one who allows it to actually get into his mind.
>
> In looking over the machinery world we find immense numbers of workers employed at peaceful work. It is a kind of work that has the minimum of strife and least personal battle for progress.
>
> The advance of an individual is frequently accompanied by a general betterment of the whole world.

Clearly, when advice is given on matters involving technology, the most critical aspect involves the human aspect. This is only natural when we consider that without the creative spirit of scientists, engineers, and inventors and, like it or not, the industrial revolution that repulses Dickensians, there would be no machines that give us comfort and pleasure. Our interaction with our machines involves an emotional investment as well as a financial one. With honest self-examination, we have to wonder how we can separate the man or woman from the machine. While the machine does not seek human companionship, it has been an eternal endeavor of humans to control their environment through invention. Through invention, and machines, we overcome our physical limitations and can apply more energy to shaping our environment.

We love our machines. We delight in the pleasures they bring us. Take one of them away, and our lives become more difficult. We would ache with deprivation if the machines that provided us with electricity suddenly went quiet. We would consider ourselves put on if our electronic spreadsheets were replaced with adding machines and typewriters, and if what is now done in seconds took days to complete. The machines that we lust for on the showroom floor, with their geometric perfection and lustrous dazzle, are designed to appeal to senses that are filled with everything but technological concepts. The color and sexy curves of modern cars appeal to a part of the brain that is ephemeral in nature and cannot be quantified.

The first automobile we owned lives forever in our minds into old age: before that, the memories of riveting your attention on the mastery your parents had of a very large machine, and dreaming of the day you could tell them to move over and let you drive. Building a machine and watching it perform its intended purpose brings memories of solutions to problems springing into one's mind in a between-sleep-and-awake moment during the night. For someone who has been inspired to create technology, an intrinsic part of his or her being is appreciating and admiring technology, created by others, on a deeper level than most. It is akin to an appreciation of difficult music, when one has struggled through hours of practice to perfect a piece, then watches someone else perform it effortlessly.

In the seventeenth century, René Descartes espoused a philosophy that changed the way we view and interpret the natural world. Separating what *can* be explained in three dimensions from that which *cannot,* Descartes's dichotomy created a model within which "spiritual technology" would seem to be a contradiction. It would most likely not even be raised as a concept or construct by those who applied his famous axiom "That which can be doubted must be rejected."

Doubt can always be cast on explanations for the underlying origins of natural phenomena. To say that the part of the brain that resonates with certain colors is "spiritual" may not be accurate. Could it be described as a part of a spiritual nature in humans? Perhaps, if spirit could be defined and its definition gain wide acceptance according to the rules set forth by Descartes. Without any certainty of the reason why my favorite color is blue, it becomes obvious that humans possess, and express, a mysterious inexplicable aspect of Nature that modern science is yet to define, even though over three hundred years of study employing the scientific method have passed.

Under the conditions that Nature requires for us to exist, therefore, the blast of the factory whistle, or ring of its bell, will summon to the factory's confines elements both mysterious and explained. Here unfolds the drama that is the interaction of humans with humans and humans with machines, to serve society and fulfill a necessary role in the future of civilization. From the outside looking in, and even from the inside being unaware, there appears to a good many observers nothing spiritual within the confines of a factory. Manufacturing work has become synonymous with drudgery.

Quite the opposite is true. The most important aspect of a manufacturing plant is its culture. Machines don't create cultures. Humans create cultures. It is human behavior within a subculture, like a manufacturing plant, that determines its success and survival. The same is true for any company, regardless of its enterprise. If, when looking at a thriving manufacturing plant, all you see is dirt, smoke, and noise, you are overlooking the beauty in its people. In "Frozen to Death," the pleading beauty of a poverty-stricken family's children in all their innocence is just a mere glimpse of the spirit and culture that intermingle with machines and, to an outsider or casual observer, this spirit is drowned out by clatter and confusion.

Even during the austere and pitiless emergence of the Industrial Revolution, the spirit of the people always shined through the darkness, and it seems that the harshest of conditions drew from sufferers the greatest expression of spirit.

An example of this could be seen in the mines of England and Wales. Coal miners emerged from anthracitic veins deep within the contoured beauty of Wales raising their voices in song. They sang to shed the harsh conditions from which they were delivered and pay tribute to the spirit of their fellow man and nation. The Welsh Male Voice Choir is legendary for its soaring, exhilarating, and inspiring effects. Similarly, brass bands throughout England formed within companies and competitions between different "Works" were part of the culture. Inspiring music has been used by cultures all over the world to lift the spirit and ease the conditions of backbreaking labor.

The evolution of manufacturing in the West has resulted in a reality of human existence that is a far cry from George Orwell's *1984,* which was published in 1949. Orwell described a world in which mankind became subservient to a government that used machines to spy on and control behavior of the people. This dystopian vision of what lies ahead for society differed significantly from the aspirations of humanity as a whole, and as such was destined to remain a fictional tale. Others of its genre, designed to give people pause and make them think about the direction they are going, have become popular science-fiction classics. *Blade Runner* and *Logan's Run,* among others, appeal to a part of our brain that can manifest as paranoia. It is a fact that when starved of information, uninformed groups will make things up. Mostly what catches fire and spreads as rumor is not very uplifting.

When ancient cultures are lauded and romanticized, the concept of "spiritual technology" seems quite at home. We are told that it existed in the past and with our pursuit of science and technology, somehow the elements have become polarized and separate. The truth is spirit and technology have always existed together and one cannot exist without the other. Spirit needs to create! Technology is the result of human endeavor fueled by an unfathomable desire for discovery in the creators of technology and by pleasure in those it is intended to please.

There has never been a time in history, nor will there be in the future, when this will not be so. Understanding that, one wonders why it is only now being discussed. Did Descartes take spirit off the agenda of scientific discourse? Some would argue yes. I would offer that he did not, and it should be recognized as being present, even if its influence doesn't enter into the discussion. Tapping into a source of inspiration is personal and private, but make no mistake, it is the fountainhead of technological progress. I would challenge anyone who has made a groundbreaking discovery or postulated a groundbreaking idea to argue otherwise.

Edward Malkowski and I agree that our existence relies on a philosophical foundation of ideas about nature and how humans react to its influence. Technology is the product of ideas and philosophy that have guided groups of people to express themselves creatively and attract others into their field of influence with their inventions. We also recognize that shifts in philosophy about nature and ourselves has a significant influence on changing the world, because it allows the exploration of new ideas and the development of new products, which, by their design and function, influence the spirit.

For instance, if a group of engineers in an enlightened manufacturing facility that made refrigerators had a philosophy of harmony with the earth, perhaps their products would reflect that philosophy. They would create devices that traditionally made annoying sounds make less annoying sounds or no sounds at all! Taking it a step further, they might make devices that create sound that would resonate in harmony with the human organism, producing health benefits. It would require a cross-discipline approach, but it would result in spiritual technology being manifested at a higher level.

For engineers who have lived most of their lives accessing left-brain data, concepts of harmony influencing design, manufacture, and function could be enhanced by a visit to Egypt. There they may learn and experience how the ancient Egyptians accomplished a marriage of right and left brain to produce perfect technological expressions of harmony, proportion, and measure that affect the spirit.

Absent a visit to Egypt, by reading *The Spiritual Technology of Ancient Egypt* you will find that Edward Malkowski makes the work of R. A. Schwaller de Lubicz accessible to those who may struggle with the scholar's original works. Malkowski's exhaustive research of modern physics and the philosophies emerging from the field of quantum physics and research into the brain find a correlation within the volumes of data painstakingly gathered by Schwaller during his thirteen years of study at the Temple of Amun-Mut-Khonsu at Luxor. Malkowski writes with a voice that modern readers will appreciate while being assured that Schwaller has been served well by Malkowski's treatment of his *The Temple of Man*.

Within the Temple of Amun-Mut-Khonsu at Luxor, there is evidence that an ancient civilization was expressing profound philosophical concepts using technology as a tool. The temple has a magnetism that attracts men and women of spiritual, philosophical, and scientific interests. Having been drawn there myself on several occasions, I can say that it is without doubt one of the most magical places on this planet. Anyone who has received training in engineering will be humbled by the expression of superior intelligence and manufacturing capacity throughout the temple. People may also consider that the inner spirit that drives them to greater engineering accomplishments today was no different for the designers, engineers, and craftsmen who were responsible for this miracle in stone.

It is an understatement to say that this temple can overpower the senses. I visited there in 2004 and was compelled to return twice in 2006. From a personal perspective, I can quite understand Schwaller's devotion to documenting its physical measures, as well as his intuitive discernment of measures that evolved from those physical measures but which lie in a realm wherein our true nature resides. These are measures of consciousness that are esoteric, intriguing, mysterious, and uniquely bestowed on you, me, and every other human. Sitting silently in the temple without interruption or chatter is the best way to access these measures. If architecture is "frozen music," as Gurdjieff said, then the Temple of Amun-Mut-Khnosu at Luxor is a movement that fills the senses with breathtaking majesty and admiration.

Through admiration for the spirit of inspired engineers in the past, a guidepost is provided for engineers today and in the future. By accessing more right-brain, intuitive data, there could be a shift in philosophy that recognizes that technology can be applied with spirit in mind, where machines will function in a harmonious manner that soothes, rather than jars, the senses.

To come back one hundred years from now and look over what the industrial landscape has become—who knows what changes Spiritual Technology will have inspired? It seems, to me, a worthwhile endeavor to consciously, and privately, draw more from one's spiritual side when preparing to implement technologies that affect the lives of the masses. In all human endeavors, both past and present, we find spirit as the motivating force. Some may call it ego. Some recognize it as "that fire in the belly" that pushes a person to succeed. Thus being, technology will continue to advance by a collective spirit that takes what has been learned in history and applies new understanding of nature to move technology further.

The greatest benefit we may learn from spiritual technology is to study how technology affects the spirit once it makes its way into society. While the right brain may drive the left brain, once activated, the left brain can drive the right brain crazy! Spiritual technology is the manifestation of spirit through inspiration, creativity, invention, and work. No greater example of this is more evident anywhere in the world than in the inspiring legacy left for us with the pyramids and temples of Egypt.

CHRISTOPHER DUNN
MAY 28, 2007

Christopher Dunn is the author of the *Giza Power Plant: Technologies of Ancient Egypt.* He began his career as an engineer and master craftsman at an engineering company in his hometown of Manchester, England. He was then recruited by an American aerospace company and immigrated to the United States in 1969.

Dunn has written numerous articles presenting his ideas that the structure of the Giza pyramid was a gigantic machine. His twenty-four years of research, involving a process of reverse engineering, has led him to theorize that civilization existing in prehistory was far more advanced than previously believed.

INTRODUCTION

The Nature of Knowledge

With all the scientific breakthroughs and the incredible technology developed in the past fifty years, we know more about biological life, the world, and the universe than ever before. DNA analysis has enabled geneticists to estimate the age of the human race. Computer modeling has shown us the likely origin of the moon, and how its gravity keeps Earth from tumbling as it orbits the sun. With the latest techniques and optics, astronomers have determined that a black hole exists at the center of most galaxies, and that this may be an important characteristic of the universe.

Through scientific investigations, many things that were once considered hypothetical are now considered fact. For the most part, science has apparently taken the mystery out of life. What can be measured can be explained, so there's a reason for everything. And what cannot be measured doesn't exist. The one exception is what you and I, and everyone else, are absolutely certain of: our conscious existence. For that, there are no measurements to analyze, no scientific explanations to offer. Although explanations exist, each of us has his or her own. *Everyone* does, without exception.

There is something very special about our conscious existence. Our ancient ancestors knew what that special something was. It is what today's scientists are discovering. This special something has been referred to as many things, but throughout the ages, it has typically been called "secret wisdom," much to the chagrin of the authoritative and skeptical. Within this secret wisdom, creationism is just as true as evolution and religious texts are just as compelling as any peer-reviewed scientific paper. Religion becomes scientific, and the scientific becomes the sacred. How such a tempestuous statement can be made has to do with knowledge. Everything—*absolutely everything*—that exists begins with knowledge.

Most everyone today knows that the physical universe is made up of matter and energy. But what is not so well known is that physical objects, regardless of whether they are natural or man-made, are useless without instruction. According to theoretical physicist John Wheeler, who coined the term *black hole* and developed the "wave function of the universe" with Bryce DeWitt, the universe is more appropriately understood as information, with matter and energy as its result.

The things we make must have information (instructions) built into them. During the industrial revolution, the method was mechanical, as with a windup toy. Today, that same toy would be battery operated and would access a microprocessor for its instructions on what to perform. The instructions stored in objects are knowledge, and in biological life, the instructions are known as DNA.

Knowledge is the essence, the fundamental nature, of life. Through the higher functions of our brain, the gathering of knowledge, its interpretation, and especially its storage are the basis for individual experience. Knowledge allows us to perceive and defines who we are and how we view the world. Knowledge is such an important and integral part of our lives that we take it for granted, especially for more subtle and mundane tasks, such as reading and tying our shoelaces. Yet like many other things, at one time reading and tying our shoelaces had to be learned. In fact, just about everything we know had to be learned at one time. How does this occur? Have you ever wondered why and how we know what we know? These questions hint at an even bigger question: What is the nature of knowledge?

One way to describe knowledge is what the individual discovers through personal experience, or what is learned through another's personal experiences. Biologically, knowledge, along with thought and all other functions of the brain, is the firing of special cells called neurons. But how do neural networks create and store the experience of perception that we as individuals so thoroughly enjoy?

Although understanding how the brain works is a fascinating subject by itself, the brain, being organic, is part of a living system, the human body, which is dependent on a greater system that encompasses the Earth's ecosystem, the solar system, and ultimately the entire universe. So understanding how the brain works also involves understanding the systems of which it is a part. In this, physics offers insight.

Physicists are generally not concerned with living systems. Rather, they are concerned with the most fundamental aspects of physical reality—atoms, electrons, and protons, and how they behave to manifest reality. Because living systems must exist as part of physical reality and are actually built from the elements that define physical reality, physics also plays an important role in the nature of knowledge.

There have been numerous breakthroughs among the scientific disciplines during the past century, many of which have led to incredible new technologies. In physics, the breakthroughs have had a profound effect on society, such as all the microelectronics we enjoy—cell phones and the Internet, for example. Even so, physics has also affected the less tangible aspects of society, particularly its cultural and philosophical views. Although it is difficult to say exactly when Western culture began to embrace the "new physics" as part of a new worldview, one milestone was the 1975 publication of Fritjof Capra's *The Tao of Physics,* an exploration into the similarities between quantum physics and the Eastern mystical tradition, which was endorsed by one of the creators of quantum physics, Nobel Prize laureate Werner Heisenberg. He was associated with Capra during the early 1970s and read the entire manuscript before its publication.[1] Another landmark book came four years later with Gary Zukav's *The Dancing Wu Li Masters: An Overview of the New Physics*. Both books have sold millions of copies and are still in print today.

In these books, and many others like them, a new worldview is offered that embraces the concept of interconnectedness between man and nature. Everything is connected, and we are all part of the same living system. And within this concept of physical interconnectedness are the concepts of mind and knowledge. In a sense, what this new worldview suggests is that physical form as biological consciousness is an expression of "mind." Birth, death, reincarnation, and evolution are its natural processes of creating a framework in order to experience. As we will find out later in the book, consciousness is a fundamental aspect of reality just as much as the three dimensions we exist in (plus time). Consciousness is viewed as eternal and the driving force for that exists. How it manifests itself in order to experience is through physical form. Thus, the consciousness we (and all animals) experience is through the brain. So, consciousness by taking physical forms creates biological consciousness. Although the mind is very much an enigmatic and debatable concept, the new physics suggests that the mind is a process of a much larger system, as opposed to being a separate entity. Such a concept brings with it complex ramifications for the definition and nature of knowledge.

Since the late 1970s, scores of books have been published concerning man, consciousness, and the new physics, some of which attempt to reconcile traditional religious views with this emerging worldview. It has also had an effect on the scientific model. According to Capra, what he wrote thirty-one years ago in the first edition of *The Tao of Physics* has successfully become a part of a new paradigm for society in general, as well as for the scientific community:[2]

> It is becoming ever more apparent that mysticism, or the "perennial philosophy," as it is sometimes called, provides a consistent philosophical background to our modern scientific theories.[3]

What is of historical importance in this "profound new insight,"[4] as Capra calls it, is that it is also a very ancient insight that I earlier referred to as "secret wisdom." A secret can survive because it resists public investigation, and its meaning remains hidden despite repeated attempts to unravel its cloak of mystery. There is no greater secret than nature's source, and this is what the new science attempts to reveal.

These new ideas about birth and death, mind and consciousness, as well as reincarnation and evolution were expressed long ago in what historians have labeled the ancient mystery schools. Although shrouded by the secrecy of the temple and rites of initiation, these schools taught through myth and symbolism an approach to understanding the world that, philosophically, rivals today's science. As did the ancient mystery schools, today's new science embraces the mystery we call life and attempts to explain our existence through the methods of science. As such, we may elect to refer to this new paradigm as "sacred science."

Few people in the English-speaking world have heard of René-Adolphe Schwaller de Lubicz (1887–1961), since his work was available only in his native French until the 1980s. Despite his relative obscurity, nearly twenty years before the publication of *The Tao of Physics,* Schwaller de Lubicz, in such books as *The Temple of Man* and *Sacred Science,* insisted that the ancient Egyptians scientifically and philosophically achieved a remarkably sophisticated understanding of the principles of nature, and that their culture has been sorely underestimated and misinterpreted by Western academics. According to Schwaller, during the earliest of times,

Egyptian civilization had achieved a technical and scientific expertise that not only made possible its massive public structures but also shaped its complex philosophy, which has been misinterpreted as religion. Such expertise is also responsible for Egypt's longevity as a nation and culture.

René Schwaller was the son of a chemist. During his late teens, he left home and moved to Paris to study under the painter Henri Matisse, who had been greatly influenced by the philosophical works of the renowned philosopher of the time Henri Bergson. Schwaller received the title of "de Lubicz" from his friend, the Lithuanian nobleman and poet Oscar Vladislas de Lubicz Milosz (1877–1939), after the First World War for his contribution to the liberation of Lithuania. While in Paris, he was active in French Theosophical groups and published articles on the philosophy of science in the journal *Le Théosophe,* which led to the publication of his first book, *A Study of Numbers,* in 1917.

Science also played a role in Schwaller's studies, particularly the new physics of the quantum world, but also the ancient science of alchemy. Later, Schwaller moved to Grasse (in the south of France), where he continued to study medieval and Arabic alchemical texts and collaborated with the mysterious man known simply as Fulcanelli, who later became famous for his 1925 book, *The Mystery of the Cathedrals.* At Grasse, fascinated by the esoteric secrets of Gothic architecture, the two attempted to re-create the alchemical red-and-blue stained glass of Chartres Cathedral, located to the southwest of Paris.

In 1937, Schwaller and his wife, Isha, visited Egypt. While in Luxor at the Temple of Amun-Mut-Khonsu, he realized that the temple architecture was a "deliberate exercise in proportion"[5]—analogous to the symbolism used by the builders of the medieval cathedrals. The more proportionate the object is, the more beautiful it is, and as a result emotions are evoked within a person's "being." So, building something with proportion in mind creates an unspoken, thereby symbolic, reference to the individual who is viewing the object. For the next twelve years, he methodically studied the temple and finally came to the conclusion that the temple itself was the Egyptians' inscription of their philosophy and science, as well as their testimony as to the true nature of Man. Deciphering the temple, its art and architecture, brought new insights into the brilliance of the ancient Egyptians. These insights led to the hypothesis that the Egyptians' knowledge was the legacy of a technical civilization of which there existed no history or knowledge in today's world.

History as Knowledge

Knowledge serves a greater purpose than being an animal concerned only with eating. It is what elevates human existence beyond what traditional science calls nature. It is what propels us to create "civilization." As the essential aspect of mankind's social organization, as well as our technical progress, knowledge is the basis for every civilization and each civilization's history. This kind of knowledge transcends the individual's direct personal experience and comes together through the discovery of fact, whether it concerns experimentation and observation of natural phenomena, or the discovery of a prehistoric settlement, or the deciphering of an ancient manuscript.

The discovered fact, however—what is observed, found, or translated—has little if any meaning outside the individual's perception and interpretation, which are based on the accumulation and amalgamation of existing knowledge. This draws attention to a basic feature of knowledge: Knowledge is not a fact. Rather, knowledge is the perception and interpretation of fact, which is the reason why there is such disagreement among individuals, be they independent researchers, scientists, or theologians. Facts are meaningless outside the personal context of perception, which includes belief and interpretation. This is part of our nature, as Ian Tattersall, anthropologist and curator of human evolution at the American Museum of Natural History in New York, so eloquently expresses:

> While every other organism we know about lives in the world presented to them by nature, human beings live in a world that they consciously symbolize and re-create in their own minds.[6]

Today, there is an enormous amount of facts through which a great amount of knowledge has been created. These facts cover everything from physics to the earliest epochs of Earth's history, and they have inspired a thirst for even more facts, which has led to the development of numerous theories that have produced many practical applications for the betterment of mankind. However useful these applications may be, any explanation put forth for just about anything is only theory, except for the certainty of our individual awareness and experience. This is particularly the case when it comes to history.

History is a chronological record of events, along with an explanation as to their causes and effects. There are a few general misconceptions associated with history, however. One of these is that history is a factual body of knowledge.

Although we certainly have factual information associated with past events that plays a prominent role in any attempt to determine what happened, history is decidedly interpretative. Despite how much we, as a society and civilization, document events, there will never be a comprehensive record of any particular past event. Furthermore, unless individuals write their memoirs, the motives and perceptions of those involved in the history of events are often difficult to know. There is also the author writing the history book based on his or her particular interpretation, including myself. In essence, we create a past based on our own memories and the memories of those who came before us, which is subject to revision by each successive generation. History is more theory than anything else, particularly in the case of ancient history. So, the question becomes "Is any certain view of the past historically accurate?" There has to be a way of verifying that events occurred, and that is best served by a comprehensive analysis of all the data, whether anomalistic or not.

Another general misconception is that prehistory was a time long ago, before mankind existed. (The term *prehistory* was coined in 1865 by Sir Daniel Wilson in his book *The Prehistory of Scotland*.) Prehistory actually refers to the period before the advent of writing, or roughly since 3000 BCE, according to the archaeological record. This does not mean that there was no history during prehistoric times. Mankind's earliest history is preserved in oral traditions and myths, which are believed to have been the currency of history long before the invention of writing. And it has been demonstrated that mankind has been around for at least one hundred thousand years, and possibly, as some research suggests, as long as two million years.

It has also been demonstrated that our planet periodically suffers global catastrophes, be they extraterrestrial, such as a comet or asteroid impact, or terrestrial, such as a massive volcanic eruption. As a consequence of these cataclysms, there have been extinctions of plant and animal life, which must have had drastic effects on any human societies that existed at those times. The last major extinction took place long after mankind had spread onto every habitable continent in the world—ten thousand years ago, at the end of the last ice age.

The concept of the ice age and our knowledge of the various ice ages serve as a good illustration of the nature of knowledge and the nature of history. Even today, the causes and effects of the ice ages are hotly debated topics. The development of the ice-age concept and its associated theories during the past two hundred years is a prime example of how we create and re-create history in our own minds.

The Ice Age: How Knowledge Creates History

Before the nineteenth century, the premises of history were very different from what they are today. Anyone claiming knowledge of an ice age that had occurred thousands of years ago would have likely been labeled as belonging to the lunatic fringe. Through the discovery of geologic facts, however, knowledge coalesced into a detailed picture of what has been called the Great Ice Age. Today, this incredible story of the power of nature is common knowledge.

According to geologists, over the course of the last two million years, at least four and possibly six periods of glaciation plowed out an assortment of rock from Canada and moved it south in a hundred-foot-high flow of ice and snow, pulverizing it against bedrock and depositing it on the midwestern plains. These ice sheets produced large quantities of gravel, sand, and silt, which mixed together and formed much of the soil in the Ohio and Upper Mississippi River Valleys.

At the leading edge of the glaciers were ice cliffs up to two hundred feet high; cold, dry winds swept down from their frozen crowns. The climatic conditions in this zone next to the ice were extremely harsh. Cold temperatures and strong winds created an arctic desert, a wasteland littered with rock debris and fine sediment. Strong winds gathered this sediment from the glaciers and deposited it in thick layers that covered much of the Midwest, extending south into Louisiana and Mississippi. Today, these deposits are the source of the rich midwestern farmlands.

Glaciers reached as far south as the Ohio and Missouri Rivers, although the most recent glacier (the Wisconsin ice sheet) stopped midway across Illinois. Local legend has it that the glacier stopped twenty miles north of Springfield, the state capital. There is a grand mound there named Elkhart Hill, more than a mile in circumference. In Logan County, of which I am a native, it is most conspicuous, towering above the surrounding landscape.

Remnants of these vast sheets of ice can still be seen today in the form of numerous lakes that pepper the Midwest's northern regions.

Retreating glaciers left large depressions in the Earth's crust that filled with meltwater, the most magnificent of which are the Great Lakes. Other, larger lakes, such as Lakes Winnipeg, Reindeer, Athabasca, Great Slave, and Great Bear in Canada, existed at one time but have since drained off and disappeared.[7]

According to geologists, it is a fact that ice has had a great impact on North America. Besides the geological evidence of moraines, kettle lakes, gouged bedrock, and erratic boulders, Greenland and Antarctic ice-core samples have demonstrated that levels of carbon dioxide have fluctuated over millions of years. Lower levels of carbon dioxide indicate cooler periods in Earth's history, but it is not yet clear whether these lower levels are the cause or the effect. Although a great amount of scientific research has been applied to the study of ice and ice ages, and much knowledge has been gained, why they occur is just as much a mystery now as it was when Joseph Adhémar published the first detailed ice-age theory in 1842 in his book *Révolutions de la mer, déluges périodiques*.

From Fact to Knowledge: Discovering the Ice Age

As early as 1787, Bernard Kuhn believed that erratic boulders in the Swiss Jura Mountains were the result of ancient glaciation. Seven years later, after visiting Jura, James Hutton arrived at the same conclusion. Until the first half of the nineteenth century, however, the prevailing model to explain the observable geologic evidence was that it was a result of the great biblical flood.

A German-born geologist, Jean de Charpentier (1786–1855), was also intrigued by erratic boulders and mounds of glacial debris (called moraines), and he developed the first theory of glaciation during the 1830s. Louis Agassiz (1807–1873) also converted to the glacier explanation of geologic curiosities, then forged ahead and integrated all these geologic facts to formulate a theory in his book *Étude sur les glaciers* (1840), that a great ice age had once gripped Earth. In a later book, *Systéme glaciare* (1847), he presented further evidence gathered from all over Europe to support his theory. In 1848, Agassiz accepted a position at Harvard and moved to America, where he discovered even more evidence of glaciation. By 1870, the theory of ancient periods of extensive ice was generally accepted by the scientific community, almost entirely through the ideas of a single man.

With a scientific consensus that the ice age had existed, the quest then became to find out what had caused it. The first theory, introduced by Joseph Adhémar, was based on the Earth's axis tilting back and forth over a twenty-six-thousand-year period, commonly referred to as the precession of the equinoxes.

As time passes, the constellations slowly change their position in the night sky when viewed on a specific date (typically measured at the vernal equinox), moving backward through the zodiac. Today, the sun rises in the constellation of Pisces at the spring equinox. Two thousand years ago at the spring equinox, it rose in Aries; beginning around 2070, it will rise in Aquarius. This tilt of the Earth's axis is called the plane of the obliquity, and it extends outward to form a great circle in the celestial plane known as the ecliptic. The angle is called the obliquity of the ecliptic and is presently inclined at 23.5 degrees to the vertical, but varies between 24.5 and 22.1 degrees. This angle of Earth's axis defines the seasons in temperate climates.

According to Adhémar's theory, whichever hemisphere had a longer winter would experience an ice age. Thus every eleven thousand years, an ice age would occur, alternately in one hemisphere and then in the other. This is due to the eccentricity of Earth's orbit. Earth's orbit goes from being nearly circular to more elliptic. When the orbit is more elliptic, the hemisphere that is tilted away from the sun, when Earth is farthest from the sun, will receive less sunlight so temperatures will be cooler. As a result, winter snow will not completely melt away and over time glaciers will build up.

James Croll, a self-taught scholar and ex-janitor at the Andersonian College and Museum in Scotland, objected to Adhémar's theory. He believed that the most plausible driving force behind climate change was variations in solar radiation, called insolation, as a result of Earth's path of orbit, which is elliptical and can vary, from being nearly circular to more elliptic, up to 5 percent over time. This eccentricity affects the amount of solar radiation that strikes Earth's surface at aphelion (our farthest point from the sun) and at perihelion (our nearest point to the sun).

According to Croll's theory, a decrease in the amount of solar radiation during the winter favors the accumulation of snow. This would result in additional loss of heat by the reflection of sunlight back into space. If winter occurs when Earth is close to the sun, winters will be naturally warmer than usual. But if winter occurs when the sun is farther away, temperatures will be colder than usual. If the polar area of a hemisphere becomes colder, trade winds will be stronger in that hemisphere, and warm equatorial ocean currents will shift toward the opposite hemisphere, further augmenting the heat loss. If Earth's orbit were circular, the slow wobble would have no effect at all on climate. Each season would occur at the same distance from the sun. Because insolation in the Northern Hemisphere is out of phase with that of the Southern Hemisphere, however, Croll believed that the ice ages would alternate from the Northern Hemisphere to the Southern.

Although the alternating ice-age theory is now believed to be incorrect, Croll's ideas laid the foundation for future ice-age theories. He was the first to recognize the importance of ocean currents, solar radiation, and the eccentricity of the Earth's orbit in building an explanatory model.

Early in the twentieth century, Milutin Milankovitch, a professor of physics, mathematics, and astronomy at the University of Belgrade, revived Croll's insolation theory and set out on the task of detailing insolation based on Ludwig Pilgrim's calculations of Earth's orbit. Milankovitch showed that insolation was dominated by a 23,000-year cycle, and concluded that ice ages would be most intense when the solar radiation dropped below a certain threshold. Since the insolation curve has an approximate 100,000-year cycle, he believed that such a cycle might be seen in the ice ages. Milankovitch also had the insight to propose that the Northern Hemisphere would dominate because it contained two-thirds of Earth's landmass. Driven by the amount of solar radiation in the north, the ice ages in both the hemispheres would be synchronized.

Milankovitch's insolation theory was abandoned when age estimates made possible by radiocarbon dating showed that the timing of his ice-age calculations was incorrect. Isotope studies of seafloor sediments, however, which focused on changes in Earth's climate, revived it during the 1960s and 1970s. According to geologists, deep-sea sediments containing the shells of small planktonlike organisms (called foraminifers) hold a history of climate change. When alive, they fix themselves to two types of oxygen atoms, the abundant and more common oxygen-16 and oxygen-18. Oxygen-18, the heavier atom, is "enriched" in ocean water; the lighter atom is found in higher concentrations of snow and ice. (When certain conditions exist, the heavier atom naturally occurs more frequently in the ocean as opposed to the air or surface of the planet.)

Whenever water is extracted from the ocean to make more ice, it leaves its calling card in the oxygen. This enrichment, from oxygen-16 to oxygen-18, is seen in the carbonate shells of the foraminifers. The carbonate precipitates from seawater, so the oxygen that builds the carbonate crystals reflects the composition of the seawater. By analyzing oxygen isotopes in foraminifers, scientists can determine when Earth produced more glaciers and the periods when ice ages occurred.

In seafloor sediments, 100,000-year, as well as 41,000-year and 23,000-year, climatic cycles have been discovered. But there are still unresolved questions. In glacial data, the 100,000-year cycle seems to dominate, with the 41,000-year cycle weaker and the 23,000-year weakest of all. In insolation theory, however, it is the reverse.[8] The 23,000-year cycle dominates, and the weakest appears to be the 100,000-year cycle.

Recent research suggests a very different approach to the fundamental cause of the ice age. Although distinct ocean currents have been known for some time, scientists have recently determined that ocean currents play a crucial role in climate and weather. New research has determined that shallow, warm-water currents from the Pacific flow westward around Africa and then northward along the African and European coasts. The flow of these waters keeps Europe balmy in contrast to its counterpart, Labrador, across the Atlantic. The warm currents provide western Europe with a third as much warmth as the sun does and are part of a global oceanic system that maintains the climatic status quo.

In the North Atlantic, the Gulf Stream carries heat in the form of warm water to the north and east. As it moves north, it evaporates and transfers its heat to the coastal areas. The warm water becomes saltier with evaporation, and when it reaches the latitude of Iceland, its density reaches a point that it sinks to the bottom. Then it becomes part of the cold-water return cycle and flows southward in the Atlantic, around Africa, and back to the Pacific.

If in some way the warm waters ceased to flow into the North Atlantic, Europe would experience a mini–ice age. Current studies suggest that is a possibility, and that this current "conveyor belt" in the North Atlantic is unpredictable. Since the end of the last ice age, the Arctic ice cap has continued to melt, allowing freshwater into the North Atlantic. If too much freshwater enters the ocean (thereby diluting its salt content and keeping it less dense), it would join the return currents at the bottom, blocking the warmer currents. As a result, the climate of Europe would change drastically.

The same type of ocean currents exists in the South Atlantic near Antarctica. There, ocean currents flow along the coast. Deep, cold currents flow back from the South Atlantic, south of Africa, and on to Australia. Cold, salty water off the Antarctic coast sinks into the depths, thereby boosting its push to the interconnected system of ocean currents. According to Wallace Broecker, of Columbia University, Antarctic waters are sinking at only a third of the rate they were a hundred years ago. But this will have a different effect. If this is what is happening, the slowdown in the Antarctic deep current that began a century ago will make the Antarctic colder and the Gulf Stream warmer. The current global warming trend began during the 1880s and received a boost from man during the industrialization of society. Broecker believes that this warming is man-made, and is working against a natural cooling trend.[9]

Another recent theory links the ice age with changes in global climate to one of the Earth's most impressive geological features: the Himalayas. According to this theory, proposed by Maureen Raymo at Boston University, as the Himalayas grew, massive amounts of rock were exposed to the elements. As monsoon rains soaked the land and combined with carbon dioxide, the face of the exposed rock eroded. This process of chemical weathering extracted so much carbon dioxide from the atmosphere that global temperatures dropped, triggering an ice age.[10] To show that this was the case, Raymo turned to the study of seafloor sediments and strontium.

There are several types of strontium isotopes, each with a different atomic mass. A heavier variety, strontium-87, is washed into the sea by the chemical weathering of rock. The lighter variety, strontium-86, is released by the spreading seafloor and comes from deep inside the earth. By comparing the amounts of the isotopes in different layers, Raymo believed that she would learn which process was more active at any point in time. Thirty-five million years ago, strontium-87 increased dramatically, coinciding with the Himalayan uplift.

With this evidence, Maureen Raymo believes she has solved the ice-age mystery. First, the uplift of the Tibetan region intensified the Indian monsoon. Then the monsoon rains eroded the mountains, stripping carbon dioxide from the air. Finally, with less carbon dioxide, the atmosphere gradually cooled.

There are other theories that explain the ice age. Although controversial, they too create a possible history in the minds of their followers. In

1950, Immanuel Velikovsky introduced the idea of catastrophism—that geologic changes were caused by cataclysmic events rather than developing gradually—to the general public with *Worlds in Collision.* His book was an instant *New York Times* bestseller. The idea of catastrophe caught on. In 1980, physicist Luis Alvarez, along with his son Walter (a geologist), discovered a layer of clay containing unusually high levels of iridium. According to the Alvarezes, the best explanation for this was a large meteor impact sixty-five million years ago. Some years later, a 125-mile-wide crater was discovered in the Gulf of Mexico off the coast of the Yucatán Peninsula and attributed to the strike sixty-five million years ago. Although the theory is still subject to debate, it has stood the test of time among the scientific community and is generally accepted as fact. Today, catastrophism is an accepted principle in both geology and paleontology.

Prior to the 1960s, the scientific community did not believe that meteors fell from space and crashed into Earth. It was the seminal work of Gene Shoemaker at Arizona's Barringer Meteor Crater in 1963 that provided the proof that asteroids did exactly that.

During the 1960s, geographer Donald Patten proposed a cosmic catastrophe as the reason for the ice age. Although his theory appears to be more theological than scientific, he suggests an interesting explanation of the effects a cometary "near miss" might have on Earth's climate. He also discusses motives and beliefs and supplies a creationist's rebuttal to geologic uniformitarianism (the idea that geologic changes take place gradually), and he provides a history of scientists and writers espousing a catastrophic approach to Earth geology. Since the 1920s, George McCready Price, Byron C. Nelson, Alfred M. Rehwinkel, Henry Morris, Charles Hapgood, Ivan Sanderson, and Dolph E. Hooker, among others, have carried the banner of a sudden-catastrophe approach to explaining ice ages.[11]

One discovery of fact that provides their proposition with some punch is the evidence of frozen mammoths, particularly in Alaska and Siberia. Although mammoths are not the sole animals that have been found frozen (rhinoceroses, sheep, horses, oxen, lions, tigers, and bison have also been found), as an extinct species, they have been at the forefront of scientific research. Their remains, sometimes whole, have been found in Siberia and Alaska by the tens of thousands. For most of the nineteenth century, they provided the world with an ongoing supply of ivory and a booming industry for Russian merchants. Between

1880 and 1900, nearly twenty thousand tusks were taken from a single island, and it has been estimated that there may be up to three million mammoths still buried in Siberia.[12] According to a *National Geographic* article, experts estimate that there are six hundred thousand tons of ivory still available for recovery.[13]

A sudden calamity, such as an asteroid impact, fares well in explaining the death of millions of animals. The precedent of asteroid or comet impact is seen in the greatest extinction of all, which happened at the end of the Cretaceous period, when a giant asteroid impact resulted in the extinction of the dinosaurs. But Patten's proposal, that a comet traveling tens of thousands of miles per second became trapped in Earth's gravitational pull, is difficult to justify. Because Earth travels at a much slower rate, approximately twenty-five miles per second, a comet moving that fast would either strike Earth or speed past it.

Nonetheless, as Patten's hypothetical comet danced in Earth's orbit, ice that had broken away from its mass through "electromagnetic defection" was deposited in vast quantities on Earth's surface. A magnetic field (such as Earth's) deflects a stream of electrons. In other words, Earth's magnetic field pulled the ice from the comet to the surface. According to Patten's theory, twelve million cubic miles of ice was dumped on our planet, six on the Northern Hemisphere and six on the Southern Hemisphere: ice whose temperature was −150° Fahrenheit. At the center of these ice dumps, the ice would have been three miles thick. The ice appeared suddenly, not over a long period. Only this, Patten claims, accounts for the sudden freezing of millions of animals.

Charles Hapgood's "wandering poles" theory is perhaps the most fascinating of all ice-age theories. It caught the attention of Albert Einstein, who believed that researching the subject was desirable and that it "would not be justified to discard the idea a priori as adventurous."[14]

Hapgood's theory began with an interest in geography and ancient maps, which led to his rediscovery of the Piri Reis map, a hand-drawn Turkish naval map that had been gathering dust since the sixteenth century. According to Hapgood's sources, the map was drawn a few years after Columbus launched his first voyage to the Americas. Admiral Piri Reis, cartographer of the map, noted that his world map was derived from very old reference maps. On close inspection, Hapgood recognized that spherical trigonometry was used in the map's layout, which required a detailed knowledge of global geography. The map also displayed the coastline of Antarctica at some remote time when it was free of ice. According to Hapgood, the map was accurate at a time when no one should have known the coastal areas of Antarctica. This prompted a search for an explanation and eventually led to his controversial theory.

According to Hapgood's theory of wandering poles, every 20,000 to 30,000 years, Earth's continental plates move as a single unit, rapidly and over great distances. This phenomenon, known as continental drift, occurs today, but at a much slower rate. If conditions arise that create an imbalance in Earth's gyroscopic rotation, Hapgood's theory stipulates that Earth's plates would move in such a manner as to return Earth to a balanced spin.[15] Geologic evidence suggesting that the poles may have been in different positions during the Pleistocene epoch (1,808,000 to 11,550 years ago) is impressive, although the physics of such a pole shift remains problematic.

Based on geomagnetic and carbon dating, Hapgood identified the locations of the four previous poles and mapped out their transitional paths. Seventeen thousand years ago, the North Pole was located in the Hudson Bay and moved to its current position over 5,000 years. Before that, the North Pole was located in the Greenland Sea (75,000 years ago), and moved southwest to the Hudson Bay. Prior to the Greenland Sea site, the pole was located in the Yukon Territory of Canada.[16]

How this movement could occur is explained by the dynamics of Earth's composition. We live on the crust, the outer surface, which comprises six main continental plates and a few smaller ones. Earth's inner core consists of solid iron surrounded by an outer core of liquid iron. Surrounding the core is the mantle, which is composed of molten rock (lower mantle) and solid rock (upper mantle). The upper mantle and crust are loosely connected and are able to slide against each other, the least effect of which is continental drift. Theoretically, each layer is capable of movement independent of other layers. According to Hapgood, the top two layers can slide, if certain forces are applied, while the core and the axis and orbit of the planet remain unchanged. But what force causes the slippage?

In Hapgood's opinion, the centrifugal momentum of ice caps, eccentric to the poles, provides this force. The weight of the ice on the poles creates an imbalance in the Earth's rotation. Eventually, this builds to a point where a change is required to correct the imbalance. Hapgood realized that the entire planet did not need to be repositioned around its axis to maintain its balance. Only the outer crust needed to move, just as the loose skin of a peeled orange can slide around the inner fruit. He envisioned a catastrophic and dramatic move of the entire crust that allowed the polar ice caps to melt in a new, warmer climate. Ice would then begin to build at the new poles, awaiting the next shift. The crust's rapid movement, of course, would create environmental mayhem. If the current level of seismic and volcanic activity was the result of plates shifting between one and four centimeters per year, a much faster rate of change would likely be apocalyptic.

After the poles shifted, regional climates everywhere would have changed dramatically. The displaced polar ice would have melted, causing incredible floods. The new polar areas would have frozen in a relatively short amount of time, quickly killing life that was accustomed to a warmer climate. Areas of climatic convergence would have shifted; deserts would have received rain and rain forests would have become deserts. Plant and animal life would have had to adapt to the new conditions or become extinct.

Still another, more recent theory explains that the global warming at the end of the ice age was a result of a cosmic phenomenon known as a galactic superwave, proposed by physicist Paul LaViolette. According to LaViolette, every 13,000 to 26,000 years the galaxy's core (the bulge where there is an immense number of stars) emits intense cosmic radiation.[17] This radiation, composed of high-energy electrons and electromagnetic radiation (from radio waves to X-rays and gamma rays), sails out from the core of the galaxy in a "superwave" traveling near the speed of light. When the superwave finally reaches our solar system, some 28,000 light-years from the galactic core, it alters the behavior of the sun.

Our solar system is shrouded in a haze of cosmic dust and debris known as the Oort cloud. When the superwave passes through this cloud, it takes with it a large amount of dust, and when the dust reaches the sun, it serves as fuel for the sun's furnace, best described as a natural fusion reactor. As a result, solar-flare activity increases and the sun's corona and photosphere increase in size, with drastic climatic effects for Earth. The increased solar activity would cause a sudden period of warming for the Earth's climate. Furthermore, cosmic particles that entered Earth's atmosphere would be captured by Earth's magnetic field and form radiation belts in the upper atmosphere. In a single day, the energy injected into Earth's atmosphere would be equivalent to a billion-ton hydrogen bomb. These climatic effects (heat and radiation) would last for several thousand years.

According to LaViolette, such a superwave was responsible for the drastic climatic changes our planet experienced between 15,300 and 14,150 years ago[18] that ultimately ended the ice age and caused the extinction of numerous species. LaViolette also believes the evidence suggests that two particularly intense solar flares occurred about 12,840 years ago and 12,730 years ago,[19] with large coronal mass ejections that overtook Earth. LaViolette believes the sudden superwarming was responsible for initiating rapid glacial melting and continental flooding.

Mystery of the Ice Age

What is the layperson with a general interest in Earth history supposed to believe? The various hypotheses explaining the Great Ice Age appear as a theoretical grab bag of pick and choose. Each is backed by scientific evidence of some kind and has its own aficionados. Each tells a story of Earth's past, and all fall into one of two general categories: those that postulate gradual climatic change and those that describe a sudden catastrophic beginning to the ice age. Hapgood, in his theory of pole shift, implies that there never was an ice age, but that is a matter of perception, geographically speaking. One might say that there is always an ice age at the North and South Poles.

Regardless of whether the ice age was the result of Earth's cyclical nature, an element of cosmic nature, or an interstellar visitor, the climate drastically altered life for those that lived during the events that caused the end of the ice age. It is a geologic fact that many species became extinct around ten thousand years ago. In North America, that extinction included the mammoths, camels, horses, ground sloths, peccaries (piglike hoofed mammals), antelopes, American elephants, rhinoceroses, giant armadillos, giant beavers, giant bisons, tapirs, and saber-toothed tigers.

Although the sudden extinction of these species remains to be proved, nowhere else is the testament of sudden catastrophe more poignant than in Alaska, as University of New Mexico archaeologist Frank Hibben (1910–2002) describes it:

> In many places, Alaskan muck is packed with animal bones and debris in trainload lots. Bones of mammoths, mastodons, several kinds of bison, horses, wolves, bears, and lions tell a story of a faunal population. . . . Within this frozen mass lie the twisted parts of animals and trees intermingled with lenses of ice and layers of peat and

> mosses. It looks as though in the midst of some cataclysmic catastrophe of ten thousand years ago the whole Alaskan world of living animals and plants was suddenly frozen in mid-motion like a grim charade. . . . Twisted and torn trees are piled in splintered masses. . . . At least four considerable layers of volcanic ash may be traced in these deposits, although they are extremely warped and distorted.[20]

The sticky asphalt of southern California's La Brea tar pits testifies to more than 565 species of animals that met their death around the same period, ten thousand years ago. During the first excavation, in 1906, scientists found a bone bed that contained more than seven hundred saber-toothed tiger skulls. Combined with wolf skulls, they averaged twenty per cubic yard.[21] There were more bones than tar, and they were discovered "broken, mashed, contorted, and mixed in a most heterogeneous mass," according to George McCready Price,[22] nearly identical to the muck of Alaska. One hundred thousand fossilized birds were also recovered, representing more than one hundred thirty-eight species, nineteen of which are extinct.

During that same period, the mammoths of Siberia were being killed in a similar fashion. John Massey Stewart, of the Smithsonian Institution, has estimated that more than five hundred thousand tons of mammoth tusks were buried along Siberia's Arctic coastline.[23] Several dozen frozen mammoth carcasses, such as the Jarkov mammoth, have been found with the flesh still intact.[24] These animals died suddenly, possibly by asphyxiation, some with still-undigested food in their stomachs, such as grass, bluebells, wild beans, and buttercups.

Whatever the cause, the climatic cataclysm ten thousand years ago also affected the climates of lower latitudes in Central and South America, as well as that of Europe. Those lands have also revealed evidence of mass extinctions. The mechanism that brought these animals to their graves is still a mystery. With various theories appealing to the assorted beliefs of those with an interest in the subject, the ice age is more speculation and theory than it is fact. Although the evidence is real, the knowledge that it produces resides in the minds of men and women, and it depends on personal beliefs and interpretation of the facts.

The obvious conclusion, and what this survey of ice-age theories has illustrated, is that history is more perception than anything else, and to reveal that *the interpretation of fact, based on belief, is the nature of knowledge.*

History as Perception

As the survey of ice-age theories demonstrates, uncovering historical truths is a difficult venture, especially when those truths require the reconciliation of current fact with events in the remote past that we know little about. This leads me to describe ancient history as perception based on our experiences, ideology, and cultural biases: a wobbling house resting on the shifting sands of evidence and interpretation. It's mostly theory and part fact, sailing uneasily in uncharted waters somewhere between science and speculation. Ancient history is more supposition than it is real, and thus there are many opinions.

All people have beliefs, a set of values and views they received through their parents and society while growing up. It is their worldview, their way of explaining the world around them. In religious-oriented homes, it also involves faith-based beliefs, called dogma by those who do not share them. This dogma can often be a round-robin affair, where one set of beliefs, viewed as truth by a particular group, is dogma to another group with different beliefs. Of course, this way of thinking and labeling works both ways. So it can be said, in general, science and religion are mankind's attempts, as they apply to the individual, to understand the mysteries of life.

In recent times, with advanced technology and passionate investigation, scientists have been able to clarify and describe in detail what this mystery is. Actually, there is a set of mysteries that include the quantum world: the human consciousness as it pertains to the animation of matter, the evolving cosmos as it pertains to the grand structure of reality, and the act of creation as it pertains to the origin of the cosmos. Everything is composed of matter. Anything living is animate matter. Anything not living, such as a rock or water, is inanimate matter. Thus, the concept of human consciousness is clearly a function of animate matter.

In this book, chapters 1 through 4 (part 1) describe these mysteries and lay the foundation for chapters 5 through 13 (part 2), which present the thesis that ancient wisdom was a product of functional scientific observation. Part 2 traces the roots of this ancient wisdom to a science-based knowledge of nature, through the genius of ancient Egypt, and shows how that knowledge was translated into religious beliefs.

Also in this book, I endeavor to explain that the science of pharaonic Egypt—sacred science—was (and still is) a qualitative approach to the investigation of nature, a way of thinking functionally by embracing the mystery of life, and that this is how the ancient Egyptians philosophically understood their existence.

This qualitative approach does not replace quantitative science. Rather, it enhances our understanding of what science is by explaining another side of nature that often perplexes us.

At the heart of the scientific quest, from a qualitative as well as a quantitative perspective, is the question of biological order and why mankind came to be. Unlike all other species, mankind has the ability to view itself as separate from nature, manipulate the environment for its benefit, and construct "civilization."

Determining what prompted mankind to organize the first civilization, or civilizations, is one of the grand goals of archaeology. Some scholars believe the building of the first civilizations was a result of warfare, a means of organizing for the common defense. While organizing people and structures into defensible cities is clearly a benefit to mankind and a reason to organize, there certainly is more to the story—scales of economy, specialization, and all the other benefits of social cooperation.

When that first civilization was organized is another matter. The traditional belief is that Mesopotamia emerged first, followed closely by civilizations in the Nile and Indus River Valleys, all around 3000 BCE. There are structural anomalies and textual curiosities, however, that have proved to be potent fodder for speculative assertions that civilization is far older than the third millennium BCE.

Even for less fervent rationalists, accepting such a notion is difficult, if not outright impossible. Skepticism (not cynicism) is healthy, but there is a significant difference between explaining and explaining away. Much has happened in the last twenty years in all the scientific fields that, when correctly applied to historical anomalies, helps explain as opposed to explain away.

There is an old cliché that truth is stranger than fiction. If so, maybe we should take those unexplainable, strange aspects of history a little more seriously and seek the answers as to why they are so. If we do, perhaps we can learn something about ourselves that will someday benefit us all.

PART I

Science Mysteries

Science is the observation of naturally occurring phenomena to discover facts about natural phenomena. The scientist's concern is to objectively understand nature, whether terrestrial or cosmic, and to create principles based on the discovery of fact that provides a theoretical explanation of those phenomena. Aside from being a method of investigation, science is also referred to as a body of knowledge that includes such disciplines as astronomy, biology, chemistry, geology, and physics. In a sense, the grand quest of science is to explain what we are and where we came from.

During the last two centuries, modern science has been responsible for a wide range of discoveries, and through the technical implementation of principles based on those discoveries, civilization has reached new heights in knowledge and lifestyle comfort. Despite these discoveries, and the ever-increasing technical know-how of the scientific arsenal, the most fundamental aspects of the natural world, particularly organic life, remain unknown.

The irony of this grand scientific quest is that in the end, the scientist is left with a mystery that requires contemplation beyond the bounds of the data set for its explanation. The effect of this curious situation is that the mystery is revealed for what it is, a fact. Physics reveals that the atom is composed chiefly of empty space, and that if there was a beginning to the universe, it erupted from an unknown substance or dimension. The most remarkable of all scientific fact is what every person every day encounters, what defines our existence. The conscious experience we call life, the number one reason why we exist in the manner that we do, is as much a mystery as it is a fact.

Such is the predicament we are in. Science objectively analyzes and categorizes nature, but it also clarifies that the mystery is not just a figment of our imagination. The mystery is real.

More than twenty years ago Carl Sagan, the great defender of science, whom I admire for his insights into the nature of the human experience, revealed to the world this mystery in the public television series *Cosmos: A Personal Journey.* The elements that make up the world as well as our bodies were created in the incredibly intense heat of stars. Through evolutionary forces, whatever they might be, we are made from the stuff of stars without any logical reason why or a precise recipe for how. We just are.

It has been said that science attempts to answer only the questions that it can and leaves the unanswerable ones to religion. Science does not profess to be a belief system, but if any doctrine fully embraced science as a means of explaining nature, mankind included, that doctrine would most assuredly be labeled as the science mystery school.

Part 1 is dedicated to the mystery today's science mystery school has so thoroughly elucidated. The principles of quantum mechanics paired with the elusive nature of consciousness—facts endowed with a distinctly mystical character—provide a cosmogony and cosmology stranger and more mysterious than any novelist could invent.

In theorizing an origin of the universe, it is significant that with the current body of scientific knowledge, we must conclude that "something came from nothing." Such an irrational notion involuntarily introduces the concept of cause without defining the nature of cause, and must be dealt with by the individual in a philosophical manner.

It is also significant that at the beginning of this modern age of science, people of learning embraced ancient alchemic and Hermetic texts in their quest for knowledge, which justifies an investigation into the source and philosophical significance of the Hermetic tradition. This is the subject matter of part 2, "The Roots of Ancient Wisdom," in which the argument is established that the Hermetic philosophy was not only based on an ancient methodology of science, but was also the inspiration for the Judaic and Christian religious traditions.

The advocate of today's new science will find common philosophical ground with the ancient wisdom of the Hermetic tradition. Any philosophical doctrine based on today's science would agree in many respects with that tradition.

1

Matter

Mystery of Quantum Reality

Atoms are mainly empty space. Matter is composed, chiefly, of nothing.

CARL SAGAN, *COSMOS: THE LIVES OF THE STARS*

Who are we? What are we? Where did we come from? Why are we here? These are the questions to which everyone wants the answers. Whether religious or mythical, during ancient times or in the present, belief systems have always been based on attempts to answer these questions. Over the course of world history, human cultures created first myth, then religion, to explain the unknown. Now, science has come to the forefront in mankind's grand quest for truth. Although based on observation and experimentation instead of subjective beliefs, the questions scientists ask are the ones our ancestors posed thousands of years ago. What constitutes the fabric of reality, and how does it relate to our existence?

There is no better discipline than physics to describe reality objectively. It is the definition of the discipline to describe why and how things are physical objects. With continually improving technology, scientific answers to who we are, what we are, where we came from, and why we are here seem to be within reach. Answers to these formidable questions are now emerging in a new scientific model. Those who are involved with or follow its developments sometimes refer it to as the "new view" or "new science." Although it is a recent movement, it originated over a hundred years ago and began with German physicist Max Planck (1858–1947) and his quest to explain heat.

Quantum Facts of Matter

Before the twentieth century, our understanding of the nature of reality was based on a discrete and concrete view of the world. Much of reality now described through the advanced knowledge of physics was a mystery. The phenomenon of heat (radiation), such as that from the glowing iron in a blacksmith's fire, had been common knowledge for thousands of years, yet no one could fully explain why or how it occurred. In classical physics, heat was theorized to be the result of gas emitted by the object or, according to an alternative theory, its oscillating atoms produced a series of waves (at the time yet to be scientifically proved). Neither theory, however, could explain the data physicists gathered through heat-emission experiments. At the time, scientists theorized that an ideal heat-absorbing material would produce infinite power when it reached a constant temperature (called *thermal equilibrium*). Yet this phenomenon, predicted by the oscillating atom theory and dubbed *ultraviolet catastrophe,* never occurred. Something was amiss in their theory.

Determined to properly describe how an object radiated heat, in 1900 the physicist Max Planck developed a mathematical constant called the *quantum of action*. He proposed an assumption that radiant energy exists only in discrete quanta and would be proportional to the radiation's frequency. Energy was known to exist as a wave function (see fig. 1.1).

Planck believed that at higher frequencies, radiation would be less likely to occur, thereby preventing ultraviolet catastrophe. If radiation frequencies in an object were quantized, its energy would be equal to a constant multiplied by the frequency. With this idea, Planck derived his famous radiation formula in which the average energy for a frequency (called a quantum) is defined as the energy of the quantum multiplied

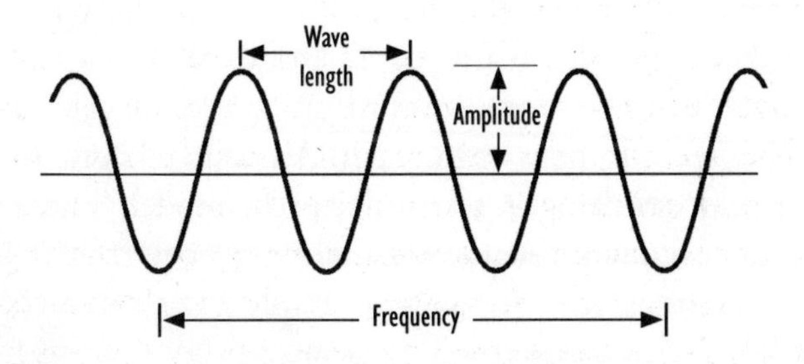

Fig. 1.1. Wave (energy) function

by the probability that such a frequency will occur. It was a simple equation where energy *(E)* equals a constant *(h)* times the frequency of oscillation *(v):*

$$E = hv$$

The constant of proportionality h is called Planck's constant, and its value is 6.625×10^{-34} Js.

What Planck was suggesting was that electromagnetic energy (heat, in the blacksmith example) is emitted and absorbed in discrete bundles, or packets, and that the energy emitted depends only on the frequency of the radiation. His breakthrough observation was that quanta, later to be known as photons, behaved like particles, not like waves as classical physicists believed.

In classical physics, particles and waves behave in completely different ways, and naturally occurring phenomena never display both. To the physicists of the time, it seemed like nonsense that Planck would describe an electromagnetic wave as a particle. But with the idea that every particle sometimes behaves as a wave and that every wave sometimes behaves as a particle, he identified a fundamental property of matter. For the observer, the physical world, at its most fundamental existence, is dualistic. Explaining this dual nature of reality became the holy grail of physics. Even today, the various explanations are a matter of contention.

From the insights of Max Planck on how heat was radiated, the complex and mysterious world of quantum physics was born. In 1918, Planck was awarded the Nobel Prize for physics after Albert Einstein and Denmark's Niels Bohr (1885–1962) successfully applied his quantum principle to the photoelectric effect and the atom. A few years later, Bohr, along with British physicist Ernest Rutherford (1871–1937), expanded on Planck's idea of the quanta and developed a model of the internal structure of the atom. According to their model, negatively charged particles (electrons) orbit a positively charged core (nucleus). It became the atomic model we are familiar with today, electrons orbiting a nucleus in the fashion of the planets circling the sun (figure 1.2).

During the late 1920s, the atomic model was enhanced further when German physicist Werner Heisenberg (1901–1976) collaborated with Bohr at Copenhagen's Institute for Theoretical Physics.

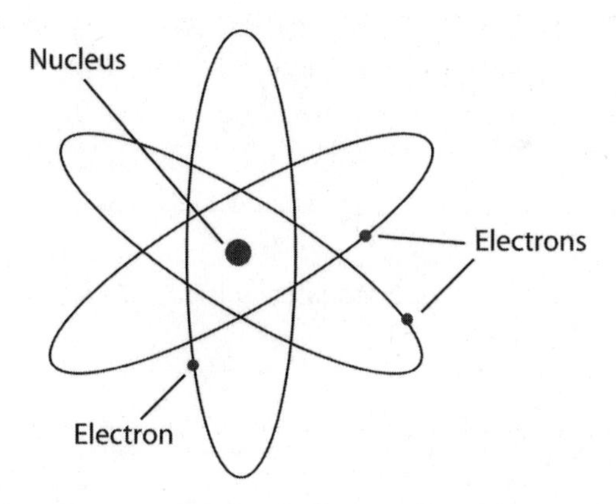

Fig. 1.2. Classic atomic model

According to Heisenberg's research, the atom's electrons do not orbit the nucleus in the way that planets orbit the sun, but exist virtually as a cloud of numerous possibilities. At the same time, another German physicist and 1933 Nobel laureate, Erwin Schrödinger (1887–1961), introduced *wave mechanics* into the atomic model to mathematically describe how electrons sometimes behave like a particle and sometimes like a wave.

$H\psi = E\psi$
H = Hamiltonian operator (kinetic and potential energies)
E = Energy eigenvalue
ψ = wave function, which represents each particle's position and time

The Schrödinger equation is a wave equation in terms of the wave function that predicts the behavior of a dynamic system—that is, the probability of events or outcome. The outcome is not a single result, but rather the distribution of a large number of possible results.

Schrödinger believed that when an electron is unobserved, it exists solely as a wave, but somehow also provides the probability of where a particle may be observed. A new "fuzzy" model of the atom was born, in which electrons do not orbit the atom's nucleus in the traditional sense but instead exist in a state of "superposition" in all possible locations within the confines of the atom's perimeter (see fig. 1.3). In other words, they exist in all places at all times until observed. They don't "almost" exist. They do exist but their position cannot be identified. It

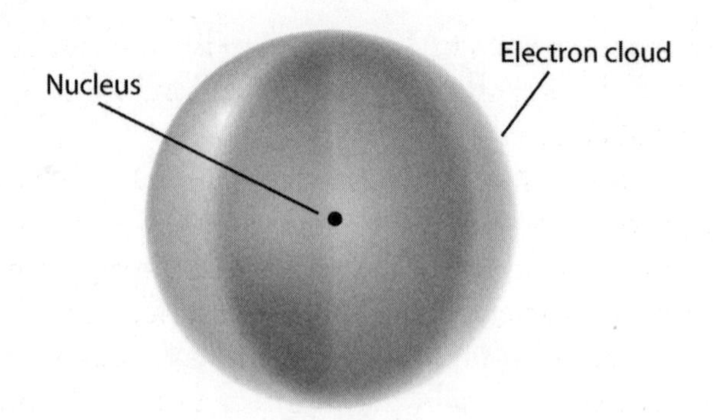

Fig. 1.3. Revised "fuzzy" atomic model (illustration by Dorothy Norton)

was discovered that the atom is a small nucleus existing in a cloud of virtual electrons. Most of the atom is simply space.

By introducing the observer performing the experiment into the equation, Schrödinger touched on a phenomenon that would later have philosophical and paradigm-breaking implications not only to the world of physics but also in science as a whole. According to Schrödinger, only at the moment of observation is the electron (a wave) seen as a particle. Thus, wave/particle duality became the mystery to be solved in the physicists' grand quest to explain reality.

Wave/Particle Duality

In the world of physics, energy is transferred through either a wave or a particle. Because all physical objects have mass and are forms of energy (we can think of an object as formed energy), mass must also have the ability to exhibit wave-like or particle-like behavior. This means that everything physical exists in a dualistic wave/particle state. The difficulty in understanding this mysterious duality is that concepts in physics, prior to Planck's idea of the quanta, were intended to describe a world of distinct and separate objects. Time, space, and all the cause/effect relationships we attach to physical reality exist only in the immediate, observable world. As physicists have discovered, however, they do not necessarily apply to the subatomic world.

The definition of a particle is derived from classical physics and based on a world of distinct and separate objects. Accordingly, a particle is a discrete, finite bit of energy that exists at a specific location with definite mass and charge. In contrast, a wave (which is not part of the immediate, observable world) can be infinite, does not occupy space, and may propagate itself to exist in all locations. How can these two behaviors that are associated with all matter exist simultaneously? It is a matter of perception!

During the 1920s, while Heisenberg and Bohr were further developing quantum theory, they realized that a new viewpoint had to be created to achieve a proper understanding of the quantum world. The classical view of a discrete world would simply not work. To accomplish this, they embraced the idea that the world is fundamentally *not* a collection of discrete objects, but an indistinct, unified world of energy where, at times, discrete objects are perceived. Heisenberg developed his wave matrix theory and Schrödinger his wave mechanics to explain their insights into the quantum world. Although slightly different in their approach, these two theories offered a more accurate description of the atom's structure than did classical physics. For their insights, both men received the Nobel Prize for physics.

What their theories state is that all matter (atoms, the "stuff" of all things) exists as a wave structure that we cannot directly see. What we do see is the localization of the wave structure with its release of energy, referred to by physicists as *state vector collapse.* The energy released is what physicists now call a photon (a particle of light). We perceive the released energy as a particle, even though it is really a wave. This occurs because that is how the human brain works.

It can be argued that the particle is not really a wave; through repeated experiments, however, physicists have established as fact that matter exists as a wave. Conversely, particle behavior has also been established as fact. The difference is that the latter is a function of and dependent on our biology, whereas the former is not. This makes perfect sense, since photons (particles) are the only things that we can see. We see the sun during the day, the stars or a streetlamp at night because they emit energy in the form of photons, which is what our retinas can detect. We see things around us, such as the moon and another person, because of an object's reflected qualities. The photon itself is what actually enters the eye as visual stimulus and provides a perception of our surroundings. So, in essence, we are always seeing a reflection (particle-like behavior) of the real phenomenon (wave-like behavior).

Proving Quantum Theory

Although initially excited about quantum physics, Albert Einstein was not enthusiastic about its direction, which was being guided by the ideas of Bohr, Heisenberg, and Schrödinger. Along with Boris Podolsky and Nathan Rosen, Einstein published an article in the May 15, 1935, issue of *Physical Review* called "Can Quantum Mechanical Description of Physical Reality Be Considered Complete?" Their argument, now commonly known as the EPR paradox, was based on a thought experiment that showed how a particle's velocity and location could both be known. It was supposed to provide the principle showing quantum theory to be incomplete.

This thought experiment envisioned a system in which two moving particles initially interacted with each other but later separated. According to the EPR paradox, even after separation, when one particle is observed, it will provide the state of the other particle *without* the other particle being observed. In other words, when the velocity of particle 1 is observed, the velocity of particle 2 will also be known. Likewise, with a second chance measurement, when the location of particle 1 is observed, the location of particle 2 will also be known. In this way, the state of particle 1 provides the state of particle 2 without disturbing particle 2. Because the velocity and the location of particle 2 can be known without disturbing it, both particle 1 and 2 must be regarded as real. Yet according to quantum theory, both particles 1 and 2 cannot be real at the same time.

What Einstein, Podolsky, and Rosen did in their thought experiment was first to establish the idea of reality. They did this by predicting with certainty an observable and physical thing (particle 2) without disturbing its state; in other words, if something real does exist, then it can be predicted, which in their example is particle 2. Accordingly, if two particles are observed together and then fly apart from a common event, both particles will behave in exactly the same way. The second particle's velocity and location can be predicted by the first. In this manner, Einstein, Podolsky, and Rosen posed a contradiction with Heisenberg's uncertainty principle, which states that the velocity and location of a particle cannot both be observed (measured) at the same time.

Not all physicists agreed with the logic behind the EPR paradox. In response to the allegations, Bohr claimed that such an argument was not allowed in quantum physics, especially the idea of reality that Einstein, Podolsky, and Rosen established as the fundamental logic of the paper. The experiment was only in thought, however, and the proof of who was correct lay thirty-seven years in the future.

In 1964, the British physicist John S. Bell (1928–1990) devised another thought experiment that could prove whether the observation (measurement) of particle 1 would affect particle 2. Bell also believed that quantum theory was incomplete and that there were hidden variables not accounted for in Heisenberg's equations. In Bell's thought experiment, two observers armed with particle detectors were to take independent measurements on a closed two-particle system. In each trial, either of the observers was free to choose a setting for his or her detector, with the assumptions that each measurement revealed an objective physical property of the system and that the measurements taken had no effect on either. Bell believed that particle measurements from either observer would limit the effect on the other particle's state after the particles were separated. In other words, each particle had its own properties before being observed (referred to as *local realism*), and changes in its properties would have a tendency not to affect the other particle.

According to quantum theory, the opposite would occur. The particle being measured would affect the other particle. Whatever choice one of the observers made on his or her particle detector setting, that particle would instantly communicate the change to the other particle. In physicists' terminology, this is called *quantum phase entanglement,* or *nonlocality*. This means that when the particles separate, even though their amplitudes also separate, their phases remain entangled. An observation on one particle will affect the phases of both particles. At the time, such a theory of reality was difficult to believe and was often referred to as "spooky action at a distance," a phrase coined by Einstein.

During the 1970s, several experiments were conducted to test Bell's idea and the EPR paradox of whether particles behave in a local or nonlocal fashion. In 1972, Stuart Freedman and John Clauser obtained results that showed Bell's idea and the EPR paradox to be false; using mercury atoms that produced violet and green photons, however, Richard Holt and Francis Pipkin obtained results in agreement with Bell and the EPR paradox. Their experiment was refined and repeated in 1976 by John Clauser at the University of California at Berkeley, and also by Edward Fry and Randall Thompson at Texas A&M University. Both experiments obtained results supporting particle nonlocality. In 1974, at Italy's University of Catania, Giuseppe Faraci, Diego Gutkowski, Salavatore Notarrigo, and Agata Pennisi devised another experiment to test Bell's idea. They found locality to be true. But in 1975, physicists at Columbia University repeated their experiment with opposite results. A year later, Mohammad Lamehi-Rachti and Wolfgang Mittig, at France's Saclay Nuclear Research Center near Paris, used pairs of protons (real matter) and found that Bell's theorem was false.[1]

Although some test results agreed with Bell and the EPR paradox, when the experiments were refined and repeated, they did not agree. As a result of this experimental data, physicists in general accept the Heisenberg/Bohr quantum model as a valid description of reality. Physical reality is nonlocal and, perhaps more important, the observer affects physical reality!

Quantum Reality

Without exception, the physical universe, including every human being, can be reduced into atoms of the ninety-two naturally occurring elements (hydrogen, oxygen, carbon, etc.). These atoms can also be reduced into electrons, protons, and a host of other subatomic particles called quarks, leptons, bosons, and several hundred other names. (Physicists have been very busy in their quest to figure out how the quantum world works and have discovered a host of subatomic particles.) Although they use such terms as these to describe the various aspects of the atom and its constituent parts, the simplest way to explain the atom's structure is that the atom is a grouping of wave functions we observe as particles. Simply put, the atom is composed of energy. I like to call it *configured energy*. The interesting question is—*What is energy?*

Energy can mean many things to different people. It can be food, gasoline, or batteries to power your MP3 player. Energy's essence, however, is movement or the possibility of creating movement. Energy can be ordered (called *mechanical energy*) or disordered (called *thermal energy*). It exists in potential (stored) and kinetic (used) forms. This concept of energy as movement is easy to grasp in the immediate, observable world. Every living organism requires energy (food) to move. What is moving is a living organism, even at rest. But in the quantum world, *the energy itself is the movement,* which creates a question and ultimately another paradox. Energy is the movement of what? There is no answer. Energy could be the movement of everything or maybe nothing, depending on how you want to look at it. Nonetheless, energy is what we are made from, since it is an established fact that everything physical is an uncountable number of atoms formed into various elements that are bonded into compounds and molecules, then integrated into biological systems.

Although subatomic particles are pure energy with a certain charge, the atom itself is not. It is energy that has somehow been arranged to exist as a three-dimensional physical object. An atom is matter. One way to think of the atom is that it is energy that has been shaped or configured. This is quickly understood by glancing at the periodic chart of elements—hydrogen has one proton, helium two, lithium three, carbon six, and so on all the way up to uranium, which has ninety-two protons. All elements are made from the same "stuff," but with varying numbers of electrons and protons. We know this is true from the nuclear weapons testing programs conducted by the United States and other countries since 1945. They illustrate Einstein's concept that energy and mass are really the same thing but exist in different states: $E = mc^2$, where energy *(E)* is mass *(m)* times the speed of light *(c)* squared.

This quantum reality—where everything exists as a wave function, which we perceive as an object with mass—creates for us perplexing mysteries. Since objects of mass are bundles of shaped energy, by what process and under what motivation were these different energies configured? What about space, the medium in which all objects of mass exist? How can something (an object with mass) exist in nothing (outer space)? What is the ultimate "cause" for matter to exist as it does?

The Zero-Point Field

Quantum physicists search for the identification and description of the most basic bit of nature in which everything that physically exists is made. According to some of the most recent research, some physicists believe they have discovered the fabric by which the vast expanse of the universe exists. At one time, it was believed that the vacuum of space was void of any energy. Now, there is compelling evidence that it is not. The evidence suggests that empty space contains an underlying sea of continuous, fluctuating energy at every point in the universe. Born through decades of research, this concept of a fluctuating sea of energy is referred to as the electromagnetic zero-point field of the quantum vacuum—*zero-point field* for short. It cannot be observed and theoretically exists even at a temperature of absolute zero (−459° Fahrenheit), where all quantum movement theoretically ceases. What *zero-point* means is that if the universe were to be reduced to a temperature of absolute zero—where all movement (energy) stops—a tiny bit of energy would still remain active.

The underlying idea of the zero-point field is from Werner Heisenberg's uncertainty principle. Heisenberg discovered that either a subatomic particle's velocity (movement) or its position (location) could be calculated with accuracy, but never both. This implies that if the velocity of the particle is known to be zero, then the possible positions of the particle have to be infinite. If such a state actually existed, in which a particle's movement became zero, the particle would disappear. Conversely, if it began to move again, it would reappear. But where did it go in between? The answer is a virtual existence.

Since time is continuous, and a forever-forward-moving aspect of physical reality, the uncertainty principle allows every real particle to be encased in an envelope of virtual particles. Although zero energy is maintained in the vacuum of space, real particles are constantly being produced, only to disappear into the virtual sea of particles from which they came. In essence, outer space is not a void but rather a vast expanse of virtual existence.

Such a concept is important because it keeps subatomic particles constantly vibrating, even those supposedly at rest. This provides a seamless existence to the areas between celestial objects of the universe, what we call outer space. Space is not a void; it is a very low-level electromagnetic field.

Evidence for the Zero-Point Field

In classical physics, electromagnetic radiation—radio waves, X-rays, and gamma rays, as well as visible light—are waves flowing through space. They are not waves of anything with substance, such as an ocean wave, but energy moving in a specific direction at a certain frequency, existing in a polarized state. For example, the microwave radiation emitted from a microwave oven moves through solid matter, causing vibration in all water molecules within its field, the area within the oven's outer case. The microwave (energy) causes friction between the water and the nonwater molecules, which results in heat. The oven's microwaves are, in fact, ripples in the state of an electromagnetic field. This is true whether the field is inside a microwave oven or the universe itself.

There are numerous wave patterns in an electromagnetic field, and because each pattern exists at the quantum level, they are also subject to the Heisenberg uncertainty principle. When an electromagnetic field is *quantized* (limited to a discrete set of values), each pattern is treated as a harmonic oscillator, vibrating at a fixed frequency and independent of amplitude (its maximum absolute value). So every pattern in the electromagnetic field must have an average minimum energy of $hf/2$ (h is Planck's constant and f is frequency). Although this is a very small amount of energy, the number of wave patterns is huge, and it increases as the square of the frequency. The end result is that the energy per pattern, although minuscule, when multiplied by its large spatial density yields a very high theoretical energy density. It is why quantum physicists believe that all space must be filled with electromagnetic zero-point fluctuations and can be thought of as a universal sea of zero-point energy.

In 1947, in a famous experiment with Ernest Rutherford, Willis Lamb discovered that the energy levels of hydrogen atoms in Schrödinger's equation were slightly lower than expected. This variance, seen as a slight shift (called the *Lamb shift*) of the electron's spectral line, was proved to be a result of vacuum fluctuations interacting with the hydrogen's electrons.

Experiments have also demonstrated that if two uncharged conducting plates are placed inside a vacuum facing each other, they will move toward each other without any apparent source of energy. This is called the *Casimir effect*. It has been a curiosity since 1948, when Dutch physicist Hendrik Casimir predicted its occurrence. In the Casimir effect, the vacuum fluctuations from the zero-point field cause two closely aligned plates to be attracted and move closer together. What moves the plates is the *radiation pressure* created by certain wavelengths of the zero-point field being excluded from between the plates. This reduces the energy density between the plates compared with that of empty space. The resulting imbalance pushes the plates together.

The God Principle

Although theories predict the zero-point energy field should exist, and experiments confirm that it does, physicists have been unable to answer the most fundamental of questions about it: Is it real or virtual? They have discovered that it is necessary for the mathematics of quantum theory, but is it required for physical reality? Furthermore, where does it come from?

Where the zero-point field came from is a point of contention. It could be the case that it was arbitrarily fixed during the creation of the universe and serves as a physical boundary. Or it could be the case that it drives the motion of all matter in the universe, where the motion of all charged particles (existing as mass throughout the universe) generates the zero-point field.

According to physicist Hal Puthoff, director of the Institute for Advanced Studies in Austin, Texas, the zero-point field is likely a result of charged particles moving throughout space. In turn, the particles themselves are moving as a result of the zero-point field. Puthoff defines it as:

> a self-regenerating grand ground state of the universe. In contrast to other particle-field interactions, the ZPF interaction constitutes an underlying, stable "bottom-rung" vacuum state that decays no further but reproduces itself on a dynamic-generation basis.[2]

Despite its mysterious origin, the field is an established fact in the physics community, with far more philosophical implications than mathematical. For some it is the *God principle.* To explain why this is so, a review of physics' second law of motion is required.

More than four hundred years ago, in *Principia mathematica philosophiae naturalis,* Isaac Newton first presented his three laws of motion. His second law defines a force to be equal to the differential change in momentum per unit of time, as developed by him in his calculus of mathematics. Momentum is defined as the mass of an object *(m)* multiplied its velocity *(v).* So the differential equation for force *(F)* is:

$$F = d(mv) / dt$$

If mass is a constant, and using the definition of acceleration *(a)* as the change in velocity with time, the second law becomes the more familiar product of a mass and an acceleration:

$$F = ma$$

Since acceleration is a change in velocity with a change in time *(t),* the equation can also be written in a third form:

$$F = m\,(v_1 - v_0) \,/\, (t_1 - t_0)$$

The above equation is the fundamental equation of motion. The important concept of the equation is that it works both ways. A force causes a change in velocity; and likewise, a change in velocity also generates a force. An object's velocity, force, acceleration, and momentum have both a magnitude and an associated direction. Scientists and mathematicians call this a *vector quantity*. The equations above are actually vector equations; they can be applied in each of the component directions. Simply put, Newton's equation states that force equals mass (or inertia) times acceleration.

Acceleration is also inversely proportional to mass for any given force. This impedes the ability to increase the speed of an object. The more massive the object, the more force is needed to get it moving. This is central to physics, and was deemed as something *not* provable but assumed to be an inarguable truth. It is also a very important and stabilizing part of the physical universe.

Alfonso Rueda, professor of electrical engineering at the University of California in Long Beach, has been researching the quantum vacuum and its applications since the 1990s. His research on the origin of inertia has received international recognition and has been commented on in *Science* and *Scientific American*. A high-level mathematician, Rueda discovered that an oscillation that is forced to accelerate through the zero-point field experiences resistance proportional to its acceleration. It was a breakthrough. Newton's second law of motion would no longer have to be assumed true. *It was true,* and Rueda proved why $F = ma$ by taking into account the zero-point field. This fundamental axiom of physical existence was reduced to electrodynamic properties of the quantum world.

In 1994, Bernard Haisch, Alfonso Rueda, and Harold Puthoff combined their research and published "Inertia as a Zero-Point Lorentz Field" in *Physical Review*.[3] By showing that inertia is a Lorentz force (a force that slows particles moving through a magnetic field), they demonstrated that inertia, which is a property possessed by all objects in the physical universe, is nothing more than resistance to being accelerated through the zero-point field. In essence, this magnetic field is a component of the field and reacts with charged particles. The more massive the object, the more particles it contains, and the greater its propensity to be held stationary by the field.

The implications of this are immense. This theory of inertia says that what we call matter is really an illusion of sorts. The background sea of energy opposes acceleration by gripping subatomic particles whenever something pushes on an object. So mass, viewed in a scientific way, is nothing more than a "bookkeeping device, a temporary place-holder for a more general quantum vacuum reaction effect," as Lynne McTaggart puts it in her book *The Field*.

The team of Haisch, Rueda, and Puthoff also realized that their research had a bearing on the most famous equation of all, Albert Einstein's $E = mc^2$—a recipe for energy, or, in reverse, a recipe for mass. They realized that matter is not really a fundamental property of the physical universe. There is only energy. Einstein's equation simply expresses the amount of energy necessary to create the *appearance* of mass. In other words, mass is not equivalent to energy. *It is energy*. Stated even more profoundly, there is really no mass, only charge. So why do objects appear to us as solid mass? The electromagnetic zero-point field of the quantum vacuum generates a force that opposes acceleration when you push on any material object. The zero-point field continuously pushes on everything and that is why matter appears to be solid and stable.

Looking at it from the quantum perspective, matter does not really possess a property of inertial mass. What possesses the property is the zero-point field, the property of resistance. So everything in the physical universe, including you and this book you are reading, is nothing more than tightly packed, configured energy reacting in and to the zero-point field. And since energy is defined simply as movement of everything (or nothing), there is really no other way to describe this quantum view of reality except by calling it the God principle.[4] It is a bold statement, but it fits, since it is the true nature of the physical universe, which possesses the characteristic that everything exists and that God is in everything. Haisch, Rueda, and Puthoff discovered a fundamental physics concept (the zero-point field) that explains the foundation of physical reality: the God principle. What's really interesting is that Einstein discovered it first, but believed it was an error in his equation.

The Human Factor

Physicists are known for their mathematical rigor in describing nature and its fundamental principles, the underlying laws governing the universe. Mathematics provides the logical structure by which theories are developed and predictions tested. Such thoroughness in physics theories has brought forth an impeccable track record of consumer applications. From polarized sunglasses to electronic watches, from nuclear bombs to television and now the cell phone, breakthroughs in physics have certainly changed modern society. As a result, physicists are the scientist's scientists. In a sense, they serve as science's high priests. They find out how things work, and all of us benefit from their research.

In the tradition of academic science, physicists have avoided speculation on how physics relates to the human experience. In doing so, they have avoided the most important factor of all, the human factor. How do we fit into the quantum model? What does quantum theory mean to the average person? These questions should not be reserved for professional philosophers and theologians, because deep down inside we are all philosophers and theologians. Everyone observes. Everyone thinks. Everyone has a perspective.

According to quantum physics, physically we are electromagnetic waves bound together to form atoms, which are then joined together to form living beings. And we exist in a giant sea of electromagnetic waves? I don't consider myself a wave, and I'm sure that you don't either. Quantum physics seems absurd.

Despite the falsification of the Einstein, Podolsky, Rosen paradox, there does seem to be an unaccounted-for variable in the quantum model. People take measurements in experiments, and in doing so they are observing what seems to be the deciding factor, the hidden factor. This is what Freedman, Clauser, Fry, Thompson, and all the other physicists discovered by testing Bell's theorem. Bell's theorem and its testing showed that the physical reality we see every day is based on the perception of the observer. Because of our brain's structure, everything around us is perceived as substantial and real, despite the fact that everything physical is nothing more than configured energy. Recognizing this leads to a profound depth of understanding, but it also leads to an eternal mystery. Why and how are we the observers? Everyone knows the answer. It is because we are conscious beings. But what is consciousness?

2

Consciousness

Mystery of the Observer

And the evidence seems to be pointing, as far as I can tell, to the conclusion that there is only one true observer in this whole universe.

PHYSICIST FRED ALAN WOLF

Our brain, like most organs, is composed of water, protein, and salt. Unlike the body, which is governed by the peripheral nervous system, the brain also includes neurons. For many decades, the primary method of information transmission in the brain has been understood to be chemical, although electrical transmission is also apparent. The best-studied aspects of psychophysiology include our two primary senses, the auditory and visual systems. Despite the well-delineated path of information through the brain, it is unclear how higher cortical centers are responsible for perception. For the psychophysiologist, it is even more mysterious how this perception is affected by previous experience and expectation, referred to as *top-down processing.*

How attention occurs, such as focusing on a task, is also a mystery to modern neuroscience. Although many brain regions have been identified as being important in tasks that require attention, it is not understood how voluntary action occurs in the brain. In fact, the only reason neuroscientists suspect that attention must have a biological basis is that "attention" is affected by some disorders, such as attention deficit hyperactive disorder (ADHD). Although no gross brain anomalies are known to exist with ADHD, subtle disturbances in neurotransmitter systems and functional differences in brain activity have been captured in laboratory experiments. Patients with frontal lobe lesions (typically stroke victims) also experience difficulties with attention.

According to University of Illinois clinical psychology Professor Dr. Greg Miller, there are difficulties in assigning behavior to specific parts of the brain. Miller explains by an example. Although a clock tells us the time of day, the inner workings of a clock do not create nor do they represent time. The clock's internal mechanisms do systematically mark seconds into minutes and then hours, but the concept of time does not depend on the mechanism. Thus, Miller asserts, it is incorrect to say that time, or even the telling of time, has a mechanical correlate. In the same way, Miller argues, behavior is not the result of a specific part of the brain. Memory and the brain suffer the same type of problem.

Although neuroscientists have been studying memory for many decades, there is still no integrated theory on how memory works. Some neuroscientists even doubt that memory exists in the brain. Nonetheless, for memory to exist in the brain there has to be a way that information is represented. The generally accepted theory is that patterns of synaptic connections create memory. However, there is a problem. Synapses, neurons, dendrites, and all the other aspects of the brain's bio-machinery of the brain continually change. Today, with evidence from new imaging techniques that examine brain function, researchers are seeing the brain in a new light. The brain is a dynamic organ, and it seems to be the case that behavior affects the brain just as much as the brain affects behavior.

Dreaming, although common and generally considered to be of no consequence, is another mystery to modern psychologists. A fair amount of research has been done on dreaming, but in the end, assumptions and a lack of definitions (consciousness, mind) prove too much for modern science. Research has revealed that certain areas of the brain are important during dreaming. Perhaps even more important, in this instance, is the lack during dreaming of brain activity in the frontal lobes, the seat of our higher brain functions. It has been suggested that the lack of logic in dreams is a result of inhibited prefrontal activity. In the dream state we still seem to be aware and conscious, but our higher brain functions, such as reasoning, are absent. This suggests that our consciousness is actually independent of our higher thought processes. For some, this may seem to validate the consciousness of animals, although the issue is more complex. It could be that thought processes do not create a conscious experience, but rather that consciousness is a priori, intangible, and ultimately a citizen of the abstract cosmos. In other words, consciousness exists whether or not physical things exist. However, consciousness can be only when there is something to identify with—the need for information from the brain, or, more accurately, the mind.

The brain/mind problem is undoubtedly the holy grail of psychology, neuroscience, physiology, physics, biology, and philosophy. Every person who has given a moment's reflection to his or her own awareness has realized the paradox in the mind/brain problem. The prevailing view among neuroscientists is that the brain and mind are epiphenomenal. In other words, the brain gives rise to the mind and does so in such a way that the result is greater than the sum of its parts. All mysteries of the brain relate to this single problem of the relationship between the mind and the brain. Even if all of the other mysteries of the brain are ultimately resolved, however, there still may not be a solution.

Describing Consciousness

Every year, millions of people receive general anesthesia before undergoing surgery. Doctors have discovered that a small percentage of surgical patients maintain awareness, despite being anesthetized, and listen in as the surgery is performed. According to the Illinois-based American Association of Nurse Anesthetists, the number of people who undergo general anesthesia and maintain some degree of consciousness could be as high as 40,000 per year.[1] This poses an interesting problem for anesthesiologists and surgeons. There seems to be no reliable way of determining whether a patient is actually unconscious. It also raises deeper physiological questions. What is consciousness, and where does it exist?

Being aware of one's surroundings is fundamental to the description of consciousness. It is the most basic characteristic that science attributes to organic life. All known animal life exhibits this feature. Human consciousness, however, encompasses more than this simple awareness. Aside from being aware of our surroundings, we have an internal perception of past, present, and future, and are also aware of our own thoughts. We are "self-aware," and we have the ability to view ourselves apart from nature, something that nearly all other animals cannot do. It is what distinguishes us as *Homo sapiens* and provides us with a unique perspective in the first, second, and third person. We can speak and think in reference to our self (first person, "I," "me"), to the person being spoken to (second person, "you"), or someone else (third person, "he/him," "she/her").

All dog owners know that their dogs are conscious and have a particular point of view. For example, my five-year-old black Lab's favorite hobby is eating (particularly sweets). Conan also likes to play tug-of-war and take the ball away, and to look out the window to watch what is going on outside. His favorite experience is having his belly scratched, and in a rudimentary way he shows emotion. When friends or family visit, he can barely hold his excitement, and when I am upset, he pulls his ears back and lowers his head. When he is hungry, he asks for food by an approach, then a soft mumble and a sideways movement that I interpret as a signal to follow him. I do, and once we are in the kitchen, he bumps the stove or refrigerator with his nose.

He also asks permission. Arriving home one day, I noticed he had something in his mouth. It was a Russell Stover crème egg, and by the look of the wrapper, he had held it in his mouth all day, but had not eaten it. He wanted my permission, and he received it.

I find Conan's behavior quite curious. He is generally a well-behaved and conscientious dog, but I did *not* directly and rigorously train him to behave in any particular manner. It seems as if he was able to take environmental cues, including my behavior, to figure things out for himself.

Whether or not Conan is self-aware—and I am unsure about this—it is apparent that some animals, with brains similar to ours, are self-aware even though they lack the language to communicate it. We humans recognize ourselves in a mirror at about two years of age. Monkeys cannot do this but chimpanzees can. More recently, it has been discovered that dolphins have the ability to recognize their reflection.

What is clearly different about humans is that we have the ability to abstract (consider abstract concepts, such as justice, apart from concrete existence), a true sense of self-awareness, and the capacity to reflect on the past as well as to contemplate the future. Every person knows that someday he or she will die. Although my dog Conan may accept that he is different from me, because he does not have the capacity for abstraction, he cannot comprehend his relationship to the world. As a result, he views himself as a part of nature and is driven more by instinct (what I prefer to call innate consciousness) than thought, and he will likely never contemplate his life.

It seems to be the case that our unique sense of awareness is not a function of brain size. With its greater focus on the auditory system, as opposed to the visual system, the dolphin brain is actually larger than a human's. Yet dolphins clearly do not have the capabilities humans do. So it must be the case either that dolphins are a lot smarter than we think they are or that something particularly interesting is going on within the human brain. In terms of production, the difference between human and animal brains far exceeds the small anatomical and physiological variations—a difference that has yet to be explained.

Difficulties in Defining Consciousness

Consciousness is obvious because each of us is the subject of experience. We enjoy perceptions and sensations, as well as suffer pain. We also entertain ideas and consciously deliberate. Consciousness is obvious. Yet how is it possible that we, as physical bodies in the physical world, experience consciousness? It is the most mysterious feature of our lives. There is no clear definition for consciousness.

In some instances, consciousness refers to an individual's waking state as opposed to the sleeping or unconscious state. In others, it is referred to as an organism's state of "being alive." In still another way, consciousness refers to a deliberate action, such as a "conscious attempt." Consciousness can mean awake or alive, or it can refer to intent.

In defining consciousness, there are also difficulties in explaining certain human abilities related to consciousness: the categorization of objects or moods, reactions to various experiences, the focus of attention, and the deliberate control of behavior. How do these aspects of the human experience—in other words, our *thought world*—emerge from a physiological state within the human brain?

An individual's thought world is easy to describe in experiential terms, but difficult to explain as a biological process. If somebody says something that makes you angry, all you have to say is, "He said something that made me angry." Everyone understands, since an event such as this, and others like it, is a part of everyday experience. But for the neuroscientist, explaining the biological mechanism of *why* we become emotional—where emotion emerges from the brain, given certain external circumstance—is an almost impossible task, aside from saying that certain neurotransmitters such as acetylcholine, serotonin, dopamine, melatonin, and the endorphins send electrochemical messages from one nerve cell to another. This is because emotional reactions are based on experience. How lifetime experiences affect the secretion and transmission of biochemicals is not easy to capture with the scientific method. It is cumulative experience that shapes how we think and act.

Experience is fundamental to consciousness. Every day we experience. We are bombarded with external stimuli (seeing, hearing, smelling, tasting, and touching), as well as internal stimuli, the things we dwell on—the great time we had last night, the upcoming game tomorrow, and ways to deal with the ongoing behavioral problems of our offspring.

Our bodies are designed to cope with various sensations. We experience visual sensations such as darkness and light, as well as color and the perspective of depth in a visual field. We also experience sound, touch, and smell: the harmonic beauty of a symphony orchestra, the smooth, soft sheets on a bed, the tantalizing aroma of a favorite meal. All these are states of experience united by the unique quality of "being" in them. We can also be in more than one state at the same time.

Experience also contains a subjective element dependent on the state of mind of the individual. For example, a group of four people attend a concert. Two of them love it, one hates it, and the other likes it. Yet all four are big fans of the performing artist, and the two that enjoy it most do so for entirely different reasons.

We experience and perceive the world around us and think in such a complex way that it would be impossible to duplicate the essence of human thought and behavior in a computer algorithm, despite the fact that computers are superb at complex computations. The eye, for example, allows light (in the form of photons) to enter through the lens (the cornea) and strike the retina, where photons interact with certain cells to create electrical impulses. These impulses are then carried through the optic nerve to the vision center (occipital cortex) of the brain, where the image is interpreted and the phenomenon known as perception is created.

When photons enter the eye, they interact with photoreceptor cells of the retina to create an action potential. (An action potential is a charge that travels down the axon of a neuron.) The action potential then travels down the optic nerve (referred to as an axon). The left and right optic nerves crisscross as they pass the optic chiasm and continue to the lateral geniculate nuclei (LGN), which are located in the lateral medial sections of the interior cerebral cortex. This is accomplished in such a way that the left vision is processed by the LGN in the right hemisphere and the right vision in the left hemisphere. Visual information is sorted out in the LGN and passed into the optical lobe, where the primary visual cortex resides. Visual information is then passed to other areas of the visual cortex and other parts of the brain. Precisely how it is turned into perception, however, remains a mystery.

Another difficulty in defining consciousness is the nature of knowledge itself, in which biological systems exist as subjects of experience. How is it explained that something exists to entertain a mental image or to experience emotion? It is an established fact that experience is derived from a physical base, but the explanation of why and how is lacking. Why should organic life produce a certain quality of inner life—or, for that matter, a lack of inner life? Scientifically, it seems unreasonable that it should, but it does.

For the human being, perhaps the biggest difficulty with defining consciousness is understanding what generates a sense of self in the act of knowing. Right now you are reading this book and are concerned with extracting meaning from these words. You are also, however, generating a pervasive thought in your mind that it is you, as opposed to someone else, who is reading. In the background of all the sensory input from our daily lives, there is a never-ending realization: "I am me." It is a presence that is so subtle we never really pay much attention to it. Yet without it, how would anyone know that his thoughts actually belong to him? Without this presence of "I am me," there would be no you.

The effect behind this problem lies in the individual perspective each of us has that is built from the proprietary knowledge we hold in our minds. Each of us is an observer, perceiver, knower, and thinker. Although the nature of consciousness is difficult to know, we can be sure that there is no homunculus, no "mini-me" who sits inside the brain in charge of knowing. Although we often perceive that there is such a creature, it is nothing more than a reflection of brain function by consciousness.

Consciousness can be understood as the occurrence or perception of experiencing by a living organism and the awareness of the simpler events an organism encounters every day. Clearly, conscious experience is a widespread phenomenon occurring at all levels of animal life, despite the fact that it is difficult to determine what provides evidence of consciousness in simpler life-forms. Regardless of the variety of life-forms, the fact that an organism has conscious experience at all means that there is a point of view for that organism. The organism observes and is an observer, as well as a participant in experience.

In humans, there is an additional element to consciousness in which perception and deliberation occurs. This is what all of us refer to as the mind.

We can describe consciousness, but it is nearly impossible to define, because it cannot be objectively demonstrated. How can we objectively describe the color blue to someone who has been blind since birth, or the sound of a passing car to someone who was been born deaf?

The Mind

It is an established fact that the brain is an organ that receives and interprets sensory input and regulates bodily functions. When we refer to the brain, we are referring to the physical organ made up of special gray-and-white tissue called *matter*. The *mind,* however, is something different, and is not a physical object that can be removed (postmortem, of course) and inspected.

Cognition, or thought, that we say occurs in the mind can mean many things. In general, it means to formulate, to reason or reflect, as in "thinking the matter through." It also means to decide, judge, regard, believe, expect, or devise. There are also the mental aspects of intention, memory, concentration, and imagination. All these descriptions—ways of thinking—are functions of the mind. When we refer to mind we are referring to the sum of all thought, perception, emotion, will, memory, and imagination. In essence, we are referring to a concept.

The mind has the capability of introspection, and is closely linked to perception. It is where all experience is internalized and becomes perception, shaped into what we call a point of view. Within the mind, a person's "inner life" is composed of the individual's unique identity and personal qualities.

This description of mind raises questions. What is the extent of the relationship between the brain and the mind? What is the relationship between consciousness and the mind?

Difficulties in the scientific quest to answer these questions are commonly known as the mind/brain problem. There are really only three choices in describing how the brain and mind interact. One is that the brain determines all behavior, as well as subjective experience. Another is that the mind (mental events) brings forth brain activity just as much as brain activity brings forth mental events. And the last is that subjectivity does not exist, and what we think is subjectivity is nothing more than a by-product of the brain's processing of external stimuli. Science, which has traditionally believed that the mind is located within the brain, has yet to provide conclusive proof as to which of these statements is correct.

The problem is that the brain is exceedingly complex, and the mind is an abstraction based on how we view ourselves. For instance, science can track the visual system from the eyes through to the cerebral cortex, but how perception occurs within the brain remains a mystery. External stimuli can be related to neuron function, but what about private thoughts that require no external stimuli? According to recent studies, the brain cannot distinguish between external stimuli and memory. Internal activity (cognition) and external stimuli appear to manifest the same type of activity within the brain. This has led to an interesting new theory describing the nature of the mind as nonlocal.[2]

Nonlocality of the Mind

In the previous chapter, we learned that according to physicists, when subatomic particles interact, they form a single state of being that remains after the particles become separated. This naturally occurring phenomenon, called *quantum phase entanglement*—popularly known as *nonlocality*—has led some scientists to believe that there are quantum characteristics to the universe that may be the foundation of the immediate, observable world. The hypothesis: The brain functions at the quantum level where nonlocality plays a vital role in how the mind and brain interact. Although new from a scientific perspective, the philosophical idea of the mind's nonlocality has existed for some time.

The French philosopher René Descartes (1596–1650) is famous for stating, "I think, therefore I am." With these words he expressed the most primitive yet profound statement concerning reality, with which any person would likely agree. Werner Heisenberg agreed with Descartes's statement and noted that the scientific method has its constraints: "In a very general sense [it is] always an empirical question how far our concepts can be applied."[3] Descartes's much quoted phrase hints that the most fundamental level of reality is thinking, and that "being" is its consequence. In other words, "being" is the result of the process of thinking. It seems counterintuitive to the common experience. How can thinking possibly be the foundation of reality?

If one assumes a priori (meaning that something is knowable without appealing to any particular experience) that thought is nothing more than biological processes in a region of the brain, then it follows that infinite thoughts are impossible. Can an infinite number of thoughts exist?

To address this question, peculiarities of experience and workings of the mind need to be addressed. For example, are memories, such as those of your childhood, still a part of your mind? If you wake up one morning from a dream but do not remember it, does the dream exist? You are reading this book a word at a time, so how can you follow what the book is about? If the universe were suddenly to disappear, would the mathematics we use to describe it also disappear? In the immediate world, "seeing is believing," so when four people view the same car in an auto show, the car is real. But if those four people have the same idea, does that mean the idea is real?

Rudy Rucker, professor of mathematics at San Jose State University (now retired) and author of *Infinity and the Mind,* views the mind in a unique way. He theorizes that the mind exists as a point, something similar to an eye, that moves about in the abstract world of all thoughts in an extradimensional (or nondimensional) area he refers to as "mental space." Just as the body shares the universe with other physical beings, the mind shares in an abstract universe of thought and is actually a nonlocal phenomenon. Just as two people can occupy the same space in the physical world (for example, when my wife and I are watching TV on the sofa we are in the same place at the same time), two different minds can also share the same position in the abstract universe, meaning that they have the same "mindset."[4]

According to Rucker's theory of the mind, concepts and ideas exist in the abstract part of the universe whether or not someone is actually thinking of them. For example, for mathematicians to discover a new principle, it has to exist prior to its discovery. Otherwise, the principle is being created, meaning that it is imaginary, as opposed to being real.

In 1931, Austrian mathematician Kurt Gödel proved that mathematical principles are real and not imaginary. He showed that in any axiomatic mathematical system, there are propositions that cannot be proved or disproved within the axioms of the system. Whatever axiomatic system calculations are based on, there will always remain true statements that lie beyond the system's reach. In other words, mathematics is not a finished object, and the abstract world is already out there, preexisting in some nonphysical state. In essence, he rewrote the rules for all of science in much the same way Einstein did in his theory of relativity for classical physics. Gödel's *incompleteness theorem* showed that nothing is certain concerning the universe, and that the rational mind may never be able to access the ultimate truth.

Although the scientific community is generally not convinced that our minds are entangled through quantum entanglement and act in ways as a single mind, quantum entanglement is actually a subtle and rare event that quickly vanishes. The notion that all thoughts are already "out there" in an abstract universe does have its evidence. Have you ever called someone, and as soon as you say hello, she remarks, "I was just thinking about calling you"? What about hunches or premonitions? Although science has yet to take such phenomena very seriously, nearly everyone has had an inexplicable bad feeling about some upcoming date or event, and subsequently found that not participating in that event was lifesaving.

The New Science of Consciousness and Mind

In the branch of psychology called cognitive science, there are two primary theories explaining consciousness and how it arises from the brain. One theory states that thoughts exist at various levels of sensory interpretation and that consciousness forms in an intermediary level between sensory input and thought. In other words, the brain gives rise to thought based on sensory input, which brings about consciousness. According to this model, the majority of thought is believed to occur at an unconscious level where, through the mind, the observer experiences only a small portion of all thoughts.[5]

Another, more prevalent theory takes a more complex approach. Called the *emergent-connectionist model,* consciousness is viewed as the result of hierarchically integrated networks that physically make up the brain. In this model, thought and consciousness exist as a result of complex biological formations that begin with the atom. Each new level emerges from the previous level: atoms build into molecules and biochemical structures, which become nerve impulses, neurons, and neural assemblies. Finally, as the aggregate of all levels, the brain emerges, and with it comes consciousness. In the emergent-connectionist model, the whole is greater than the sum of its parts and exerts control over subordinate structures.[6]

Both these models explain that consciousness arises from the brain. Their shortcoming is that they do not describe the process by which consciousness occurs. Certain characteristics of consciousness exist, most notably perception, which is difficult to understand and explain through conventional neuroscience models. A new view on how the brain works would have to be put forth for researchers to continue investigating the process of brain-emergent consciousness. Referred to by some scientists as *quantum consciousness,* this new area of scientific exploration has attracted the interest of a small but growing number of scientists.

One new model of quantum consciousness, called *orchestrated objective reduction of quantum coherence in brain microtubules* (orchestrated OR for short), theorizes that quantum states inside the neurons give rise to thought and perception. Events that lead to consciousness occur in binary fashion inside *microtubules,* a neuron's substructure.

Microtubules, once believed to be without an internal structure, are hollow, crystalline cylinders twenty-five nanometers in diameter. They are composed of hexagonal lattices of proteins, known as tubulin. All tubulin molecules contain an electron that exists in one of two states (which can be thought of as analogous to the 0 or 1 state of a data bit in today's computers). The alternating quantum states (phases) of a single tubulin form a "bit" of information; the aggregate of quantum tubulin, in effect, forms consciousness.

Stuart Hameroff, professor of anesthesiology and director of the Center for Consciousness Studies at the University of Arizona in Tucson, cocreator of the orchestrated OR model, believes that the quantum world and the immediate observable world are in some way bridged by consciousness.

To develop the orchestrated OR model, Hameroff teamed up with the theoretical physicist Roger Penrose during the 1990s. At the time, Penrose was working on a theory of quantum gravity in an attempt to explain the relationship between the quantum and the classical worlds. For Penrose, it was possible that quantum gravity was the force that collapses the state vector of the wave function—the phenomenon that transforms an energy wave into a particle for our perception. Penrose called this self-collapse wave "objective reduction" and theorized that it occurs when a critical threshold of quantum gravity is attained.

With this theory, Hameroff and Penrose proposed that objective reduction is based not on mathematics but on biology—specifically, the brain—and that stimulus is transformed into perception through the processing of consciousness in microtubules and associated structures within neurons.[7]

According to Hameroff, "Consciousness is a self-organizing process on the edge between the quantum world and the classical world, and a connection between biological systems and the fundamental level of the universe. Orchestrated OR is consistent not only with neurobiology and physics, but with spiritual traditions such as Buddhism, Hinduism, and Kabbalah."[8]

You are thinking that this sounds more like philosophy than science. It is, to a point, but with any new theory there must be an insight, a suspicion, a philosophical foundation, to form a hypothesis to investigate. In essence, the rest of science is catching up with the physics breakthroughs of the 1930s by adopting the philosophy of those who created quantum theory. Although they have been remembered for their quantum mechanics, Planck, Heisenberg, Schrödinger, and Bohr were just as much philosophers as they were physicists.

Philosophical Foundations of the New Science

These creators of quantum physics were well aware of the impact their work in physics would have on the rest of science—indeed, the rest of the world. In 1933, Max Planck, discoverer of the quanta, wrote *Where Is Science Going?* In writing *My View of the World,* Erwin Schrödinger turned to the Eastern philosophical tradition for his insights—consciousness and monism: that is, everything that exists is a single entity being expressed through the physical universe. In 1934, Niels Bohr wrote *Atomic Theory and the Description of Nature,* explaining the philosophical consequences of the emerging new science and how quantum principles describe nature. Werner Heisenberg published *Philosophical Problems of Quantum Physics* and *Physics and Philosophy: The Revolution in Modern Science.* The latter hinted that ancient principles of "natural philosophy" were more relevant to modern physics than was Newton's classical physics. These great scientists, all Nobel Prize laureates, were saying something profound and philosophical about the world we observe.

Perhaps the most celebrated (and mystical) statement ever uttered about quantum physics is Heisenberg's "Atoms are not things"—which implies that the objects we see are also not really things, since all matter is composed of atoms. What Heisenberg was alluding to is that quantum theory denies an objective existence for the atom. Real objects have a location in space *and* a meaningful momentum. According to his uncertainty principle, atoms can have one or the other but never both. The strangest aspect of what Heisenberg was describing is that human observation plays a role in defining reality. This is what Hameroff and Penrose are now trying to describe occurring within the brain's neural microtubules. Bohr agreed with Heisenberg. Only the intervention of the observer makes the "fuzzy" existence of the atom concrete. With the idea that the observer is an integral part of the scientific process,

Heisenberg concluded that nature is not quite what scientists conclude it is: "This again emphasizes a subjective element in the description of atomic events, since the measuring device has been constructed by the observer, and we have to remember that what we observe is not nature itself but nature exposed to our method of questioning."[9]

> The conception of the objective reality of the elementary particles has thus evaporated not into the cloud of some obscure new reality concept, but into the transparent clarity of a mathematics that represents no longer the behavior of the particle but rather our knowledge of this behavior.[10]

The renowned British theoretical physicist Stephen Hawking also agrees:

> The theory of quantum mechanics is based on an entirely new type of mathematics that no longer describes the real world in terms of particles and waves; it is only the observations of the world that may be described in those terms.[11]

So what is an atom (of which everything is made)? "In the absence of observation, the atom is a ghost," writes theoretical physicist Paul Davies in *God and the New Physics*.[12] Atoms are the way we observe and perceive the world. They are ideas, or concepts, or maybe just a bit of information or a bit of knowledge. Nature, it seems, is something else, something beyond the grasp of reason. The way quantum mechanics works prompted Bohr to proclaim, "Anyone who is not shocked by quantum theory has not understood it."[13]

At that time, science lacked the proper conceptual framework to describe the empirical evidence and how it relates to human experience. Heisenberg believed something was missing: quantum theory "does not introduce the mind of the physicist."[14] So, he implied, to take into account the processes of thought, another concept would have to be introduced.

It took another seventy years after Bohr, Heisenberg, and Schrödinger created quantum mechanics for science and society in general to catch up. The new concept that Heisenberg was talking about is what science has always insisted on removing from the method—the way the human brain/mind works and its effects: consciousness.

The "Reality Quest" Explosion

Today, it seems that society is beginning to understand the philosophical implications of quantum mechanics. A great number of books exist on the subject, such as *The Quantum World: Quantum Physics for Everyone* by Kenneth Ford and Paul Hewitt, *Quantum Reality: Beyond the New Physics* by Nick Herbert, and *Taking the Quantum Leap: The New Physics for Nonscientists* by Fred Alan Wolf—and too many others to list. Amazon.com has 5,366 titles that contain the word *quantum*.

Books on consciousness are of a similar nature—*Consciousness and Experience* by William G. Lycan, *Consciousness Explained* by Daniel C. Dennett, and *The Quest for Consciousness: A Neurobiological Approach* by Christof Koch, to name a few. Amazon.com lists 4,275 books with *consciousness* in the title.

One of the more notable researchers into consciousness is Francis Crick, codiscoverer with James Watson of DNA's double helix structure. In 1994, Crick published *The Astonishing Hypothesis: The Scientific Search for the Soul* (Simon & Schuster: New York). In it, Crick seeks to understand the nature of consciousness through visual awareness, and how brain activity combines discrete elements into meaningful images.

There is also the popular movie *What the Bleep Do We Know!?* (2004) starring Marlee Matlin, with its mixture of biology, religion, and quantum physics. In this docudrama, scientists discuss how reality is subjective, and based on how we think as opposed to what we see. The filmmakers took their inspiration from the new scientific view and suggest that this view is "in agreement with what mystics have been saying for centuries."[15]

The quest for all these researchers and authors is to discover what lies behind our reality and the defining core of the human experience—the mystery behind the observers that we are. In other words, it is a quest for the fabric of reality and how it involves consciousness, mind, and brain.

Some scientists believe consciousness may be just as fundamental as matter and energy and that it extends beyond the individual body and brain. Yet others believe that it arises solely from the brain as a result of electrochemical activity. Marilyn Schlitz, anthropologist and research director of the Institute of Noetic Sciences in Petaluma, California, believes there is compelling evidence for consciousness to include something beyond individual awareness, and that it is impossible for all the religious mystics throughout history to be wrong.

Theoretical physicist and author Fred Alan Wolf suspects that consciousness may be the true reality and that matter, the physical universe, is like an illusion. Physical materialism (there is only the physical) and dualism (there exists the physical and the spiritual) have been the models of choice in the Western world. Wolf believes it is possible that both these models are incorrect and that a dimension (or nondimensional state, in my opinion) exists where all of reality, abstract as well as concrete, occurs—an indescribable realm out of which the physical world pops into existence.

For those who believe consciousness occurs on the level of the quantum wave function collapsing into a single state, the universe in every place is "conscious." But it also seems likely that consciousness can be reduced to brain states, since consciousness exists *only* from the viewpoint of the organism that experiences it, which includes the objectivity of brain states but also the subjective state of mind. There is an even deeper mystery, however.

Mystery of the Observer

Consciousness and mind are so integral to the human experience that we take them for granted. Apart from scientific inquiry, rarely do we question their significance. This leads me to the conclusion that consciousness and the phenomenon of experience are a result of brain activity. Mind, on the other hand, although associated with brain activity, is not so obvious a deduction. Some scientists believe the mind's nature is nonlocal; others believe it, like consciousness, is a direct result of the brain. The difficulty in ascertaining where the mind exists stems from a lack of understanding on how the brain becomes the mind—how vision and hearing generate perception. This is a mystery. But an obvious deduction about the mind and how it is associated with the "self" is that it relies on memory. Without memory, not knowing what happened yesterday, or any other day, for that matter, we would live the life of animals. Self-awareness and thought beyond the current moment would be impossible.

Your first memory can be considered your first thought and can be aligned with the emergence of self-consciousness. My first memory was formed in the summer of 1963, at the age of two, when I discovered that roses have thorns. After that, I began to remember consistently even the more mundane things of life, such as playing in the backyard with my sister and watching *Batman* on TV. For the next five or six years, although I remember events, I have no recollection of thinking in any particular way. In a sense, I was oblivious to my own thought life. I was naturally, instinctively doing things rather than thinking. All that changed somewhere around the age of seven or eight.

I remember asking myself, *Why am I me?* and *Why am I not someone else?* There was no desire to be someone else. I was simply curious. While contemplating the question, I also remember feeling strange, bewildered almost, because it was (and is) one of those unanswerable questions in life. It was as if I realized for the first time that my perspective, thought life and all, was unique to me. I was an observer of my own thoughts just as much as an observer of the world around me. It was (and still is) a mystery.

Ask yourself the question *Who am I?* If you answer with your occupation, you are not answering who you are but what you do. If you answer with what you look like, you are describing the physical characteristics of your body. If you answer, *I am a person,* then you are answering *what* you are as opposed to who you are. The best answer to the

question of who you are sounds silly, but it is correct: *I am me*. Meaning, you observe the world from a unique and exclusive perspective and are aware of the fact that you do. Everyone does, and in a sense everyone really is "me." (We will return to this concept later.) Such thoughts make for strange logic. Nonetheless, apart from our perspectives, none of us really knows who we are! Why this should be the case is a mystery. *It is the mystery of the observer.*

In psychology, there is a concept called the *central executive* who runs the show, enabling us to think and make decisions. The concept of the central executive is linked to working memory. (Without memory, it would be impossible to make any decisions at all.) Exactly where this "executive" is who deliberates and makes decisions, however, is unknown. All that can be said with any confidence is that the central executive is the conscious self. But who or what could the self be? With this question, philosophy must come forward to present a hypothesis that will later be investigated from a physicist's point of view. Is the nature of "the self" (the observer) abstract or physical?

We exist in a physical reality and have bodies that must be sustained. So we eat, sleep, and endeavor to keep ourselves free of disease as a matter of survival. The brain is fundamental to our physical existence. Everyone knows this. So does the brain create the self, and could it simply be the size of the human brain that gives us our ability to think in ways far different from those of animals? If it were, dolphins would be a higher order of being than we are, and they are not. It is the mind and the workings of the central executive, not brain size, that distinguishes us as sapient beings—the functioning and organization of which has engineered the building of civilization. In this way, the nature of the self can be viewed as abstract, and is the human consciousness that involves thought, perception, emotion, will, memory, and imagination.

In this mystery of the observer, although all people must share a global physical reality, each person perceives reality in his or her own way. We all see the same things (visual stimulis), but that does not mean we perceive (interpret) them the same way. This directly affects our idea of truth. For example, let's say Joan and John are watching antiwar protesters in New York City. Joan strongly believes in the war in question, so she disapproves of the protesters even to the point of calling them treasonous. John, in contrast, hates the war and perceives the same protesters to be patriotic heroes. Joan's and John's eyes do not see the world differently; *the observer does through the mind,* and in a sense each person creates his or her own reality through interpretation. The story of Nobel laureate and mathematician John Nash, although extreme, is a case in point.

In the Ron Howard film *A Beautiful Mind* (2001), a fictional drama inspired by a scholarly biography of Dr. Nash, the mathematician suffers from schizophrenia, a mental disease that often alters the perception of reality. He sees and talks with people who, according to everyone else, do not exist. When the movie begins, you, as the observer, are convinced that what Nash is experiencing is real, but it is not. Thirty minutes into the film, Nash's wife, Alicia, has him forcibly hospitalized. For me, it took another ten minutes to be convinced that, according to the film, he was truly ill.

Although Nash was never cured of his condition, he learned to distinguish *his reality* from the realities of others, which enabled him not only to function, but also to succeed in his chosen academic field. His story is a testament not only to the difficulties associated with the brain/mind problem and the mystery of the observer but also to the courage and will of the human spirit.

There are various theories about what causes schizophrenia. Most medical researchers believe it has a physical basis. Chemical neurotransmitters within the brain are what provide us with mood, emotion, and other mental processes. Studies have shown that schizophrenics exhibit subtle abnormalities in brain structure, such as the enlargement of ventricles (fluid-filled cavities), a decreased size of certain brain regions, and decreased metabolic activity in certain brain regions. According to the National Institute of Mental Health, the brain formation that causes schizophrenia may occur during fetal development.[16]

Again, this leads to the conclusion that the seat of consciousness is in the brain. Yet how can neurons firing in the brain create such a complexity of thought in the mind? How can something such as abstract thought be contained in a physical organ like the brain? It is an established fact that the brain processes external stimuli, and it is believed that electrical signals, through a configuration of neural networks, compose abstract thought. What is the mechanism? Attempting to understand the relationship between the brain and mind leads into a vicious circle between philosophy and materialistic science.

The bigger problem is the nature of "me," the observer. Ultimately the "self" interprets the electrical signals in our brains, but science has yet to find out what this "self" really is.

Brain Physics

Physics has an impeccable track record of discovering fundamental principles of nature and applying those principles for the benefit, and sometimes detriment, of mankind. The application of quantum theory is not only responsible for nuclear fission but also the reason for the technology society enjoys today: integrated circuits, transistors, computers, televisions, and so forth. Since quantum mechanics agrees perfectly with experimentation, the scientific community has generally accepted it. According to Stephen Hawking, "It has been an outstandingly successful theory and underlies nearly all of modern science and technology . . . and is also the basis of modern chemistry and biology."[17]

Given such a record of accomplishments, I have to ask: What role does physics play in the quest to solve the mind/brain problem and the identification of the observer? Can physics tell us something more about the human condition that has remained the territory of biology and the more esoteric areas of religion and philosophy? Niels Bohr believed it could. In his book *Atomic Theory and the Description of Nature* (University Press, Cambridge, England, 1934), he suggested that mechanisms of the brain are so sensitive and delicately balanced that they should be described by quantum theory. Bohr also believed such an approach would help describe what we like to call "free will."

The new physics that Heisenberg, Schrödinger, and Bohr developed also created a revolution in thought for a new generation of physicists. By the early 1970s, nonlocality had passed the experimentation phase, which helped muster support for the general idea of quantum entanglement, the so-called Copenhagen interpretation of quantum theory. As a result, the observer and the observer's systems of measurement became an issue, as did the conscious act of observation. How we observe and are conscious was beginning to be taken seriously as a physical phenomenon worthy of investigation.

In the Hameroff/Penrose model of consciousness (orchestrated OR), the microtubule ground state is tied to state vector collapse, from which consciousness emerges as a function of the brain. According to Penrose, gravitational differences among various potential states force the system to collapse into a single state when observed. For the model to work mathematically, Penrose had to modify the Schrödinger equation slightly by adding a term to bring about state vector collapse. Although equation manipulation is a common practice in the world of theoretical physics, according to physicist Evan Harris Walker, past attempts to modify the Schrödinger equation in such a way have typically suffered from incorrectly predicted results.[18] With the elusive, unknown disturbance that brings about state vector collapse, Schrödinger's equation seems to have stood the test of time. Despite its mathematical difficulty, the Hameroff/Penrose model with its research into microtubules has made a "quantum leap" into new vistas of how the brain works.

Other physicists, such as Henry Stapp of the University of California's Lawrence Berkeley Laboratory, have also proposed ideas about the quantum brain. In *Mind, Matter, and Quantum Mechanics* (Springer, 2004), Stapp also discusses the idea that consciousness is related to state vector collapse within the brain. In other words, perception (our observable reality) occurs as a consequence of consciousness collapsing wave functions (the ultimate reduced state of the universe), which gives rise to events. He agrees with Heisenberg that only waves and events exist, and it is the brain that brings the universe "alive" to us.

What drives these "events" that we call perception? In his model, Stapp builds on the works of Schrödinger, Heisenberg, and Dirac. He blends Schrödinger's possibility set for a system, Heisenberg's idea of choice—that we can know something only when we inquire of nature—and Dirac's process of getting an answer—the state of the universe changes, which corresponds to our brain states. For Stapp, the quantum wave dispersion of calcium ions likely plays a direct role in synaptic functions, which form the brain states attributable to consciousness. Together, all synaptic connections of the brain work in a holistic manner, continuously making choices—continually creating the observable reality and the forever-forward movement of time.

Quantum theory was originally developed to explain the entanglement of the observer's mental properties with the physical properties of what was being observed. Now, scientific investigators of consciousness are including the observer's brain. In effect, the brain has become part of the system as that which turns waveform energy into our observable reality. Exactly how this may work is a fascinating story.

One Scientist's Quest to Understand Consciousness

Long before today's reality-quest explosion, a University of Maryland physicist named Evan Harris Walker was studying the nature of consciousness. In 1970, Walker published a paper in *Mathematical Biosciences* entitled "The Nature of Consciousness," to consider "the framework of current scientific philosophy . . . [concluding] that consciousness is a real, nonphysical entity, as these terms apply in modern physics."[19] After three decades of research, he still believes that consciousness is something in its own right, apart from the physical fabric of nature we call atoms and molecules. Just as gravity is a fundamental feature of reality, consciousness is also,[20] and interacts with matter to animate it.

How and where this interaction occurs has been Walker's quest to understand the brain and consciousness. He proposes a model in which quantum mechanics serves as the basis for brain activity from which consciousness arises. Specifically, electrons, through quantum-mechanical tunneling at the neuron's synaptic cleft, initiate the transfer of information from one neuron to another.

Quantum-mechanical tunneling is a phenomenon in which an energy wave moves from its source to its destination, bypassing any barrier that lies in its way. The barrier (cell tissue) is sufficiently thin that some of the wave moves through the barrier and continues on the other side. Although this appears to be akin to a ghost's fabled ability to move through walls, its functioning has been verified in such naturally occurring phenomena as the thermonuclear reactions on the sun, for which it serves as the initial step.

A neuron (nerve cell) is a unique type of cell found in the brain and body that specializes in transmitting information. In the brain, there are about one hundred billion neurons that pass information to other neurons through appendages of the cell called dendrites and axons. Dendrites bring information to the cell body, while axons send information away from the cell body. A small gap exists between the dendrite of one cell and the axon of another cell, referred to as a synapse. (The sending neuron is referred to as the *presynaptic neuron;* the receiving neuron is referred to as the *postsynaptic neuron.*)

Although the flow of information through the brain is electrical, the passing of information between neurons is chemical. When a packet of information, called an action potential, moves down the axon and reaches a synapse, pores in the cell membrane open and allow an influx of calcium ions (positively charged calcium atoms) into the presynaptic end of the neuron. This releases a small amount of chemicals (neurotransmitters) into a small gap (referred to as the *synaptic cleft*) between the two cells. The neurotransmitter then spreads across the synaptic cleft, interacting with receptors (specialized proteins that serve as ion channels) that are part of the postsynaptic membrane. Pores are opened in the tip of the receiving neuron's dendrites, allowing ions to move through and onto the body of the receiving neuron.

Based on his analysis of experimental data, Walker believes that an electrical field is created within the synaptic cleft and that electron tunneling serves as the catalyst for the neural exchange of information. He also states that the design of the synapse is tailor-made for this quantum mechanical process.

On each neuron, patches of a particular material face each other across the synaptic cleft. One side of the cleft (called postsynaptic) serves as the source for tunneling electrons, while the other side (called presynaptic) acts as a receiver and provides the mechanism where chemicals (neurotransmitters) are released. The synaptic cleft becomes electrically polarized when an impulse arrives. This allows vesicles carrying neurotransmitters to bind with the vesicle gate molecules and calcium ions (electrically charged atoms) to flow through the calcium gates in the receiving neuron's membrane, linking with the vesicle gate molecules. Walker explains:

> The result is that the cleft itself becomes strongly polarized between the postsynaptic elements and the gate molecules. This large electric field causes the electrons to tunnel across to the macromolecules that make up the vesicle gates on the presynaptic side of the cleft.[21]

The interaction between the electrons and the calcium ions (now attached to gate molecules) changes the shape of the gate macromolecules. Then, gate molecules open a channel to the cleft and into the interior of the neuron. At the same time, they open the vesicle, allowing neurotransmitter chemicals (its contents) to flow into the cleft. Neurotransmitters then spread into the cleft and on to the postsynaptic membrane, where they open its ion gates. Excess positive ions flow through, and a new neural impulse continues to the next neuron. Walker writes:

> All this complex activity actually works out very smoothly. Each step is necessary, and the whole thing functions to provide the great flexibility that chemical transmitters give to the brain. But more than this, this synaptic structure seems to be just capable of letting the fluctuations of quantum uncertainties—the dispersion of quantum mechanics, with the attendant need for observer effects—come into play in the brain. The synapse—the whole exotic machine—is poised ready to fire in the flickering passage of a thought.[22]

Aiding electrons in their process of tunneling from one neuron to another are large organic molecules of soluble ribonucleic acid (RNA). They serve as stepping-stones (messenger molecules) for the synaptic electrons. "They 'take orders' from the cell's DNA," says Walker, "and then control the production of all the cell's proteins."[23] According to Walker, these RNA molecules are roughly the same size, and structurally similar to, molecules where synaptic electrons originate on the synaptic interface. RNA molecules also provide an easy way for electrons to land on a neuron's surface regardless of the electrons' initial energy. Amino acids in the RNA molecule are also likely to act as information encoders "possibly serving as part of a molecular memory mechanism in the brain."[24] Furthermore, RNA molecules, by their very nature, are designed to carry the information of genes, which are, in fact, a type of memory, the genetic code. (Walker is referring to the creation of proteins, since the information that genes contain are the instructions to create proteins. Proteins are the implementers of biological function.)

In this process, Walker calculates that a single electron could travel across the brain (ten centimeters) in about 0.084 millisecond, which is a third of the time it takes for a single synapse to fire. The result is that there is enough time for a single electron to travel to remote areas of the brain, making it possible to join all quantum events in the brain into a unified and holistic process. For Walker, "this holistic process brings the mysteries of quantum mechanical uncertainty into play in the functioning of the brain. It links 'thought'—that little fire of the fleeting, jumping electrons—and the specter of the 'observer' of quantum mechanics into one orchestration of mental phenomena."[25]

According to Walker, how synaptic connections operate through quantum tunneling not only explains how quantum mechanics works in the brain, but also explains the brain's form, structure, and behavior. It is in this complex world of living cells that the secret of the mind/brain interaction hides.[26]

The Brain's Quantum Interface

We know that billions of neurons are associated with each other in the brain to form neural networks, and that the firing of neurons in these networks is the essence of brain activity. However, *why* a neuron fires is a neuroscience mystery.

Current scientific theory for *how* a neuron fires states that when a nerve impulse reaches the presynaptic terminal of the sending neuron, neurotransmitters are released and spread across the synaptic cleft on to the postsynaptic terminal, which causes the next neuron to fire. How these neurotransmitters are released in each cell is believed to be through the diffusion of calcium ions within the body of the cell.

According to Walker, the calcium hypothesis is correct, but incomplete when explaining the process of a neuron firing. What is lacking is a source of energy to open the vesicles and their gates, allowing the neurotransmitter chemical to flood into the synaptic cleft. That source is electrons from adjacent neurons that tunnel their way across the synaptic cleft.

In Walker's theory, calcium ions attach themselves to the molecules of the gate vesicle and neutralize existing electrons. (Each vesicle is composed of six to nine molecules.) This neutralizing effect enables the molecules of the vesicles to accept electrons that are attempting to tunnel their way across the synaptic cleft from the postsynaptic terminal of an adjacent neuron. When the electron tunnels its way into the gate vesicle, its energy (seventy millivolts) is just enough to open up the gates, thus allowing the neuron to fire.

The source of the tunneling is the postsynaptic terminal on the neuron's dendrite. In essence, the receiving neuron requests information from the sending neuron via electrons tunneling across the synaptic cleft. In this way, the quantum mechanics of electron tunneling serve as the trigger or catalyst for firing of neurons.

Walker's Model of Consciousness

The previous chapter described how Erwin Schrödinger's wave equation encapsulated the mystery of wave/particle duality—why a particle sometimes behaves like a wave and a wave sometimes behaves like a particle. All objects in the universe are fundamentally electromagnetic waves. But for us, according to Schrödinger, at the moment of observation, the wave is seen as a particle. How does it happen that what is really a wave we see as a particle?

According to Schrödinger's equation, the system (physical reality) is defined as a multistate state vector (of wave functions), where each state is a potential and observable configuration of the system. How we experience what we perceive as reality is a function of how the brain works. This in itself is nothing new—that the brain is important to sensory processing—but the mechanics of how it happens and how it creates consciousness have never before been tackled by science.

Everyone can agree that consciousness is the primary element of our lives. Consciousness is real, and it is also a deep mystery. It is what we lose when we fall asleep and regain when we wake up. Walker's quest to solve the mystery of consciousness began with Albert Einstein's theory of relativity. Since time and its passing are dependent on an observing person, meaning that space and time are relative rather than absolute, Einstein discovered the effect of the observer.[27]

The "observer" effect of relativity (the concept that relativity requires an observer) led physicist Evan Harris Walker to hypothesize that the brain is a device that encapsulates the mechanics of the Schrödinger wave equation and produces our physical reality, individually and aggregately, turning electromagnetic waves into the particles that make up atoms, which in turn make up everything we see.

According to Walker, through consciousness, which is a local phenomenon and bounded by the brain, consciousness sets up state vectors. Through the collapse of the state vector, a nonlocal and time-independent phenomenon, we experience physical reality (the system of electromagnetic waves that make up the universe). Within the brain, consciousness arises from a quantum mechanical process involving synaptic connections between neurons. Electrons tunnel between synapses causing neurons to fire, thereby releasing neural transmitters. This electron-tunneling process within the brain becomes self-sustaining, and an ongoing process. Connecting neurons, as a result of the electron tunneling its way across the entire brain, create consciousness. It occurs so quickly that the signals throughout the brain become a single grand, unified process.[28]

Walker calculated that for the mechanism of electron tunneling to become self-sustaining, synapses in the brain must satisfy a minimum condition for the rate of firing—

$$f_{min} = Mt/nNt^2$$

where M is the number of molecules in the 31 grams of RNA, t is the time for the electron to make one quantum jump, n is the number of such electrons that travel out from one synapse, N is how many synapses are in the brain, and t is how long a synapse usually pauses before firing.[29]

This equation mathematically represents the physical conditions for consciousness to emerge from the brain.

Walker's theory is that the self, the observer, resides "somewhere within the brain and in the consciousness of that brain."[30] Interestingly, this interface between the quantum world and the immediate, observable world boils down to a single electron tunneling itself across the expanse of the brain. "The Great Oz is but one electron," Walker writes.[31]

Sensory perception is a cycle of will (state selection) and observation (state vector collapse) that manifests itself as consciousness and

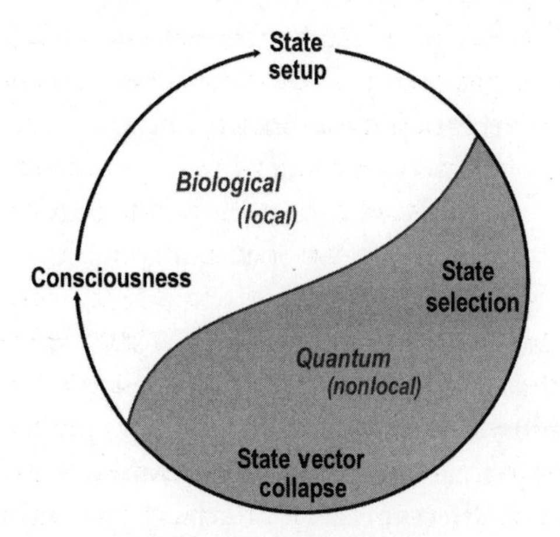

Fig. 2.1. Walker's model of consciousness

the availability of potential actions (state setup) continuously moving through the quantum realm into the immediate, observable world.

Individually and collectively, we—by what can be referred to as "will"—select a state from an array of possible states. In other words, we decide on a behavior. As a result, a particular synapse fires from a selection of consciously coupled interacting synapses. For Walker, this suggests that the "brain-willed" state vector collapse is tied to the external event of state selection. Both are nonlocal. Therefore, consciousness is a reciprocal relationship between the quantum and the observable world, and a continuous loop of information gathering and analysis.

At a rate of one hundred million bits a second, sensory data is channeled through comparison loops by the brain bringing about state vector collapse. In effect, the brain is a measuring device for the system we call physical reality. Every measurement of every kind, performed by the brain, will always become the perception of the observer. Of course, these measurements are really chemical changes in the brain associated with the firing of neurons. Yet to our consciousness, these measurements are seeing, hearing, and feeling. Ultimately, they become our perception. Continuous looping links the observer and the measurement event in the loop. "These loops require state vector collapse to occur," Walker points out, "and at the same time, they carry with them the state vector collapse of the events that we observe."[32] It all happens so fast that for us it is a single, continuous event that lasts from the time we awake in the morning until we fall asleep at night.

By viewing the brain as a measuring device and observation as consciousness, Walker's model appears to solve the long-standing paradox of quantum physics. Reality is matter consisting of particles, but it is also matter consisting of energy waves. The brain, which is part of the system, forces state vector collapse on the entire system, turning the waves into particles.[33] Walker concludes that the "'observer' of quantum mechanics is also really a quantum system himself"[34] and that "consciousness is the bringing about of this ongoing state vector of possibilities that runs through the brain."[35]

If this is the case—that individuals select a state at will—then why does everyone generally experience the same thing? Shouldn't we all experience something different based on our own inner thought life? Those who are unfortunate enough to have contracted a brain disease often do experience a different reality, but that is not the norm.

The answer is that Schrödinger's equation does not allow for different final states of perception for various individuals. Every animal's brain is designed with a constraint to collapse in a coordinated manner. In other words, all observers experience the same reality, which means that "will" is collective as much as it is individualized. The information contained in state vector collapse is the same information that goes into "will."

Walker's Conclusions

We see objects in space that move and change over time, and we refer to this as the *space-time continuum*. According to Walker, however, this is what we perceive, as opposed to what actually is. What we call "space" does not really exist; only observational events exist. "Space is merely the ordering of those things that could have happened but did not happen," Walker says. For observation to occur, a four-dimensional space-time matrix is required, in which one time coordinate is associated with the events that actually occur.[36] Stars, planets, the vastness of space—everything around us—everything we see becomes reality as a result of the brain!

In 1928, the British physicist Paul A. M. Dirac (1902–1984) introduced special relativity into the Schrödinger wave equation, making it "relativistic." Special relativity states that the laws of physics are the same for any inertial frame of reference, regardless of location or speed, and that space and time are not absolute. Space and time are relative, so only relative positions and velocities between objects have meaning.

In his equation, Dirac provided a description of spin-½ particles (particles are characterized by their spin) that was consistent with quantum mechanics as well as special relativity. It was a monumental triumph that predicted the existence of antiparticles, specifically the positron—the antiparticle of the electron. (In 1932, American physicist Carl Anderson discovered the positron. Its existence was proved in 1933 by Patrick Blackett and Giuseppe Occhialini.) Dirac's wave equation is important because special relativity and quantum mechanics seemed to be at odds with each another. Dirac showed that they are not.

Now, Walker shows that by replacing the mass term (electron or positron) in the Dirac equation with his "information term" (a term resulting from the process of consciousness), the information term becomes the source of mass (particles)! In rewriting the equation with only the space-time term and the information term that causes state vector collapse, Walker found that the result is, again, the Dirac equation.

Here lies the mathematics for describing reality, including the vastness of space, as the observer observing. Walker writes:

> If we assume only that conscious observation exists, that alone is enough to let us understand where space-time and matter come from. There is no space as such, no matter as such. There exists only the observer, consciously experiencing his or her complement. And in doing so, the observer weaves the illusion of space-time, and matter falls like snow from the conscious loops of the mind.[37]

In other words, space and matter do not exist as an absolute reality, and the death of a living organism occurs only in that which was created by its own mind.[38] Einstein himself foreshadowed such a statement when he claimed, "Reality is an illusion, albeit a very persistent one." Even the big bang theory, with its irrational beginnings, points to such an existence. Cosmologists theorize that the universe as we see it was in the beginning a singularity, an infinitesimally small point. All matter, space, and time existed as a single quantum state. Perhaps the big bang theory is just a physical perspective of our quantum selves.

Walker finds that the quantum state and the mind are the same, that quantum fluctuations are the objects of consciousness and will. In doing so, he concludes scientifically, "In the beginning, there was the Quantum Mind, a first cause, itself time-independent and non-local that created space-time and matter/energy."[39] And "our consciousness, our mind, and the will of God are the same mind."[40] Everything (space, time, matter, and energy) depends on state vector collapse, the brain, and the perception of the observer—not specific to any particular individual, but an amalgamation of us all.[41] Yet, inundated by a sea of visual and auditory stimuli, we find ourselves tied by the constant processing of consciousness in a physical world, and the need to keep the body going.

The Conscious Mental Field Theory

Evan Harris Walker is not alone in his new approach to understanding consciousness. According to longtime consciousness researcher and University of California professor of physiology Benjamin Libet, the unity of human consciousness poses difficulties with traditional (reductionist) approaches to the brain/mind problem.

After a long and distinguished career in neuroscience, Libet concludes that neural networks within the brain offer little insight into how subjective experience occurs. According to Libet, "A knowledge of nerve cell structures and functions can never, in itself, explain or describe conscious subjective experience."[42] There is nothing in the brain suggesting that subjective experience—the role of the observer—is actually occurring. Yet through complex neural functions, a unified image is somehow produced for that individual where thought occurs.[43] As does Walker, Libet believes that the conscious, subjective experience is "another unique fundamental property in nature."[44]

Libet explains that during the late 1970s, John Eccles proposed that "experienced unity comes not from a neurophysiological synthesis but from the proposed integrating character of the self-conscious mind"[45]—a separate nonphysical mind must exist that detects and is integrated with brain activity. According to Libet, other neuroscientists, such as Roger Sperry and Robert Doty, have come to similar conclusions, and in general believe there is a growing consensus that no single cell or group of cells is responsible for conscious experience. Rather, it is characteristic of a more globalized function of the brain.[46]

During the early 1990s, Charles Gray and Wolf Singer discovered that there is a widespread synchronization of oscillating neuronal responses in relation to certain visual configurations. This led to the hypothesis that the synchronization of electrical oscillations could produce a unified subjective image, although explaining how the subjective experience is unified in a holistic manner remains elusive.[47]

In explaining how subjective experience (the mind) arises from the brain, Libet has proposed that a "conscious mental field" exists as a result of neural activity. It is in this field, which is at this time of an unknown nature (and not electromagnetic), where subjective experience occurs and, in turn, alters some neuronal functions.[48] Libet does not claim that all cognitive functions, such as information storage, learning, and memory, are mediated by the conscious mental field but rather that only conscious subjective experience is.[49]

Whatever the case may be as to the fundamental source of our conscious existence, our struggle manifests in the immediate, observable world. Attempts to reconcile the quantum view of reality with the classical view have always had their problems and critics. Even the creator of the equation describing state vector collapse viewed quantum mechanics as a paradox.

Schrödinger's Cat

In 1935, Erwin Schrödinger contrived a thought experiment, now quite famous, to show how quantum mechanics contradicts our perception of reality. The experiment entailed a box, a cat, a vial of poisonous gas, and a radioactive atom with a 50 percent probability of decaying in an hour. The cat, the gas, and the radioactive atom were placed in the box. When the atom decayed, it would release the poisonous gas.

A paradox arises in this situation because the atom, being ultra-microscopic, must be described by quantum mechanics. After an hour, and before the atom is observed, it exists in a "superposition" state. In other words, it simultaneously exists as an intact atom as well as a decayed one. Yet, if quantum mechanics is a complete, universal theory, it must describe the whole system. Since the state of the cat is correlated with the state of the atom, the cat must also be in a superposition of being dead and alive. This, of course, contradicts our everyday experience, since we know that a cat, or any other living creature, for that matter, cannot be both dead and alive.

Although Schrödinger's experiment was only a mental exercise, it gets across the idea that we live in a single reality, not two realities of the quantum and the immediate, observable world. It also gets across the absurdity of quantum theory as it relates to our view of the world, which has been a criticism of the skeptics of the quantum mind theory. These critics argue that the participation of human consciousness in the world is little more than our subjective perception of whatever reality is, and that viewing the brain as a device for state vector collapse is irrational and runs against the grain of what science is supposed to be about. They put the quantum mind into the category of myth and fable, citing the fact that if we could really determine our own reality through our conscious minds, the world would be a very different place. The evidence is that the world is not the place of our dreams. Or is it? Isn't life the aggregate of what each and every individual on the planet makes of it?

Skeptics claim that if quantum realism is true, individuals should have the ability to create their own reality. They pose examples, such as jumping off a building and changing the concrete pavement into a feather pillow, as the evidence required to prove the existence of the quantum mind. Such a proposition is clearly out of context for any scientist or science enthusiast who espouses the so-called new view of physics. The theory of quantum mind is rooted solidly in the tradition of the scientific method, and it does not make any of the extraordinary claims its critics suggest it does. It is my belief that the quantum approach to explaining and understanding life does a remarkable job of reconciling the opposing constituents of creation and evolution, which we'll address in the next chapter.

Quantum Weirdness

When we view physical reality through the concepts of quantum theory, it would be correct to say that physical objects exist in all possible states. This means that all objects are simultaneously all colors and shapes. Only when we observe an object do we create our own instance of reality by forcing the physical world to conform to a series of quantum properties through state vector collapse. Since all brains perform the same task, everyone sees the same thing. Thus, a single reality is continuously created. It sounds absurd that "constant creation" keeps everything appearing as it should. But quantum physics, as an experimental as well as a theoretical science, seems determined about the quantum nature of reality.

At the most reductionist level, we are atoms: electron clouds bound to an infinitesimally small nucleus, electromagnetic waves with certain probabilities of existence until a photon reflects off us and the electron is observed. But we are more than electrons. We are also highly complex physical bodies, and in considering the delicate balance required for the human body to exist, the idea that particles are constantly popping in and out of existence seems ludicrous. But if you lost a single cell, how much would you care? Imagine, then, that the cell reappeared and another cell disappeared. As a person with approximately seventy-eight trillion cells, you probably would not miss even a million cells. Also consider the great number of atoms per cell, and then an average of ten electrons per atom. Even a vast amount of electrons popping in and out of existence would not matter—those that are created are replacing those destroyed.

We could argue that since the phenomenon of particle appearance and disappearance occurs only at the quantum level, it is irrelevant for the immediate world. Reasoning such as this, although quite practical, is misleading when addressing the fundamental issue of reality. Everything occurring in the observable world is derived from the quantum world. So is everything constantly popping in and out of existence, including humans? It is a madcap thought, for the world does appear seamless and fluid. Scientific fact, however, suggests it is indeed the case. Matter, it seems, is a matter of perception. But our senses are permanently configured to the immediate world, so we never realize what is really happening.

Suppose a car can move only in increments of two-thousandths of an inch and always skips the distance between the first and second points. In other words, the car appears at one spot, disappears, and then reappears at another spot. If the car made 250,000 jumps per second, an observer would say that the car is moving at a constant speed of five hundred inches per second. Yet the car is actually starting and stopping. It is the same principle with motion photography. The brain, particularly the cerebellum, allows for smooth movement. We watch a film and to our eyes it appears as constant motion, but we know that it is really millions of stills moving exceedingly fast. It is in this manner that both observable and quantum realities are true. So quantum events really do provide causality for everything occurring in the visible world.

In the sense that the immediate world is a result of the quantum world, all matter everywhere is constantly being created. It is not far from the truth that we live in a bubbling sea of virtual particles, a quantum matrix, so to speak. Furthermore, it must also be the case that when an object is unobserved in either world (quantum and immediate), it exists in all possible states. Sounds too weird, but it is what Erwin Schrödinger was implying with his thought experiment.

All this quantum weirdness becomes understandable through observation. It is what creates the forward-moving path of the arrow, what we see as continuous existence. Wherever there are humans, there are also observation and perception, the fundamental elements of consciousness that provide the seamless picture of physical reality. Perhaps reality can exist only where there is consciousness. The observer (consciousness) has created the cosmos to experience and through evolution created the being (man) where the experience is most dramatic. This has to be viewed in an esoteric fashion. Everything that exists is actually a single entity that exists in a nondimensional state.

3

Cosmos

Mystery of Life

We are describing a universe, which may be infinite. It may go on forever, and already the mind has a hard time conceiving of things that just go on endlessly.

And, to make it even stranger, it originated from a period in which things were infinitely close together, from which everything was infinitely, densely packed, and yet the universe was still infinite, perhaps. The mind has such a hard time comprehending this sort of concept that the name of the game is just to understand a little bit from this angle, and a little bit from this angle, and try to see if you can make sense out of it all. And you don't expect, really, to understand it, in the full sense that we think we understand the immediate world around us every day. But that little sense of that glimpse of understanding, I think is very exciting and makes you feel like you are in touch with something magical.

PHYSICIST SAUL PERLMUTTER, IN THE BBC DOCUMENTARY *FROM HERE TO INFINITY*

For the past seventy years, astronomers and physicists have been confident not only that the universe was expanding but also that it was slowing down in its rate of expansion. As the force of the big bang loses its momentum, the theory goes, the universe will eventually reverse its direction and come crashing back together into a primordial singularity, the way it all began. Out of curiosity, physicist Saul Perlmutter, at the University of California's Lawrence Berkeley Laboratory, was determined to find out when, in the future, this "big crunch" would happen. By looking deep into space, he identified a specific type of supernova (type Ia) that could be used to approximate the universe's rate of expansion. According to theory, these exploding stars should be decelerating in their journey away from Earth. Likewise, all stars at or near the distance of the supernova would also be moving at the same rate. As a result, the overall rate of the universe's expansion could be determined. Measured over time, the universe's deceleration would provide a way to calculate when the big crunch would happen.

The expanding cosmos physically stretches not only the distance between galaxies, but also the wavelengths of light (photons) as it travels. By analyzing the spectrum of light given off by a star or galaxy, scientists can tell whether it is moving toward or away from Earth, and how fast. A shift in the starlight's wavelength into the blue end of the light spectrum means that the source of the light is approaching; a shift into the red end means it is traveling away.

Perlmutter's results were shocking. In fact, they were paradigm breaking. The expansion of the universe was not slowing down; rather, it was speeding up. At first the scientific community believed his data were incorrect. After further research, however, Perlmutter and his Supernovae Cosmology Project team determined that the data were indeed correct. The explanation was that "dark energy"—what might be referred to by other scientists as the zero-point energy field—dominates large, empty areas of the universe and is likely responsible for the force that keeps pushing galaxies farther away.[1]

When new theories based on groundbreaking research are introduced, they are often difficult to accept and require further study. The idea that everywhere there is a force created by particles appearing and disappearing, in what used to be considered true nothingness, was not an exception. Nonetheless, it is compelling evidence for *the magical,* as Perlmutter so honestly puts it—not charms or spells, but a distinctive quality that produces effects in the cosmos that are inexplicable and unaccountable.

The Solar System

Notwithstanding the possibility that a quantum field exists in the emptiness of space, there is an abundance of reasons why mankind has always pondered its place in the universe and why scientists today continue to do so. Why we are here, on Earth, at the edge of the Milky Way, requires a set of unique circumstances that is a mystery in itself.

According to the available evidence, our solar system and planet came together from the whirling debris of some cosmic catastrophe in an area away from the high-energy dangers of the Milky Way's galactic core. Rich in metals and removed from active gamma radiation, our solar system is something special. It is not a binary. Nor is it near a globular cluster or other dangerous cosmic event, such as a pulsar or magnetar. Our sun is neither too small nor too large, and it will not exhaust its fuel anytime soon.

The entire solar system is housed in an outer shell of asteroids and comets called the Oort cloud. This is an immense spherical shroud approximately three light-years in diameter, thirty trillion kilometers from the sun, and at the edge of the sun's physical and gravitational influence. Inside the Oort cloud, just outside Pluto—the planet farthest from the sun—is a doughnut-shaped ring of asteroids and comets called the Kuiper belt. These objects are believed to be the primitive remnants of the solar system's violent beginnings. Astronomers suspect that the Kuiper belt is the source of short-period comets, acting as a reservoir for these bodies in the same way the Oort cloud acts for long-period

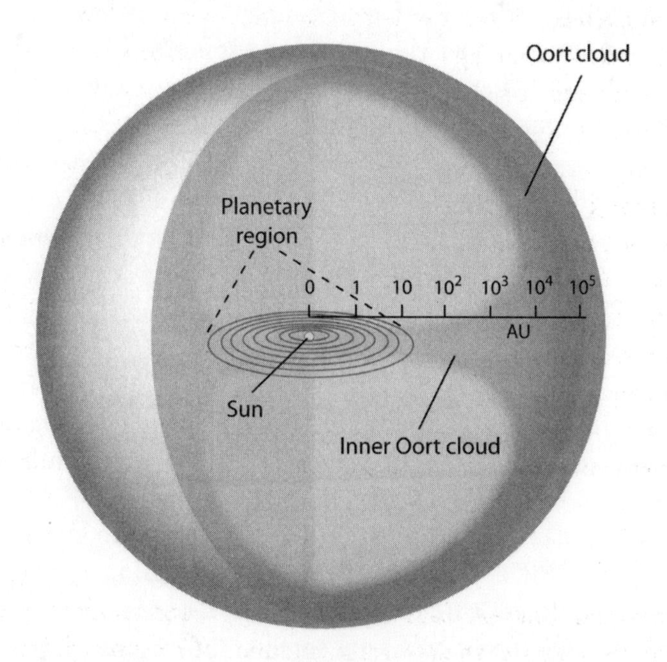

Fig. 3.1. The solar system (illustration by Dorothy Norton)

comets. ("Short" and "long" period is in reference to the length of time of a comet's orbit around the sun.)

As our solar system developed, comets and asteroids bombarded Earth. But this rain of rock and ice was never intense enough to annihilate Earth, thanks to the presence of Jupiter, the solar system's gas giant. Jupiter served, and still serves, as a cosmic vacuum cleaner. Its immense gravity attracts almost everything that comes its way. Comets that did get past Jupiter and collided with Earth brought with them the elements of life—hydrogen, oxygen, carbon, and nitrogen.

Earth's early, violent history proved to be what was needed for life to emerge. In its beginning, Earth was a spinning, molten sphere of liquid metals. Over time, Earth's surface cooled, but its center remained a swirling molten core. As a result of the core's electrical circulating currents, our planet is shrouded in a magnetic shield. Without this protective covering, Earth could not support life as we know it. But what makes a planet like Earth extremely rare is its ability to sustain plant and animal life.

Unlike planets in other solar systems so far discovered, Earth's nearly circular obit and distance from the sun provide stability, allowing global temperatures sufficient for liquid water to exist. Our moon, with its gravitational pull, larger than any of the moons of the other planets in the solar system, keeps our world upright and spinning like a top. Without such a large moon, Earth would tumble in space as a lifeless planet. No stable weather systems could be sustained anywhere on the surface.

According to the latest scientific findings, the elements responsible for organic life were constructed as a result of a stellar explosion billions of years ago. In a sense, we really are children of the stars. How is it that all the correct conditions for life came about so precisely?

Life *Extraordinaire*

With the invention and refinement of the telescope, together with modern society's technical progression, for the UFO enthusiast and scientist alike it has become the curiosity of curiosities to know if we are alone in the universe. Billions of galaxies exist with billions or more planets composing each galaxy. It is a mind-boggling struggle imagining how insignificant a single planet is compared to the universe.

Because life evolved on Earth, surely it did so elsewhere in the universe, so goes the argument. For the last thirty years, astronomers of the Search for Extraterrestrial Intelligence, or SETI, have been scanning the sky searching for a systematized transmission, the anticipated trademark of an extraterrestrial civilization that has attained some level of technology equivalent to or greater than our own. So far, nothing has been received.

There is also a hunt for planets in other solar systems. To date, a fair number have been found, but none seems to be like our Earth. The typical solar system appears to have planets moving in orbits more elliptical than that of Earth. Some solar systems have gas giants orbiting at such a high speed and so close to their star that one complete revolution occurs every three days. From what scientists have discovered so far, it appears that Earth-like planets are very rare. Given the many unique qualities of our solar system, could our ordinary existence really be extraordinary?

Cosmic Uniqueness

Although we take life for granted, there is also a certain uniqueness about the universe for it to exist at all. In *Our Improbable Universe,* former physics professor Michael Mallary lists fourteen points that make the universe unique and ideal.[2] There is one fact Mallary believes cannot be ignored: energy, as a result of the big bang, somehow became people who find joy—a joy simply in being. How could a statistician calculate the odds for such an outcome? The physical existence of the universe seems to have a uniqueness and rarity in and of itself.

To understand the rarity of the immediate world in which we live, we must embrace the complexity of why it does exist. Around fourteen billion years ago, a tiny point, an abstraction called a singularity, a non-dimensional quality, exploded into the three-dimensional universe we now see. This is a theory very few scientists dispute. What is curious is how matter (any matter) survived after the first three minutes of the big bang. It is one of the great unsolved mysteries of physics. Even a slight change in the amount of energy and matter that was produced would likely have resulted in a universe without galaxies or stars. Nevertheless, the "bang" was perfect, and it went on to develop at least one stable solar system where life could spontaneously emerge. According to Mallary, there are at least thirteen other properties of the universe required for the big bang to produce life as we know it.

For any matter (and mass) to be produced by the big bang, the universe had to be capable of producing at least *six kinds of quarks*—subatomic particles having an electrical charge one-third or two-thirds of that of the electron.[3] Nearly all of our body mass (99.95 percent) is made of this nuclear matter. These six varieties of quarks allow for a subtle asymmetry in how matter and antimatter behave, referred to as *charge conjugation-parity symmetry violation*, or just *CP asymmetry*.[4]

Without CP asymmetry, the big bang would have produced the same amounts of matter and antimatter, resulting in mutual annihilation. The universe would have been nothing more than an expanding ball of light and inert particles called neutrinos. No matter would have existed to coalesce into galaxies and planets. But CP asymmetry and the six varieties of quarks allowed an extra particle of matter to exist for every billion particles of matter and antimatter. These extra particles make up the universe we see today. So, early in the universe's life, there had to be a time when energy and matter existed in a state of imbalance. There also had to be a mechanism for the transmutation of quarks and antielectrons into each other.

When the universe was only the size of a grapefruit, just under one-trillionth of a second after the big bang there was *just enough energy and matter to be of importance*.[5] The density of mass and energy was just right in relation to the universe's rate of expansion. If the density had been higher, a contraction of everything would have occurred too soon; if it had been lower, stars would never have formed. The universe would have turned out to be a diffused ball of isolated atoms. It did not, and the universe continued to expand with *just enough lumpiness* to initiate the process of galactic formation.[6] If it had been too smooth, matter would not have formed into stars. Or, if it were too lumpy, monstrous black holes would have formed, swallowing everything.

At some point early in the life of the universe, four forces emerged to define the complex behavior of matter.[7] Of these forces, gravity is the most familiar. Galaxies, stars, solar systems, and planets would never have formed without it. At an atomic level, electromagnetism causes electrons to orbit the nuclei of atoms, forming the basis of everything that has concrete existence.[8] Two more forces, referred to as "strong" and "weak," operate only at the subatomic level, gluing protons and neutrons together to make complex nuclei. This makes it possible for stars to burn hydrogen in a slow and steady way for billions of years. The weak force converts protons into neutrons, allowing complex atoms, such as carbon and oxygen, to form. It also permits old stars to explode, spreading the elements of life into space.

The nuclear force between two protons is exactly what is needed for stars, including our sun, to have a very long life. If it were any stronger, protons would stick together permanently, forming a helium-2 nucleus—two protons stuck together without any neutrons—and resulting in the emission of large amounts of radiation. The consequence would be short lives for stars, which would burn up their supplies of hydrogen in less than a hundred million years. If the force were weaker, hydrogen would burn only in giant, short-lived stars. But since protons do not quite stick together, a star burns a very long time.[9] The fact that helium nuclei do not quite stick together prevents a helium nucleus from turning into carbon, which would result in a massive explosion. But if they did not stick at all, helium would never turn into carbon and other elements necessary for life.

Carbon, the element, is amazingly suited for the complex chemistry needed for life. When a third helium nucleus joins a pair of other (stuck together) helium nuclei, the resulting carbon atom attains an excited state of energy, enabling it to stay together. This results in an enhanced rate of carbon production and allows carbon to outpace its conversion into oxygen. Without the excitability of carbon and the calmness of oxygen, there would be very little carbon in the universe—and life, as we know it, would not exist.[10]

Large stars explode at the end of their lives, spewing out complex elements created in the stars' interior.[11] These elements include those necessary for Earthlike planets. According to astronomers, this is precisely how the Earth came to be five billion years ago.

Because of *heavyweight neutrons,* stars have a long life.[12] The big bang would have created an abundance of very heavy elements if a neutron's mass were less than the combined mass of a proton and an electron. Although there would be no lack of chemicals if neutrons were lighter, all hydrogen would be in such a form that it would burn rapidly in stars.

According to particle physics, protons and neutrons have a long life, even though they eventually decompose.[13] Their rate of decay, really a theoretical prediction, is so slow it is yet to be observed. Experimentation suggests that the decomposition of protons and neutrons will not be a problem for another ten billion trillion trillion years.

We exist in a three-dimensional reality, a universe where objects have height, width, and depth. Although a two-dimensional universe would suffice to allow stable orbits for planets, only a three-dimensional universe supports the neural networks required for the animal brain.[14] In a two-dimensional universe, neural fibers would be required to jump over each other to make a complex connection. That would be impossible, because a neural fiber would have to cross over another and lift off the plane's surface. Adding a fourth spatial dimension creates instability in the rotation of a planet around its host star. Like a Phillips screwdriver standing upright on its point, any disturbance whatsoever would force the planet from its orbit, causing it to spiral into the sun. Complex chemistry would likely not exist in anything less than a three-dimensional universe. Carbon atoms form bonds in multiple directions, a vital element for the building blocks of life to occur.

The wave nature of matter, as discussed in the previous chapters, is crucial in that quantum waves resist being squeezed into small volumes.[15] The smaller the space that a quantum wave occupies, the greater the energy it contains and the more force it extends outward. Without such resistance, the universe would be nothing more than a collection of black holes. Interestingly, the interplay of an extraordinarily large number of waves (matter reduced to its simplest form) is the foundation for our physical reality.

According to Wolfgang Pauli's exclusion principle, formulated in 1925, particles are reclusive.[16] For physicists, it is one of the most basic observations of nature: particles of half-integer spin must have antisymmetric wave functions and particles of integer spin must have symmetric wave functions. Spin is the angular momentum that is built in to the particle. It is the fundamental property of that particle. Other elementary particles have 1 (photon) and 2 spins (graviton). What this means is that it is impossible for two electrons to occupy the same state, a principle that all stable particles with mass obey. Without it, there would be no complex chemical bonds, only metals. As a result, particles would group together in extremely dense gravitationally bonded clusters and quickly collapse into black holes.

Although there are likely more than Mallary's fourteen crucial ingredients for a universe, his list illustrates the precision and complexity built in to physical reality, what the ancient Greeks referred to as the cosmos, *the Order*. It also shows the uniqueness of our Earth and of the solar system we live in, as well as of the universe.

Consciousness and the Cosmos

It is nearly impossible to imagine how all the aspects of consciousness could exist without complex organized matter. Perhaps consciousness cannot, which raises an interesting question. How does energy organize into matter in the first place? In an evolutionary approach, the idea of basic quantum entanglement and its associated nonlocal quantum correlations is the most fundamental form of awareness. (Quantum systems in an entangled state, such as microtubules in the brain, serve as an information channel to perform computational and cryptographic tasks.) Proponents of this theory envision a self-organizing system that learns in a nonlinear quantum holographic manner and that, in principle, could arrange itself into what we observe in the universe, where consciousness itself is intertwined with the fabric of the cosmos, the zero-point field.

Although many aspects of the cosmos remain a mystery, there can be little doubt that the concept of a self-organizing, regenerating universe is as viable a theory as any other. It is observable and provides some level of causality in a scientific way. Even some ancient cultures, such as the Hindus and their Yuga system, recognized that life operates in regenerative cycles. Given the evidence, it also seems plausible that consciousness may somehow interface with the quantum world (the zero-point field), as Evan Harris Walker theorizes, which facilitates the explanation of mystical and paranormal encounters millions of people experience. Yet if the universe did organize itself into organic life and evolve into the consciousness we recognize today, by what biological necessity did it occur? While intellect can be argued as an important element for the evolution of humans for reasons of survival, the primary attribute of consciousness and self-awareness has little to do with intelligence and more to do with biology.

The Conscious Body

Since the discovery of DNA in 1953 by Francis Crick and James Watson, the biological sciences have developed a paradigm that everything about us is built into our genes, a concept known as *genetic determinacy*. Its essence is that we are a product of our genes, and through genetic changes, life has evolved. During the last twenty years, however, new research is challenging this orthodox view

Evolutionary biologist Bruce Lipton, former professor of anatomy at the University of Wisconsin's School of Medicine and a proponent of the new science, has had profound insights into cellular biology and the role of consciousness. According to Lipton, in his book *The Biology of Belief,* consciousness is a function not necessarily of the brain but of the entire body. Every cell in the body has its own awareness and looks to the environment for clues on how to behave. The immediate world, as well as internal physiology and our *perception* of the environment, directly controls cell behavior. Thoughts, attitudes, and beliefs create the conditions of our body and the external world, resulting in body-wide molecular communication through awareness of vibration and resonance. In other words, consciousness at a microscopic level operates the body.

Science has known for some time that proteins provide an organism's structure and function, but why proteins occur has remained a mystery. The actions of random proteins simply do not create life. In theorizing how protein replacement (protein production in our bodies) allows for complex behavior, in the past biologists have looked to the memory factors of proteins and how they provide for heredity. When DNA was discovered, it became the answer, since it encodes the proteins' structures, the character of the organism, and allows protein synthesis. This led to the assumption that DNA controls life, with the nucleus of the cell as its brain. Science refers to this as the *primacy of DNA,* and it is considered to be the answer to the question of causality.

Through experimentation, however, Lipton discovered that the primacy of DNA has been a false assumption and that the cell nucleus, with its genetic material, is not the brain of the cell. In all organisms, removal of the brain results in immediate death; so, if the nucleus is the cell's brain, its removal will result in the cell's immediate death. In Lipton's experiments, denucleated cells did not die, and, in fact, some cells lived on for months. If the cell nucleus is not the brain, then what is? The only organized part of the cell is the membrane, which is responsible for containing the cytoplasm, digestion, respiration, and excretion. Lipton believes that the membrane of the cell is the brain, and the nucleus is nothing more than the cell's gonads, to be used as a chemical blueprint for reproduction.[17]

In the traditional view of the world, cell receptors respond only to matter (molecules). The latest in cell research, however, suggests that energy, as defined by quantum physics, also invokes receptor response. It has been shown through experimentation that pulsed electromagnetic fields regulate nearly all cell functions, including DNA synthesis, RNA synthesis, protein synthesis, cell division, and differentiation. In other words, biological behavior, including thought, can be controlled by energy.

Lipton concludes that cells read their environment, assess it, and then select appropriate behavior to survive. Receptors, called *integral membrane proteins* (IMPs), built into the cell membrane continually scan the environment, while effector IMPs create the response, translating environmental signals into behavior. These receptor IMPs are aware of their surroundings, and their behavior controls gene expression. Another way to describe what is going on is that the cell, aware of its surroundings through sensation, is engaging in perception! According to Lipton, "Each receptor-effector protein complex collectively constitutes a 'unit of perception,'" which exists between the cell's surroundings and its expression.[18] Accurate cell perception results in life-enhancing behavior, whereas inaccurate perception leads to poor health.

In multicelled organisms, constituent cells have additional receptors to provide integration as well as a communal identity. These respond to hormones to coordinate and regulate the cell community. Another, group of receptors bestows identity, so that all cells of a like kind act collectively in accordance with a central command. This is one of the main problems physicians face in organ transplants. Self-receptors of the immune system distinguish "self" from other foreign organisms, resulting in rejection of the new organ.

Lipton believes he has discovered the fundamental aspects of consciousness—how life is animated. Although an individual is a single entity, he or she is also a community of fifty trillion cells working together for the common good. Each cell, Lipton explains, is the functional miniature of a human being with its own digestive, respiratory, nervous, and reproductive systems. Every single cell has it own awareness through cellular (IMP) perception units and receives signals from the environment to alter proteins to be used in bodily function. Since cells are in harmony with the environment, in the aggregate, the whole body is in harmony with its surroundings, with the brain and its complex arrangement of cells, and with awareness, acting as a governor.[19] The end result is animated life.

Evolution in Perspective

According to a February 2001 Gallup poll, only 12 percent of those surveyed believe that man evolved by chance, without any influence from a supreme intelligence or being.[20] But 37 percent stated that they do believe mankind developed from a lower life-form, trusting that divine providence, in some way, played a crucial role; they viewed evolution as the mechanics of creation, attributable to an intelligent design.

Those who believed that God created man "as is" sometime within the last ten thousand years accounted for 45 percent of the people surveyed. Another Gallup poll focusing solely on the issue of God's existence found that 86 percent of Americans believe a personal God exists and answers prayers.[21]

According to a 1997 survey in *Nature* magazine, 40 percent of U.S. scientists stated that they believe not just in a creator, but in a God to whom one can pray with the expectation of an answer.[22] Stefan Lovgren, author of the *National Geographic* article that reported on the *Nature* magazine poll, noted that the number of scientific deists might have been higher if the question had simply concerned God's existence.

Americans, scientists included, seem to be convinced God exists, so why all the fuss about evolution versus creationism? The trouble lies in the details and the dogma, for which both sides deserve some blame. The role of science is to maintain a theory or explanation until another, better one comes along. Religion's role is to give purpose and understanding to life, a belief system—an important and fundamental aspect of society. When attempting to describe man's origins, however, religion falls short of a rational, mechanical explanation, which science offers through evolutionary theory. From the beginning of the universe to the formation of our solar system, evolution mechanically explains how we got here. There is also a fair amount of evidence in its support. On the other hand, it is anything but free of controversy.

Although *evolution* is the common term for the theory that all life evolved, it should be made clear that the theory that all life evolved from a single-celled organism is more appropriately called *biological evolution*. It refers to the cumulative genetic changes of living organisms that occur over a long period. As an organism's genes mutate (recombine) in various ways, new traits are passed on to the next generation through reproduction. As a result, through some serendipitous event, some organisms inherit a new characteristic that may provide an advantage for survival and reproduction. These characteristics tend to increase in frequency in animal populations, since that specific organism outlives other organisms with less advantageous traits. This propensity for an organism to live longer and reproduce more often as a result of a preferred genetic mutation is a process known as *natural selection*.

According to its proponents, the Darwinian theory of evolution has withstood the test of time, as well as thousands of scientific experiments, particularly with simpler life-forms such as the fruit fly. They claim that evolution has not been disproved since it was first proposed during the 1850s. Not all scientists agree, however, and a number are skeptical of the validity of evolutionary concepts.

In 2001, the Public Broadcasting Service (PBS) aired a seven-part documentary series entitled *Evolution,* coproduced by the WGBH/NOVA Science Unit and Paul Allen's Clear Blue Sky Productions. It was a traditional Darwinian approach to man's origins. In response to the TV series, the Seattle-based Discovery Institute, which describes itself as a nonprofit, nonpartisan public policy center for national and international affairs, declared that one hundred scientists had banded together voicing their skepticism of evolution. Each scientist endorsed a statement announcing:

> I am skeptical of claims for the ability of random mutation and natural selection to account for the complexity of life. Careful examination of the evidence for Darwinian theory should be encouraged.[23]

Of the scientists signing this notice of dissent, twenty-eight are chemists, twenty-one biologists, fourteen physicists and mathematicians, and two psychologists. Interestingly, those signing the proclamation also included two anthropologists and a geologist, disciplines that favor evolution theories.

In marking the beginning of life on Earth, evolution theorists point to a period five hundred million years ago, when most of the major animal groups, such as brachiopods (marine invertebrates resembling bivalve mollusks), trilobites (extinct marine arthropods), mollusks, echinoderms, and other odd animals of extinct lineages, suddenly appear in the fossil record. For these theorists, this is hard evidence favoring evolution. Life seemed to burst forth from a world where only soft-bodied animals, algae, and bacteria had existed, an event known as the Cambrian explosion.

There are perplexing problems with such a conclusion, however. Those who are skeptical of evolutionary theory point out that there is no adequate explanation for an origin of life from lifeless chemicals. Even the simplest forms of life are extremely complex.

Evolutionary theorists themselves point out other difficulties with their theory. Harvard University's Stephen Jay Gould (1941–2002) argued that a completely different evolutionary course could have developed, one that does not require the existence of mankind.

According to Gould, there is no reason for any evolutionary event to lead to the development of the human body. The traditional concept of evolution claims that our planet's environment created selection pressures that produced similar forms of life to those around us, humans included. Gould, who was a professor of geology and zoology, believed that major evolutionary changes are likely to occur in sudden bursts as opposed to a gradual process.[24]

Another difficulty facing evolution's credibility is the interpretation of skeletal remains, particularly those of primates, and whether these primates represent transitional human forms. Since all types of primates seem to have been fully formed by the time their bones/skeletons appear in the archaeological record, it is just as likely, one can argue, that a number of species always existed and later became extinct, as opposed to one or several species evolving into higher life-forms. Alternatively, it is difficult to believe our planet suddenly appeared intact without any type of process to shape its surface features, atmosphere, and ultimately its biological life.

Amid all the debate, there lies a fundamental misconception of evolution by its opponents. Evolution is not a substitute theory for the physical creation of the universe. *Evolution is a scientific theory of process, and does not address the creation of matter.* Nor is evolution concerned with proving whether God does or does not exist. Evolution is concerned only with the scientific description of a process believed to govern the development of the universe and life on Earth. And as with all scientific theories, the theory of evolution undergoes change as new evidence becomes available.

New Developments in Evolutionary Theory

According to the current evolutionary model, modern humans emerged about one hundred fifty thousand years ago and began moving out of Africa fifty thousand years later. Genetics scientist Spencer Wells, however, in his book *The Journey of Man: A Genetic Odyssey,* suggests that the migration occurred much later, around sixty thousand years ago. Wells's date for this "out of Africa" journey is based on DNA samples (the Y-chromosome of males) taken from peoples around the world. Since the Y-chromosome does not recombine with other parts of the human genome but passes from father to son intact, it serves as a marker in genetic studies. In studying this Y-chromosome marker, Wells traces mankind's genetic origin back to a single man in Africa.

He also discovered in his study that African populations display more genetic diversity than any other peoples of the world, which means they are older than all other populations.

Evidence such as Wells's worldwide DNA analysis makes the story of how mankind came to be much less clear, since it conflicts with the established archaeological model. According to the archaeological record, two different types of humans existed during prehistory. Skeletal remains from what have been called "archaic" humans go back two million years, but anatomically modern humans have a much more recent origin, approximately one hundred fifty thousand years ago. How archaic humans—*Homo erectus*—became "modern" man is a mystery. Could there have been more than a single species of humans, between which interbreeding gradually replaced the archaic species? Or did the new, modern type of human evolve in isolation, out-reproducing all other *sapiens* types, and win a long-term economic and cultural war of attrition? According to Wells, the data he gathered strongly suggest that interbreeding did not occur, and that anatomically modern humans arose somehow from a unique change in DNA.

Whatever the case for the emergence of modern humans, it seems that Wells's genetic model of origins and migrations supports what others have previously postulated, although his time frame produces a forty-thousand-year discrepancy. Modern humans arose in Africa and then left in several migratory waves to inhabit the world. Modern humans gradually replaced existing *Homo* populations as new territories were settled.

One avenue in explaining Wells's discrepancy is to view behavior and anatomy as mutually exclusive. Although anatomically modern one hundred fifty thousand years ago, *Homo sapiens* did not display modern behavior but developed it much later on as a consequence of genetic mutation. Stanford University anthropologist Richard Klein believes it is possible that the ability to think enabled modern humans to develop technology that enabled them to craft not only sophisticated tools, but also the means to travel to remote locations.[25]

It appears that the concept that mankind evolved sequentially over millions of years no longer holds much weight in evolutionary thinking. Rather, there were a number of species of *Homo* that had to compete socially and economically for territory. Of those various species, paleoanthropologists generally believe that *Homo erectus* was the ancestor of anatomically modern humans, whose origins were in Africa, but expanded his territorial homeland into parts of Europe and Asia beginning close to two million years ago. In Africa, skeletal remains of *Homo erectus* have been discovered that suggest that these individuals were more than six feet tall. Nicknamed "Goliath" by the writers of *National Geographic*'s documentary *Search for the Ultimate Survivor,* this type of human, anthropologists Lee Berger and Steve Churchill claim, did not make it, as a result of a changing climate in Africa. Goliath (whose anthropological name is *Homo rhodesiensis*) was too big for his own good and could not adapt to the warming climes of Africa. The evidence of Goliath's existence disappears in the fossil record.[26]

When anthropologists talk about a certain human type becoming extinct, however, they are not always insinuating that a whole new type of human was somehow created to take its place. What they are saying is that those who were of a smaller frame, and thereby better fitted for a hot climate, had a tendency to survive longer and pass on their traits to the next generation. So Goliath himself didn't really die off; his large build did, as an adaptation to the changing environment—a statement that, ironically, agrees with some ancient texts, biblical as well as nonbiblical, particularly the Book of Enoch and the introductory passages of the account in Genesis of Noah's flood. According to Berger:

> What the fossil evidence shows us now is that our ancestors were giants. They were creatures that were much bigger in their mass and stature than humans are today, all except the very rarest, the very biggest of all humans.[27]

It seems the more anthropologists and archaeologists discover, the more science recognizes that a greater number of pieces to the puzzle are missing than previously thought, and that we really do not know how we became human. Nor do we know how life originally began. Exactly what happened in terms of human development and extinction between two million and fifty thousand years ago we will likely never know. What we do know is that by about 30,000 BCE, *Homo sapiens sapiens* was the sole surviving species of hominids.

Today, a very different picture is being painted of man's origins. According to Ian Tattersall, curator of human evolution at New York's American Museum of Natural History, there is no compelling evidence that a gradual transition took place of one species into another.[28] There are too many gaps in the fossil record. For Tattersall, if these gaps are telling us something, it is that evolution is more a matter of sporadic innovation,[29] and that the march from monkey to man as a slow, incremental process over tens of millions of years is no longer valid.

Tattersall's case in point is the modern human body. Primitive upright walking "archaic" hominids (short legs and long arms) lived in Africa for millions of years before the modern human body structure emerged. The change of body form from archaic to modern, which involved modifications throughout the skeletal structure, simply happened in one fell swoop:

> Throughout this long time the basic archaic body structure appears to have remained essentially unchanged, even as new species came and went; and when modern body form appeared, it did so suddenly, and pretty much out of the blue.[30]

Despite the problems associated with evolution, there is no other way to mechanically describe how life, as we know it, came to be, other than having magically appeared "as is." Still, this mechanical approach to life is limited given the complexity of life on Earth. For cosmologists, there remains a philosophical complexity in explaining why our universe, solar system, and planet exist, and how mankind came to be in its present form. Science is not perfect, but it is all we have in our quest to understand why we are what we are.

For University of California astrophysicist Joel Primack, the sciences may actually enlarge the concept of God. "The search for scientific truth can be a form of guidance. It is as divine as any other. The foundation-building revolution that modern cosmology is undergoing today, as it seeks a verifiable description of the origin of the universe, requires that we transcend previous notions of space, time, and reality."[31] By viewing science for what it is—a description of processes and the existing relationship between cause and effect—we can be confident that it is an established fact that we live in an extraordinary universe, as well as on an extraordinary planet.

Conscious Evolution and Canine Domestication

In his research, Lipton lays the foundation for a new paradigm in biological evolution. Lipton's "biology of belief" is a holistic approach that moves beyond the scope of modern science and embraces creation as well as evolution. Although Lipton may be the first biologist to publicly embrace a creation/evolution marriage, statistics show that such an idea is not new to science or the layperson, as we have seen in the Gallup polls previously discussed.

If, as Lipton suggests, the random, motivating force behind the evolution of species is not the gene but rather consciousness, then how does a species alter itself? The domestication of the dog, I believe, provides a good example.

From archaeological evidence, we know that the relationship between dogs and humans goes back at least fourteen thousand years. Before that time, there were only wolves, coyotes, and jackals—all of which were potential ancestors of the dog. Although some studies, focusing on mitochondrial DNA sequences, conclude that the dog was actually domesticated more than once,[32] according to recent DNA studies, the wolf appears to be the canine's closest wild relative.

Exactly what series of events led up to humans and dogs becoming best friends we will never know. But a likely scenario is that wolves were attracted to human campsites and settlements in hopes of finding discarded bones and scraps of food, since they are predominantly scavengers. As a consequence, perhaps the wolves kept away other predators. Humans, seeing that this was a good thing, offered the wolves parcels of food as a friendly gesture. For the wolves, it was a deal. Although the first few generations were likely not pets as we know them, as time went on, wolves and humans became closer. Later, the wolf became the scout of the hunting party and was given a place inside the camp's perimeter. Before long, puppies were being born that had a natural liking and affection for humans, an innate sense that humans were the true leaders of the pack.

How this works, according to Lipton's theory of biological consciousness, I believe, is that when a living organism perceives something as beneficial (or hazardous, for that matter), that perception becomes a part of its "being," or its thought processes. Ultimately, this perception and its associated behavior (via neural net configuration of the brain) are stored and saved in its DNA so the next generation will inherit the same tendencies and ideas.

What happened to the wolves is that, first, they began to perceive humans as friends. As the wolves continued to live alongside humans for a number of generations, this notion became stronger, and at some point caused a change in the canines' DNA. As a result, every subsequent wolf was born with a predisposition to consider the human as a friend. But it didn't stop there. The longer and the more closely the wolf, now officially a dog, lived with humans, the more it decisively changed its behavior to better accommodate people and ensure its own survival. Humans likely helped out by killing dogs that were unfriendly. Nonetheless, the conscious behavior of the dog was stored in its DNA and passed on to future generations.

Today dogs are trained for a variety of official duties, from guide dogs for the blind to police dogs and personal guards—not to mention the family pets that perform a number of functions, such as getting the paper and alerting the household that someone is at the door.

Although dogs cannot speak, they try very hard to do so with different types of barks, whines, and growls. Everyone who has ever had a dog knows this. They can also process some verbal communication from us humans. My boyhood dog achieved a vocabulary of nearly twenty words, not by rigorous training, but simply through daily living. My sister and I, when we were fourteen and thirteen, tested her. Each time she was asked to retrieve a specific toy, such as the carrot, Frisbee, ball, or chew-rag, she would find it and bring it to us, never confused by what toy we were asking for. She also knew our names and would take the toy to whomever she was asked to take it to.

The history of the dog tells not of an accidental genetic mutation but of a conscious recognition, albeit survival-based, that it would be beneficial to befriend the human. Perhaps this is the case for all life-forms, that the effects of brain function (consciousness) are the impetus for organic life and its evolution of form into human. This, of course, leads to deeper questions, such as Where did we originally come from? And how can matter organize itself, seemingly out of nowhere, into biological organisms that become self-animating?

Explaining this deep mystery seems to be beyond science, and perhaps even beyond our ability to comprehend. So far, no one has offered a good explanation, despite numerous attempts by various cults and religions over the millennia. The best that can be hoped for in anything close to an objective manner is an approximation through the principles of mathematics, through number.

4

Cosmogony

The Origin Mystery

Number is the purest and most perfect expression of esoteric knowledge, and we must return to those bare bones.

R. A. Schwaller de Lubicz

Mathematics might be prized or ignored, but it is equally true everywhere—independent of ethnicity, culture, language, religion, ideology.

Carl Sagan, *The Demon-Haunted World*

According to big bang theorists, the creation of all matter and energy in the universe occurred from a singularity—a single infinitesimally small point—some time between twenty and ten billion years ago. Our three-dimensional reality emerged from a nondimensional reality, or, put another way, the concrete appeared from within the abstract. Big bang theorists also say that the cosmos began in an indescribable explosion of immeasurable heat, which was responsible for creating not only the fundamental subatomic particles of which everything is made, but also space itself. For cosmologists, the big bang explains why remote galaxies are speeding away from us, and why there is background radiation in all areas of the universe.

The big bang, as an explanation for the creation of the universe, is the consensus of the scientific community—what technical skill and expertise tells us. But such a theory—or any theory suggesting that

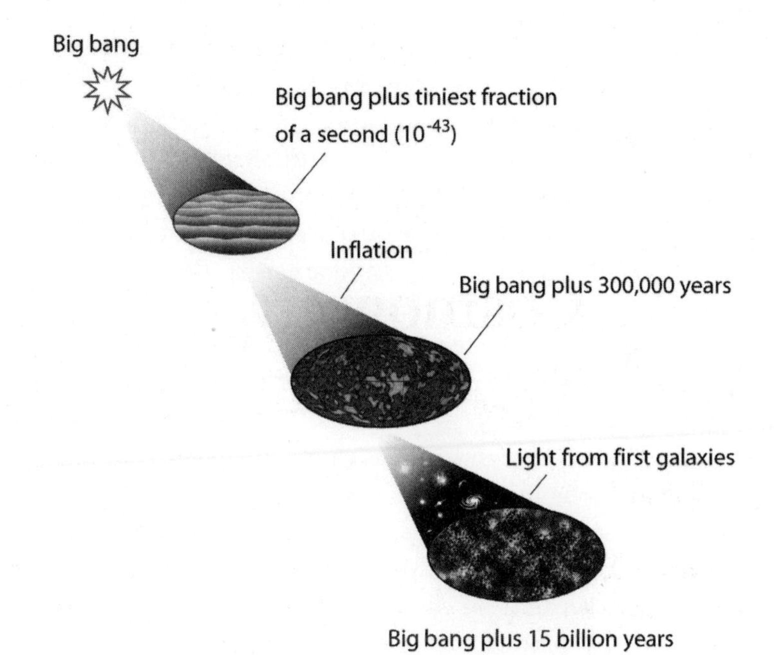

Fig. 4.1. The big bang (illustration by Dorothy Norton)

something was created out of nothing—is irrational. Of course, it is also a possibility that our universe was created out of something; we just don't know what that something was or is. Whatever the case, science and the scientific method have objectively brought us to the edge of the abyss where we must conclude that the origins of the universe are truly unexplainable.

Scientifically, there is no way to observe or measure such phenomena as the big bang. This creates a predicament and a great scientific assumption: *The universe exists.* Pragmatically, it does not matter how it was created. Cosmologically, there is plenty to study about the universe's structure and how it has evolved since its inception. Our interests, however, are cosmogonic—how the universe emerged from a nondimensional state—which brings us to an obvious difficulty: how to explain the unexplainable.

How does science, or anyone else, for that matter, reconcile the creation of all matter and energy from a singularity, an abstraction, and a single infinitesimally small point? How do three dimensions emerge from the nondimensional? On a more materialistic level, why does the proton attract the electron?

Science has a tendency to avoid explaining such questions simply because the answers fall into an area of immeasurability and subjectivity. Science is supposed to be objective. Reality, existence, and nature, however, taken to the farthest end of scientific investigation, become subjective. Science gets around this inherent subjectivity by limiting the scope of investigation. In doing so, objectivity is attained, which is quite useful in deducing relationships about nature. But this dodges the essence of true science, as Perlmutter so honestly puts it, to be in touch with "something magical." Life *is* magical, and it is the ultimate quest of science to want to discover how the universe and everything in the universe began and where they may end.

For our ancestors, questions such as those posed here were answered through myth and religious beliefs. Today's explanation comes through science, of which math and physics are the most important elements. They provide the ways and means of measurement and analysis. Mathematics, of course, is based on the concept of number. We take numbers for granted as a part of culture, but, it must be asked, What are numbers? What is the nature of numbers?

Number in Nature

Today, when the typical person thinks of number, calculation in a mathematical sense automatically comes to mind, such as figuring the monthly payments for a new automobile, the profit-and-loss statement for the owner of the car dealership, or even all the complex calculations the engineer made when designing the new car. Although number is abstract—you cannot hold the number one in your hand—number is almost always associated with a quantity of physical objects or the description of a physical object. In this method of thinking, it appears that number has nothing to do with the natural world.

If number has nothing to do with the natural world, then it must be that number is a concept created by mankind. While it is clear man created mathematics as a measurement system to benefit society as well as describe the natural world, number is not so clearly defined as a man-made concept. Although the measurement system is conceptual and can be anything from cubits to meters, the value derived from a measurement is based on the actual properties of the physical object being measured.

We see objects, and objects have form. The brain, particularly its visual ability, is designed to distinguish form automatically. This occurs through comparisons. So if all natural phenomena were invariant, we would be unable to see. We do see, so we know that variations in nature, which define form, are real. And when measurements are taken, it is the variations in the object that are being measured. These variations are assigned a value we call number.

In nature, number is disguised as form and exists behind the scenes as an integral part of the system, so much so that number is a part of our natural mental faculties we take for granted. We never notice it. But it is these variations (numbers), and how they relate to each other, that mathematics and physics use to describe reality.

For the mathematician and the physicist, these relationships are equations. The equation is man's observation and description of a given phenomenon or object. Number is what provides the physically detectable variations in nature that provide a way to compare and perceive. An equation states a relationship, but it is through number that the equation has meaning, because number is the source of the equation.

Could nature be described without number? Yes, it could, but its description would be confined to words such as *big, small, many,* and *few* or *bright* and *dark.*

As a perception, number is intrinsically tied to quantity, and quantity is inherent in nature. We see a multitude of objects—quantities—every single day: people, cars, houses, trees, and so forth. It is our mode of navigating the world, built into the brain through the visual cortex.

Number is also built into hearing. For blind people, hearing becomes acute. They learn to navigate through the tapping of a cane and the sounds of the environment; the greater the number of sounds, the more descriptive the environment. But sound is a mechanical wave (vibration) and exists without physical form.

What is it about nature that number is an integral part of it?

Nature is active. In fact, everything from subatomic particles to galaxies is in constant motion. But what is the nature of activity? The nature of activity is function. Every function of the natural world or the universe returns some consequence (value) that we perceive. The activity of subatomic particles results in atoms. The activity of atoms results in elements and compounds. The activity of elements and compounds, under another function called gravity, results in celestial bodies. What is being described is a function.

Any function has a force, a method, and an outcome. In nature, the force—the underlying impetus for all that is—is unknown and constant. (Consistency is required from the impetus for continuity—that is, we experience a continued existence, and so the source has to be constant.) The outcome, however, which is the result of the function, will vary depending on the method of the function. Thus, it is the method within the function that determines the variability in outcome. If there is only one method, then it must be the intensity of that method that allows for various outcomes. If there were a single intensity to the function, there would be only one outcome. Nature, the cosmos, would have no diversity. This variation in method can only be described as our understanding of quantity, which is what we label as number. Thus, number creates nature.

We understand nature as number only mechanically, but there are variations (numbers), just as the periodic chart of elements indicates. All elements—their atoms—are composed of protons, electrons, and neutrons. If variation (number) did not exist, nature would consist exclusively of hydrogen, which is composed of a single proton and a single electron. An increase in the quantum intensity results in helium, which is composed of two protons and electrons, and so on. Therefore, the intensity of charge within the atom is what creates variability through the number of electrons and protons. It is with this understanding of number that Schwaller states:

> We are led to this conclusion: There is a unique impulse, original and constant, and number reveals its modes to us through the varieties that make up nature. Number is therefore the essential—but also the last—word explaining the Universe. It is not a question here of calculation, but of the esoterism of number.[1]

The function at the foundation of reality Albert Einstein defined as $E = mc^2$—which from our point of view is $m = E/c^2$—is imposed on the knowable universe. Although $E = mc^2$ is well understood, its inverse, $m = E/c^2$, is esoteric. How does energy manipulate itself into mass? Hidden from our perception, with its secret origin, this function defines mass and ultimately defines us. Energy (the source), as we have discovered, is the movement of an unknown substance, of unknown origin.

Such a statement is irrational. Nonetheless, it is fact, and it demonstrates that numbers, and the relationship of one number to another, have a real existence. How could mankind develop a body of knowledge (mathematics) on the basis of something that does not exist? According to Schwaller, "The functional character of numbers is not relative or accidental, it is cosmic, the conscious revelation of our innate knowledge."[2]

The number One best describes the impetus of the function that is nature and the foundation of physical reality: an absolute existence without variation. Self-multiplied, absolute existence reveals itself through the number Two. This original movement in creating Two from One is the birth of enumeration and the polarization of energy in which physical reality is based. The entire universe, the construct of reality—number—is imposed on us as knowledge a priori. Otherwise, everything that comes naturally to us would have to be learned. In turn, this a priori knowledge becomes a part of our psychological or mental consciousness, from which we interpret and describe nature through number.

Those who object to this Pythagorean style of thought and label it as mystical, occult, or numerological fail to grasp its meaning. The true meaning of Pythagoreanism, Schwaller suggests, is that "there is an esoterism of number that proceeds from numbers seen as functional symbols . . . which is to say that the behavior of numbers with respect to each other reveals, in their simplest expression, the functions that we see acting throughout all of nature."[3] In nature, number is function, and any calculative approach to number, whatever that approach may be, ceases to be a function and becomes a description. In functional thinking, number is active. In rationalistic, calculative thinking, number is the halting of activity and a finality.

For example, *phi* is believed to be an irrational number with the approximate value of 1.618, also known as the golden mean. This is based on rational thought, as it exists in mathematics. In nature, however, *phi* is the function that governs the growth and replication of many organisms, from an artichoke to the proportioning of the human body. In nature, it is the standard unit relative to the organism.

When viewing nature for what it really is, numbers cease being a notation specifying a quantity and become the expression of life itself. From the atom to the cosmos, the function of number exists on all scales.

Numbers are very closely related to how we observe and perceive the world. Through the comparison of the things we see, we continually observe number in that a *quantity* always exists. Quantity and, therefore, number are somehow tied to reality. To investigate this further, let us engage in a thought experiment and follow the universe back in time to its creation.

At the present, we see a universe that is composed of an enormous number of objects. There are billions (to the billionth power) of objects such as stars, planets, asteroids, comets, and galaxies in our universe—more than the total number of grains of sand on all the world's beaches. But as we move back in time in our thought experiment, we eventually encounter a universe of a billion objects, then a million. As we keep moving back, there are a thousand, then a hundred objects. So far the universe, in our thought experiment, is easy to visualize. Eventually, however, we will arrive at a point when there is only a single object in the universe, which is much more difficult to picture, but not impossible. Yet, according to science, we cannot stop there, since the universe, in its primordial state, was a singularity.

At the moment just prior to the universe's creation, we enter a realm without time and number, the realm of the abstract. How can you picture a universe that is infinitesimally small and existing with no dimensions? As you can see in our thought experiment, a world without quantity—without number—is an inexpressible world for us. The concept of number and quantity is fundamental to our perception of physical reality.

Another analogy illustrates the special relationship that visual perception has with number. Almost everyone has a personal computer. The visual interface software companies have developed for their operating systems provides the user with a virtual "desktop," displayed with colors and icons (picture symbols) customizable to the user's whim. The general idea of the visual interface—referred to as a graphical user interface, or GUI—is to provide a friendly way of allowing the user to access the system and its programs by mimicking the way people naturally see the observable world. Whether a person uses a computer for gain or games, instead of viewing a text-based command prompt (the original computer interface), the user sees a host of colorful objects to choose from to initiate an application.

Today's computer games are very realistic, allowing the user to perceive events in the game as if the user were actually inside the game. (Military and commercial pilot training programs have been using this technology for some time in flight simulators.) Behind it all is complicated mathematics that not only can process numbers up to four billion in a single clock cycle (for 32-bit processors) but also display scenery and motion with extraordinary precision.

The interesting part of this analogy is that we interact with the computer's operating system and applications as if they were real.

Although the computer is real (a physical object), the software that allows the computer to perform its duties—what it displays and what we interact with—is not. *Computer software is abstract.* It exists only as the processor loads source code—which is compiled or interpreted into machine language—into memory. A form of the software exists that developers manipulate, in a source-code editor, as written text, but this is of little use unless it is compiled (or interpreted) into an executable file that the computer understands. (Some languages, such as Visual Basic, are interpreted as opposed to being compiled.) In the immediate, observable world, the printed source code is of no use.

In cyberspace (the electronic world of the computer), reality is brought to our eyes from the abstract through number. We rely on this cyberspace for a variety of things, such as reading a topic of interest found by a search engine, e-mailing a family member, chatting with friends, checking our bank account, purchasing a book. Some grocery stores now offer online shopping, with the goods delivered to your doorstep. The computer and the world of the Internet have created instantaneous communication and knowledge, which, in a sense, is an extension of our minds. It has changed the world we live in. Yet underneath this cyberspace, there is nothing more than an extremely complex manipulation of 0's and 1's.

In essence, I believe the development of the computer, with its visual properties, has touched on a fundamental principle of reality: Quality relates to the abstract as quantity relates to the concrete, and the connection between them is number. With the application of quantum physics—the fundamental principles of nature—we have created an abstracted extension of reality.

Whatever qualities may or may not exist in the abstract world, they cannot be expressed as a physical object without number. In this way, numbers can be viewed as a universal symbol of expression. Perhaps number is the very essence and nature of knowledge; and perhaps it is how the abstract quality of life relates to the manifest quantity in which we perceive that life, in that number, symbolizes the defining principles inherent in nature. It is how we understand the world around us.

Number as Symbols of Expression

The symbolic way in which we understand the world and communicate with each other through number was the life's work of twentieth-century

French philosopher René A. Schwaller de Lubicz, on whose work this chapter is based.[4] In his earliest book, *A Study in Numbers,* first published in 1917, Schwaller viewed numbers as symbols of expression, an innate part of the universe and the purest form in which to convey a cosmogony.

The principal difficulty with current cosmogonic thinking is that at the moment immediately prior to the big bang, the known laws of physics are irrelevant. It is the same condition just outside the expanding edge of the universe. Quantifiable reality does not exist. Only ideas, concepts, and principles—the arena of the abstract—exist. This creates a problem in understanding the universe, since we know that the universe is expanding. What is the universe expanding into? By definition, the physical universe has to exist in something or else there would be no expansion to observe. At this point, all that can be said is that the universe exists in the nonphysical. By definition, the universe encompasses all that is physical, so it is not possible for the universe to exist *in* something physical. The question is: What is it—this nonphysical arena—in which the universe exists?

We can think of this nonphysical arena as an absolute state, and by definition, its existence precedes all else, even the creation of energy and matter. Therefore, we can say that the Absolute exists by itself in the highest degree of completeness. Defining precisely what this state is, however, is not easy and can be expressed only as a concept. Even Harris Walker, whom we met in the previous chapter, chooses to refer to this state as Will, the *quality* inherent in the individual that he refers to as the observer, or the Self. And since all of physical reality is contained within it, in essence it is the vital animating life force of nature—what makes the proton and electron attract and what produces life at the instant of conception. Without the attachment of any significance, this abstract state is more commonly known as *infinity.*

Abstract Origins

Infinity *is* before and after the creation of the universe. It is an absolute world of abstraction, where quantity does not exist. Everything that may exist is part of everything else, and can be thought of as quality—the nature of all that is possible. In quantum theory, it is the totality of state vectors before state vector collapse. In number, it is represented by One, since it is the single source that existed before all else. The cosmogonic mystery: How does this absolute, abstract potentiality—this quality—become the cosmos and the multitude of physical objects that we observe? It occurs through polarization such as we see in the atom, with its subatomic particles of opposite charge, the electron and proton.

Think of an atom of gold (the element), which contains seventy-nine protons in its nucleus and seventy-nine electrons in its electron cloud. Gold, as does every other element, has certain properties (qualities). Gold is unique in that it displays a brilliant color and is durable and malleable. What defines gold's quality is the specific arrangement of electrons and protons, a specific polarization we have discovered that we refer to as gold's *atomic number*. Add an extra electron and proton to the gold atom, and you have mercury, a liquid. Remove one, and you have platinum. Although science understands that gold, as well as other heavier elements, was created through the immense heat of stars, where the simpler elements, such as hydrogen (one electron and one proton) and helium (two electrons and two protons), originated is unknown. Somehow, their qualities were harnessed into physical matter in the initial phase of the big bang.

Quality without quantity exists in a holistic state and can be viewed as an uninterrupted flow. But through polarization, it can exist as something other than a holistic state, something other than One. Schwaller illustrates this principle with the analogy of an electrical circuit. In an absolute state, the energy within the circuit continues forever in a circle, uninterrupted. But if resistance of some type breaks the circular current, the energy will naturally seek out a stable state. This break specifies a quantity of energy in two qualities of the circuit, and creates the moment of arrival, as well as departure, for the current. In other words, it creates two poles and a line of force. In this manner, resistance in the current resolves a *quantity,* thereby completing a unique and indefinite *quality* (the inherent, distinguishing characteristic) of the current. Herein lies the principle by which we perceive the universe. Any quantity abstracted is an indeterminate quality of an absolute state of existence.[5]

The concept of quantity is an easy enough principle to grasp, since it refers to objects. A person holds four apples, two in each hand. So the quantity of apples is four. Quality is a little more difficult concept. In our example of the apples, what would be the inherent quality (the abstraction) of the apples? Aside from minor differences, the answer is a specific taste, color, shape, odor, and texture, all of which apples share. What determines the quality of the apple is the experience, which is derived from seeing, feeling, smelling, and tasting the apple. So quality, per se, is an aspect of mind as our physical senses relate to it. Quality is the intrinsic characteristics of objects that we know as a result of experience. Where this quality comes from is another matter entirely, and is likely somehow tied to cosmogony.

For Schwaller, consciousness can be thought of as a result of the relationship between quantities and the absolute state, where we can conceive of absoluteness only in relation to quantity. This is why any phenomenon occurring by the opposition of quantity to pure quality can also be thought of as resistance to activity. When quantity and quality (resistance and activity) are equal, any phenomenon achieves a harmonic state, although limited by time and mass. The phenomenon is then said to be observable and *real.* The abstract quality, which is indefinite, appears as quantity, which is definite.

To overcome the limitations of time and mass, the process of resistance to activity must occur on a continual basis. When the moment of harmony is again interrupted, the phenomenon itself becomes a cause working against a resistance, which in turn is greater than its precedent. This cycle of harmony-disharmony-harmony, the opposition of a quantity to indefinite quality, continues, resulting in continuous creation. Since the size of quantity is always momentarily equal to that of the quality from which it came, the process of creation emerges from duality. In other words, our perception results from the contemplation of an active, indefinite, absolute state with respect to a passive, quantitative state.

This rational approach aids in the understanding of how our three-dimensional universe emerged from, and is expanding into, a nondimensional state. The very first instance of disharmony—the initial break in the absolute state (the uninterrupted flow)—becomes a new cause that contains the information of its future form. It establishes a pole, the first pole, and also a line of force whose quantity and quality vary with its nature. In essence, the pole and its associated line of force are the cause that defines form, and in it, duality becomes a unity, taking possession of similar unities to fix the form in the three directions we recognize as height, width, and length. In nature, we see it is the cyclical characteristic of formative growth. Schwaller illustrates the process with a salt solution as an example.

Sea salt grows, or crystallizes, on three regular axes, limited by six passive poles around an active central pole that results in the form of a cube. In a beaker containing a saturated salt solution, as soon as the conditions of perfect equilibrium exist, the first instance of polarization occurs and the solution begins to crystallize (or grow). Crystallization first occurs at the most accommodating location, such as the beaker walls, a foreign body, or a crystal of the same salt artificially introduced. Salt molecules naturally and automatically fix themselves along the traced axes and form a crystal. For the salt crystal, it is the birth of form.[6]

Everything that exists physically must undergo this birth of form from the abstract. We might say that the abstract is self-forming into the concrete, physical world. We simply call it nature. More appropriately, it can be referred to as first cause, which originates with the concept of number, in that quantity is a function of quality perceived—the cycle of harmonic/disharmonic succession of an initial, undefined (absolute) state. So form is mass as it pertains to quality, where quality resides in a single, indivisible state. Consequently, the indefinable absolute state (quality) contains all forms (quantities).

Since all that exists in this absolute state are possibilities or potentials, it is a single quantity with no dimensions. Numerically, this undifferentiated state is One. All else is an addition from the absolute state, then the multiplication of its first unity with absolute state. This process, taken to its farthest end, results in quantities manifesting as subatomic particles that organize themselves into new unities, which serve as a new cause. They organize and grow into atoms, then into elements and compounds, as well as solar systems and galaxies. In effect, they present to us the existing cosmos as we observe it.

The successions of poles, identical reproductions of the original pole created from the continuous equilibrium-to-disequilibrium process, serve as the construct for phenomena occurring in the cosmos. Here there is a qualitative relation (the succession of poles) that serves as a quantitative relationship (the phenomenon occurring in the cosmos). This self-sustaining and self-organizing process constitutes all that we experience in the cosmos, including plants, animals, and human beings.

For Schwaller, such reasoning is sufficient to claim that two cosmos exist in a single universe.[7] One cosmos is absolute (the macrocosm) and the other is harmonic (the microcosm), or quality fragmented and organized into quantity. Such thinking mirrors the state of physics where there are the immediate and quantum worlds—two very different worlds that exist simultaneously.

Through cycles from polarization, ideation, to formation, the physical is created. We see this in everything from the quantum world (electron/

proton, matter/antimatter) to the immediate world around us, which cosmically is light/dark and biologically male/female. In essence, I believe this conceptualized process of events is the "magical" to which Perlmutter refers, the simultaneous manifestation of the universe from the abstract to the concrete, from the quantum to the classical. It is movement (energy) measured by us as moments of time.

Symbolically, the manifestation of the universe can also be expressed through numbers, proportion, and hierarchy. In tracing back all that exists into quantities of indeterminate quality, there is only number. Although there are other ways we might seek an understanding of the universe, number is efficient and effective.

From the Abstract to the Concrete

We take numbers for granted, with the understanding that the concept, taught to us at an early age, has always been around to perform useful calculations. Have you ever considered why the concept of number exists at all? One possibility is that long ago mankind saw that it was useful to keep track of things: children, or the number of animals that must be killed in the hunt to feed everyone in the clan. Also, through the observation of nature, it would be obvious that a multitude of objects exists—plants, animals, rocks, and people. Why are there so many things?

Man must also have observed through his own experience that all of nature reproduces, and that any given population would continue to grow unless checked through predation or by some natural catastrophe. As soon as the human brain took its modern form, reflection and introspection must have come into play in mankind's traditions. Rationalizing life is a natural human characteristic. Man must have wondered where everything came from, just as we do today. Through reflection, it would long ago have been an obvious deduction that all people eventually reduce to two. This poses a more mysterious question. Where did the first man and woman come from? Or, for that matter, what about the first plant and the first animal? These early humans would not have been able to conceive of a world before Two. Nor can we.

If life was not dualistic, if there was no Two, there would no process for multiplication. With this biological fact, there is no rational way to arrive at a single person, plant, or animal. Yet it must be explainable how we, as well as all of life, are here. Deductively speaking, it must be the case that the abstract absolute state, what Schwaller calls the irreducible One, possesses a dual nature.[8] In other words, this duality is a prime quality of the cosmos. *Why* this is so cannot be addressed. What can be affirmed, though, is that a progression occurred from an absolute state to a cosmos containing a multitude of objects. As a result, we experience quantity as well as quality, the concrete and the abstract. So it also can be deduced that the absolute state's dual nature, which is both passive (feminine) and active (masculine), must manifest itself. In this, Schwaller finds pure causality for the act of polarization, a desire to separate.

Although the nature of the absolute state, of One, is unfathomable, it is possible to recognize through Two its manifested dual nature. And Two, which is a unity of opposites (proton and electron form a unity known as an atom, for example), is the manifest nature of the irreducible One. But there are really three distinct natures: the irreducible One that is potentially active as well as passive, passive (feminine), and active (masculine). From this moment on, when there is motion (energy), there already exists opposition to it.

As in the example of the electrical circuit, the infinite flow of energy is broken, creating the first polarization. The active nature is the moment of the current's arrival. The passive nature is the moment of the current's departure, or the part that is left. Since the circuit is now empty, it can receive, and since it opposes activity, it determines a *quantity from a quality*. Schwaller explains that this abstract process can be thought of as the fertilized, productive nature that creates through the "quantitative determination of vital activity."[9] Therefore, quality is the number whereby quantity is the measurement that fixes the quality. This is the mysterious act of creation that occurs at the subatomic level, the process of the abstract becoming concrete. It is why we experience the abstract as a function of the concrete.

In our example, however, the feminine and masculine natures still have not become individualized and organized. They are still interdependent. (Procreation has yet to occur.)

In figure 4.2 there are two passive (feminine) natures and two

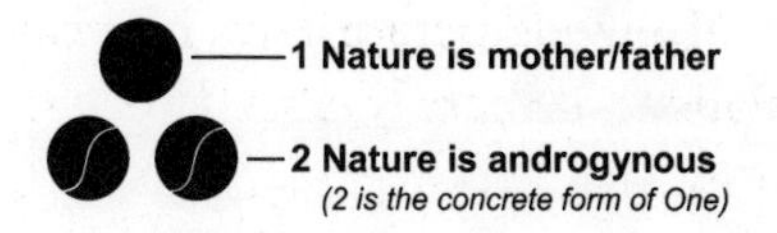

Fig. 4.2. The number Two as Creative Unity

active (masculine) natures, plus a nature that cannot be defined except by the term *father/mother,* indicating that both natures exist simultaneously in One. But since Two must include One, it contains all five natures. Yet the union of One and Two is Three, which can be called a unity, and two times this unity (which is Three) is six. Schwaller refers to this as the Creative Unity.

In figure 4.3 are the five natures of the Creative Unity, plus three dual natures of unity manifested. So in Three there are eleven new and creative natures: $5 + (3 \times 2) = 11$. Although Three is a harmony in itself (three masculine/three feminine natures), it will be able to procreate only by the fecundating action of the Creative Unity (the first triangle). As a result, Three is unity perfectly manifested.[10]

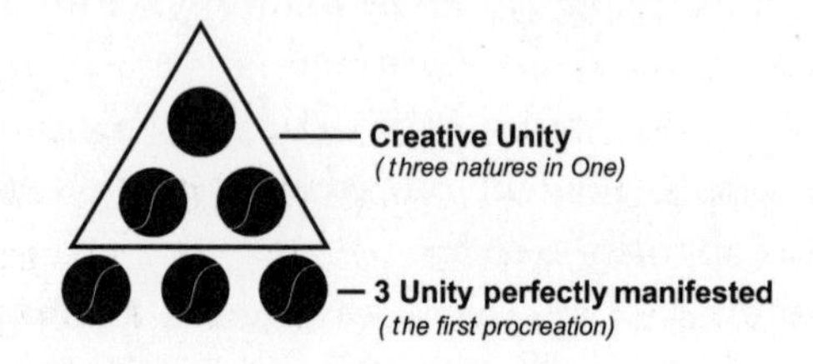

Fig. 4.3. The number Three as unity, perfectly manifested

It is noteworthy that creation is produced internally from within the absolute state and is purely a *qualitative* disequilibrium, whereas quantity is the shaping receptacle of activity. On the other hand, procreation requires activity acting on an independent resistance of the same progression.

In figure 4.4, there are the eight natures (4×2) in Four, the first number to be procreated by what has previously been procreated (Three). Four contains eight natures (4×2) plus the triune (three-in-

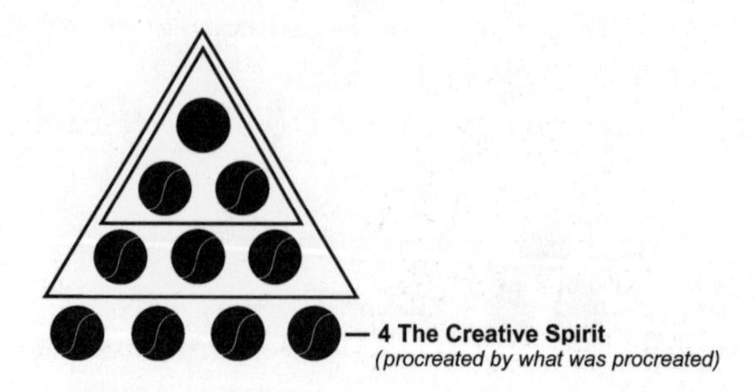

Fig. 4.4. The number Four as the Creative Spirit

one) nature of the Creative Unity. So in all, there are eleven natures: 8 + 3 = 11. In this final abstraction exists the potential manifestation of all eleven natures. Four also includes the creative power, 11, so it has the ability to procreate. As number, however, it is 10: 1 + 2 + 3 + 4 = 10. Nine principles surround it, as well as being contained by One, the source of fecundation.[11] Four occurs through procreation by what was already procreated, and is creation relative to concrete multiplicity.

Four can be thought of as the most fundamental source of creation—the Creative Spirit or the active principle—that fecundates and maintains the cosmos and all life.[12] It is the first power of the One. What this means is that the energy of One (the absolute state), once manifested as the first vital quality in number ten, is the process of the abstract (nonphysical) becoming concrete (physical), which is quality being quantized.

In figure 4.5, this process of the absolute state acting upon itself is advanced one more step in that the "becoming" of the abstract into what we perceive as the concrete is complete, meaning that matter exists as the physical universe. *Five is quality fully quantified.*

For all conscious life, however—the ultimate product of quality formed as quantity—this process appears as a mystery, since we have no recollection of it. This is because as the Self (the observer) descends from the abstract to the concrete in the process of quality being fixed through quantity, there yet exists a physical organ able to provide a memory, which is a necessity for higher thought. There are two characteristics of the physical world (whether person, animal, plant, or rock): quantity and quality. Quantity is easy to understand since it is the most obvious concept we experience. Not only can we see quantity but we can also feel it, such as a blind man feeling the number of apples on a

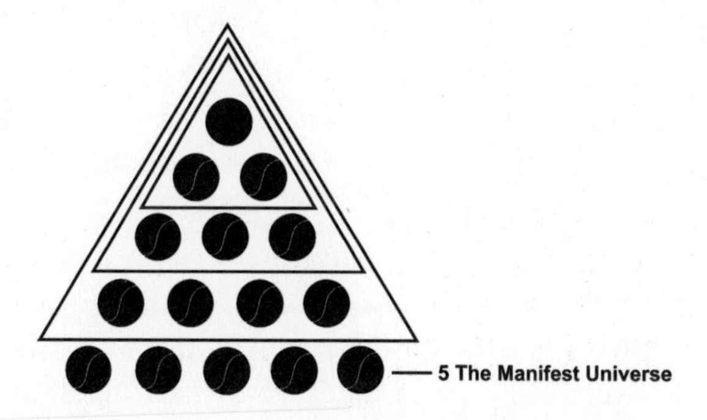

Fig. 4.5. The number Five as the Manifest Universe

table. Quality, however, is not so easy to understand. We understand it because it is fundamental to our biology. Quality is abstract and simply exists. What Schwaller is expressing in his philosophy is that quality exists in a nondimensional state and that quality is the essence of the core of our being. In order for quality to be experienced, it must become (fragment) or descend into the physical reality.

For the Self, the abstract world is real as long as there is nothing concrete about it. As soon as the Self enters the concrete world, however, when the Self is appropriated to height, width, and depth, it can no longer conceive the abstract through its physical representation. Yet there is a residual memory of the abstract embedded in the awareness that all life has. This memory—how to interpret and recognize shapes as abstract qualities—is what produces a unique identity and perception for all individuals. Schwaller refers to this residual memory as sympathetic consciousness or "innate consciousness."

This innate consciousness is inscribed into the nature of all organisms. By necessity, the cyclical principle of harmony, which serves as the quantizing force, establishes the function of the organism,[13] where the function is the actual principle of living nature and is used in this context like an equation $a \times b = c$, where a is distance and b is speed. Organisms are functions of nature. (For example, digestion is a function of all living things, as is respiration.) For Schwaller, "This principle is a reality beyond corporeal matter, but it assumes a body; it incarnates by means of the harmony of the ambient elements."[14] Innate consciousness is the "experience of life" and "the specific character of each thing," and "the quality that is pure spirit in Nature."[15]

Unity and the Role of Unification

According to the big bang theory, the universe emerged from a nondimensional state into three dimensions. Such an idea, involving the emergence of the concrete from the abstract, requires a conceptual process of unification. In order to understand that the concrete emerges from the abstract, one has to be able to conceptualize that everything exists, at the most fundamental level, in a unified state.

Quality's descent (movement upon itself) into quantity, expressed through number, can be a difficult concept to grasp, since we relate primarily to the final sequence in the quantification of nature (the universe as we experience it). Here is an easy way to think of number as the quantification of quality. All elements are made of atoms. Gold has a specific number of electrons, as do silver and all the other elements in the periodic table. All atoms are made of the same thing, energy. What distinguishes gold from silver is the *number* of protons. The energy is the quality and by itself has no dimension. When it is bundled into an atom, it takes form, becomes physical. Numbers are abstractions and can never physically exist, yet there exists a multiplicity of people and trees, as well as planets and stars, each of which has and displays certain intrinsic qualities. These qualities are based on molecular structures, a specific arrangement of one or more elements, which serve as the basis for various characteristics we observe. In this way, everything is of the same nature existing in a single universe formed with a common set of elements. Some elements are simple, such as helium and hydrogen. Others, like carbon, are more complex.

It is only through a sequence of a unifying principle (unity) and the role of unification applied to movement (energy) that a multiplicity of objects emerges. The sequence is the long story of cosmic evolution that began with an abstraction and through many sequential events ends up with us. Gravity, which is a mysterious phenomenon itself, can be thought of as a unifying principle. So can chemical bonds and the strong nuclear force of quantum physics. From the quantum level up, unity occurs at each level from elements into molecules, and then into compounds that take on a distinct form. Another way to envision unity is that each abstraction of the next number, working upon its predecessor, creates a new union, and so the complexity we need to experience in three dimensions occurs. Before there can be three dimensions there has to be two, and before two, one, and before one, a nondimensional state exists. These dimensions (number) emerge out of the previous dimension. Think of it as geometry. There is a blank paper, then a dot appears, then a line, then a geometric shape (square or triangle), then the paper has to fold in order to truly be three dimensions.

Absolute Unity

To describe the knowable universe, three dimensions are required, plus time (perception of movement) as well as quality as it relates to quantity (the specific and intrinsic characteristics of an object). So five factors are naturally involved in describing the universe, as well as the absolute state (infinity) in which it exists.

In an absolute or unchanging state, quality exists without quantity. So quality exists as a single entity and contains all possibilities of what the known universe may be. Quality is something as opposed to nothing. Numerically, One (the number 1) represents the absolute state.

For quality absolute to manifest in a way that we recognize, four events must occur. First, the universe must be created from the absolute through polarization that occurs as an abstraction from One. The universe will be abstract (without form), represented by a single infinitely small point. Now, there are *two* qualities, since the absolute state differentiates the universe from itself. As a result there exists Two, the number 2.

For the universe to take a shape we recognize, quality must differentiate three more times. First, a second point is created from which a line can be drawn, represented by Three, the number 3. Then it must differentiate into a plane (Four, the number 4), and finally into a volume (Five, the number 5).

Since each state, represented by number, is the product of the previous state, it retains the quality of the previous states. So Five contains 1, 2, 3, and 4, the sum of which is 10. Therefore, the absolute state, in its manifestation as the concrete world represented by number, takes form when it reaches 10. So it can be said that the first power of 1 is 10, the second power being 100.[16] In other words, Five, the physical universe (matter) as we observe it, is a quantity through the *quality of 10* creative principles. Through these ten principles, Absolute Unity exists among quality differentiated, giving rise to quantity. Put another way, how we use the value of 1 in the concrete world by referring to a single item (a quantity) is really an abstraction that requires 10 principles or qualities to achieve.

Notice the number of widgets in figure 4.6. There are ten of them within the outermost triangle in the number Four.

In figure 4.6, the triangle of Creative Unity (Two) contains its creative association (in its passive nature) in 20. Likewise, the perfectly created (as opposed to procreated) number, Three, is in manifestation 30. The procreative triangle of the creative spirit, which is a generative disequilibrium, is found in 40—a number, by the way, that is found in numerous ancient mythologies and religions. For example, before Jesus began his ministry, he fasted in the desert for forty days. Forty represents the generative impetus, the internal drive, for him to begin preaching the message of the Gospel. It also rained for forty days and nights in the story of Noah's flood, signifying that the deluge was a result of not normal weather but the creative forces of the Hebrew God. According to the story, the old world and civilization were destroyed and new ones created.

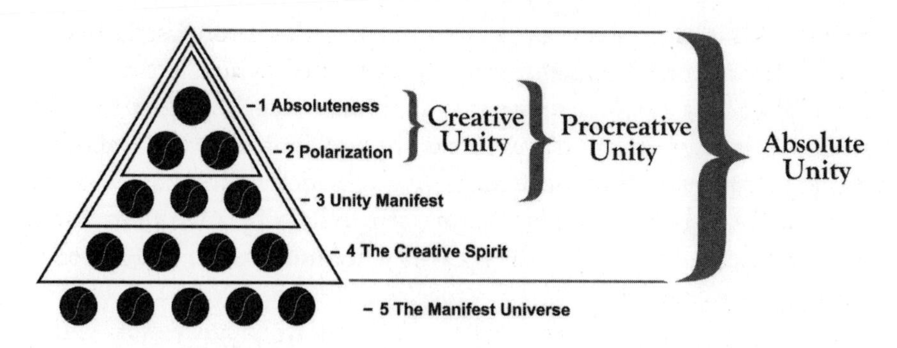

Fig. 4.6. Absolute Unity

It may seem a little confusing that 1 is 10, and 2 is 20, and so on. Accordingly, it must be recognized that numbers do not simply designate quantities (values such as 1), but are also concrete definitions of formative energetic qualities (principles) of nature. The numeral 1 is a value associated with quantity, a single physical item. For that quantity of One to exist, however, ten principles must come into play to quantize quality.

This act of the abstract becoming concrete is founded on the idea of harmony (unity) and disharmony. Each successive unity is a triple power and a double nature, in that one nature is active and the other passive. The active and passive natures are coupled to each other by mutual action and reaction. Furthermore, the duality of nature exists in everything created, since it is of the creative and defining source, and will exist in all subsequent procreations.

In the world of the abstract, only 1 through 9 exist as qualities only, although we have yet to mention 6 through 9. The two natures, masculine and feminine (active and passive), result from the relation of numbers to each other, and separate out. The numbers of a masculine nature are called odd, and are 1, 3, 5, 7, and 9; even are 1, 2, 4, 6, and 8.[17] Since 1 contains the masculine as well as the feminine nature, it is considered both odd and even. Even numbers, which are equally divisible, are harmonic and receptive. In other words, they are not creative by themselves. Masculine numbers are disharmonic, cannot be equally divided, and are therefore active and creative. They can be divided only by themselves and by 1. This mutual action and reaction of the masculine and feminine through polarization, ideation, and formation produces the entire universe, the intangible aspects as well as the tangible.

The constant action and reaction of these two natures are best represented in the concept of *pi* (π), a number that is irrational, since it can be defined only by what it is not. It cannot be represented as a ratio of any two whole-number integers, and consequently it does not fall into a repeating pattern of any sort when written in decimal notation. In *pi,* Schwaller finds that the relation of the circle's diameter to its circumference is the same as the relation of 1 to 9. In his own words, *pi* is the equivalent of saying "Nothing exists without the life of the Spirit."

Four as the Idea of a Physical World

It is from the principle of unity that numbers develop in the abstract world. One and Three are particularly attached to the cycle of polarization from which other numbers have their origin. After the first procreation (Three, a result of Procreative Unity), all other numbers are derived from procreation.

After polarization, creativity occurs as a result of idea (thought), which is the image of the form to be generated. During polarization, *generative poles* are selected and arrange their positions to fix the shape. Complex lines of force, determined by the generative poles, establish the frame of the physical form. Polarization led to creation by addition. Now, procreation occurs by reproduction (multiplication) with the numbers 1 through 9, and their powers 10, 20, 30, and so forth, as the energetic foundations of physical forms.

The absolute, One, is the point that generates lines by addition, since it cannot manifest itself any other way. ($1 \times 1 = 1$; and 1 times a point is always a point.) The point becomes a line by adding to itself another point, in much the same way a cell adds to itself. (Division and addition are really the same effect.)

This occurs through the triple nature of the absolute, the One. This is the quality that allows the concept of the procreative function to exist, during which division is simultaneously addition. Addition is the consequence of division, and division generates multiplication.

From a point called One, Two is another point generated by division. A third point (Three) simultaneously occurs, which is the solution, provided by the function of creativity. Three is engendered by the first multiplication of Two, and provides the first procreative state. Four, then, signifies both the last term of creation and the first term of existence manifest. It is a square, and the first stable figure embodying the creative essence of the first procreation.

The point creates only by addition (division). The resulting line can only procreate, and never create. A line adding to itself remains a line. As soon as it is multiplied by itself, however, it becomes a face. With the emergence of idea from a single point, the first term is created and the last term procreated.

The numbers 1, 3, 5, and 7 are creative numbers, since they do not contain a procreated form. They are prime numbers, meaning that they are not composed of any other numbers except 1. As prime, they generate all figures and forms by addition. In the concrete world, they play the same role as does the absolute One in the abstract world. Through addition, creative numbers produce (procreate) other prime numbers, such as 11, the first representation in the concrete world by the abstract Two. Eleven can be produced only by addition through 1, 3, 5, and 7. It cannot be produced by multiplication.

Conceptually, the numbers 1, 3, 5, and 7 provide the perfect transition from the abstract to the concrete world by multiplication. Multiplication of a unity, not itself engendered by multiplication, presides over this cycle of ideation. Its natural consequence is the creation of the face.

Although the square is not the first face, it is the first stable face, meaning that no complex effects are produced that wander from the natural consequence of the first function. The first face is the triangle, since three vertices are adequate to establish a face. As soon as the triangle is multiplied by itself, however, it immediately produces a square. This is because the triangle's face is not engendered by multiplication. The triangle results from the first addition.

The first perfect triangle needs to fulfill the following conditions:[18]

To be the first face
To be constituted by three vertices
To engender a stable face by multiplying itself
To be constituted by a stable face or contain its potential

The first triangle (Two, or Creative Unity) fulfills all conditions except the first and the fourth. There can be no face that comprises only a single unit of height and two units of length. (This constitutes a line, not a face.) Nor does it contain a stable face, as the fourth point requires. The second triangle (Three, or Procreative Unity) fulfills the first, second, and third conditions. The fourth condition is met beginning with the third triangle (Four, or Absolute Unity).

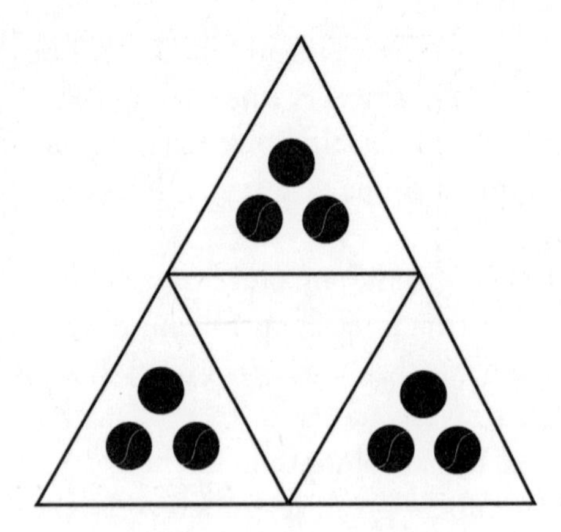

Fig. 4.7. The quaternary triangle

Four, or Absolute Unity, constitutes the tetractys or quaternary, which (1) is the first possible face; (2) is triangular, and by this fact made up of three vertices; (3) is engendered by multiplication, with itself being a stable face, which is 9; and (4) potentially contains the first square, which is 4.[19]

Since the first triangle (Two, or Creative Unity) is, as we know, considered to be the first reducible unity, we must regard the quaternary triangle as potentially being the first triangle constituted by three equal angles, which we can show (fig. 4.7).

This determines a triangle whose height is half as great as its base, and clearly frames Absolute Unity, which is, according to Schwaller, "the vital breath in the things of this world."[20] Therefore, contained in the quaternary triangle are actually two distinct triangles. These are the triple creative nature in the One, and this triple nature as manifested in the thrice, or the three unities of the first triangle.

This figure, which is the first manifest triangle, fulfills the above conditions and engenders the perfect square, which is 9, for it contains all the numbers. In this way all squares become divisible into two triangles, since 4 is 1 and 3; 9 is 3 and 6; and so forth. The square, beginning with 10, becomes compounds (fig. 4.8).

The uneven (positive, masculine) numbers form squares where the numbers being represented form two constituent triangles, one negative and the other positive (even and uneven). For the moment, the even numbers give the square whose triangles are either both negative or both positive.

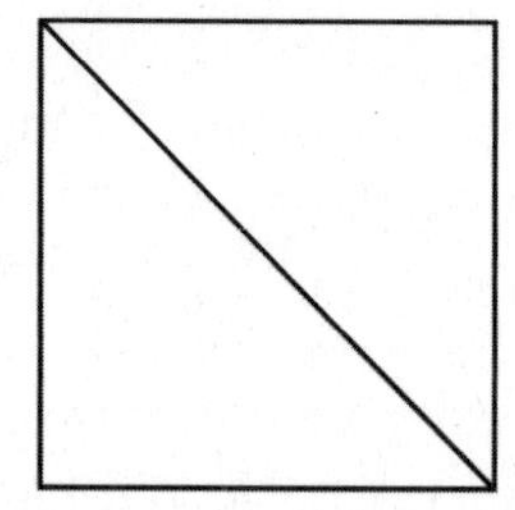

Fig. 4.8. The square, divisible into two triangles

Although there is no reasonable explanation for this fact, Schwaller contends, it is sufficient to indicate that the even numbers are all compounds—2 equals 2 times 1, 4 equals 2 times 2, 6 equals 2 times 3, 8 equals 2 times 4, and so forth—and this function determines the intermediate or variant forms of the square, such as the figures or the parallelograms with two unequal sides.[21]

Creative numbers 1, 3, 5, and 7 establish the following principles and are found everywhere at the base of the two natures' creation. The multiplication of any quantity by itself (procreation) results from the addition of a positive, odd (masculine) nature to a negative, even (feminine) nature. Other figures—pentagons, hexagons, and polygons—are the effects of an addition of different triangular and square figures, just as the numbers greater than 10 result from the addition of the numbers 1 to 9, to 10.[22]

Formation and Form

Physicists have been in search of the smallest indivisible subatomic particle for some time. Some scientists believe, in theory, that particles can be split indefinitely, and that the whole substructure of physical reality is infinitely small. The underlying truth of the universe may be field-like, however, as opposed to particle-like. According to this theory, if particles are split indefinitely, *there is only form and no content,* a point that most physicists agree is a good place to start. The various features of the universe, then, are explained in geometric terms of various fields, and particles that seem to exist are explained as features of the flow.

First there was a point. Addition (by division) of that point created a line. Multiplication of the line then created a face, the square. From the face, passing through all the phases of polarization and ideation results in the square's multiplication into the cube. Polarization and ideation is how Schwaller expresses creation of the physical world from a single source. The source had to act upon itself in order to create something different. The square (one dimension), being the point of departure for the form, proceeds to the second dimension to become the third dimension. The first cause is the point, the second the line, and the third the face. Recall that each of the three cycles is subdivided into three cycles of activity, inert reaction, and effect, whereas only the cause varies, and with it the effect as well.

At the base of all creation is movement (energy), where addition is the first movement and is linear. The second movement is multiplication, which is planar. Multiplication of the planar is the third movement, which is a solid.

For a solid, movement is composed of three principal directions. If a body moves individually and is not in relation to another body, then the rotary movement is about itself. The first movement, which is linear, constitutes the axis. The second movement, which is planar, constitutes the equator. The third movement results from the mutual action of the first two. As one increases, the others diminish, producing centrifugal force.[23]

As soon as the face exists, there exists also a reason for immediate movement to a form. Similarly, there is also reason for the existence of rotary movement as soon as an object exists, which is an effect from the line to form, because at its base, there is always the existence of a formative axis or a first line of force. Procreation begins with the function of a linear movement, followed by two circular movements. (Remember that the first face is the triangle, whereas the perfect face is the square.)

The first perfect triangle is Three (Procreated Unity) and the first perfect square is the quaternary, Four (Absolute Unity). The first perfect form will therefore be that which is made up of these two triangles—that is, of 16 units. The first square, which is 4, gives the first perfect form: the cube, which is 16.[24]

The triangle alone cannot give rise to form through the procreating function (multiplication), because a triangle times a triangle gives a face, which is 9. Therefore, the first form exists by the addition of two triangles (which gives the square) and the second, by the multiplication of this square.

In principle, the first triangle is Creative Unity, the second being Procreative Unity. These two added together are the Absolute Unity or the Creative Spirit, or 4, and when multiplied is 16. The result is the first cube. Although the cube is the first perfect form, it is not the first form.

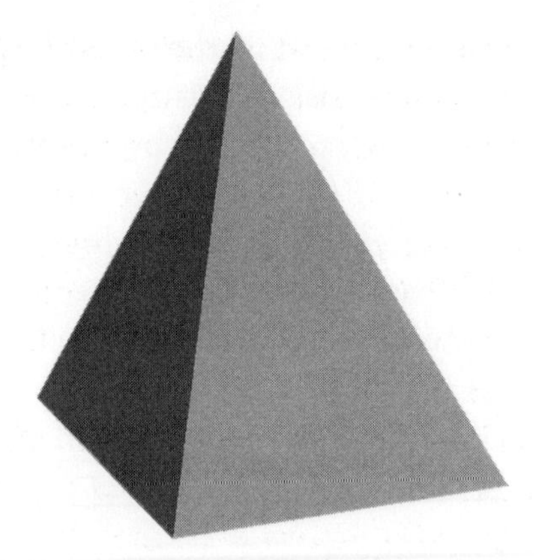

Fig. 4.9. The tetrahedron

The triangle is actually the first form, having an effect only by addition, but it cannot enjoy the procreating qualities of the perfect form. This means that the triangle is the creative or causal form whose ideal form is the cube. So, by necessity, the first form is triangular and engendered by the addition of triangles. This form (see figure 4.9) is called the *tetrahedron*.[25]

Four times the triangle gives the tetrahedron. From the face of the three sides, the fourth (the base) is established. Therefore, in principle it is made up of 12 units. The fact is, however, that it is composed of 9; so nine is the cause of the final cube, manifest by 12. From the moment the base is constituted by the fourth triangle (when it has established the twelfth point), it creates the reason for a form, whose multiplication by 12 becomes the cube (fig. 4.10). It is made up of 24 times the

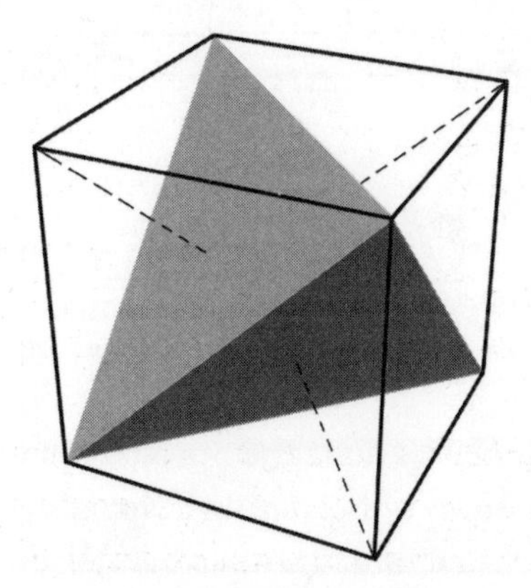

Fig. 4.10. The tetrahedron creates the cube.

tetrahedron, whose base is formed by the triangle defined by the diagonal of the face of the cube whose apex is the central point of the latter. Therefore, all other forms may be derived by addition from the cube and the tetrahedron.

According to Schwaller, beyond the third dimension no other dimensions can exist.[26] Although time is often referred to as the fourth dimension, time is really the first nondimensional state—quality, causal space, or, in absolute terms, everything or One. More appropriately, perhaps, time is a concept to express the physical changes we perceive in the immediate, observable world. Time is our perception of movement within the universe.

Proportions of Nature

Everything existing in the cosmos—quantities of the universe—is based on, and limited to, the proportions of number. This is why mathematics is so successful in describing the world around us. Through observation, and by analogy, if the immediate cause of a phenomenon is investigated using the first proportions, its immediate cause and effects can be determined. In other words, the intrinsic nature of the phenomenon can be understood. This provides a basis for conceptualizing the cosmos as the manifestation of absolute quality into definitive quantities. For the construct of the physical world to occur, three events in each of three cycles must be complete: those of polarization, ideation, and formation.

TABLE 4.1. THREE CYCLES OF THE PHYSICAL WORLD

Cycle	Polarization	Ideation	Formation
Event	Creation	1st Procreation	2nd Procreation
Activity	3	5*	7
Inert reaction	4	6	8
Effect	5	7	9†

*1 active and 1 passive, thus 1 androgyne, a new active cause.
†By scission, this last cycle produces the two distinct natures that will procreate in what follows.[27]

In the cycle of polarization, the absolute One becomes Two. During the process (addition) of becoming Two, five principal natures (two feminine, two masculine, and another being both mother and father) are created. This also creates the first disharmonic state, which becomes the entry (activity) point of the next cycle. In the cycle of ideation, the One fecundates the Two, thereby producing Three, which in effect is the reducible One whose nature is equally active and passive. This is the fundamental energetic form and the first quantity, Five. Here, the fecundated nature absorbs the seed, establishing lines of force for the next stage, formation.

In the cycle of formation, the internal separation of the fertile seed's two natures occurs in a way that one predominates, constituting growth of the new seed to maturity, in quantity and in form. That is, Three, through movement (multiplication), becomes Four.

Seven is the seed (what has been referred to as the face) that, through formative growth, becomes the manifest object. What results is the succession of numbers as proportion and measure of the cosmos. This process occurs through activity, inert reaction, and effect—three events within three cycles. Seven terms (factors) constitute the cosmos in appearance, whereas nine terms (functions) constitute it as fact.[28] Within the cycles, the effect becomes cause and *creation occurs continuously,* although procreation is the only creative form we have rational knowledge of. True creation appears to us as magical, and we tend to attribute it to God.

Schwaller contends that constant creation (what is hidden from us through our own mortality—that is to say, *perception*) is the normal, active state of the cosmos developing through polarization, ideation, and formation. All that exists in the universe is generally passive relative to the absolute. In the absolute's (One) manifestation as a multiplicity of forms, the first activity, what Schwaller refers to as the Primordial Scission, alone produces the effects. *These effects are the physical cosmos.*[29]

Constant Creation

There is difficulty in relating these effects (the cosmos) to modern scientific theory in that the initial act of creation, the big bang, requires the existence of an intelligence or creator. Science has yet to discover any evidence that an individual creator or intelligence exists. But how would we know of a creator if one did indeed exist? More importantly perhaps, there are difficulties reconciling the big bang theory with quantum theory. Specifically, and this is a big problem, the laws of physics break down when the universe must be contained in a singularity.

Such difficulties have led some physicists to develop alternative cosmogonic theories. The big bang, once the darling theory of the scientific intelligentsia, now has its doubters. One of those doubters is renowned physicist Stephen Hawking, who is "now trying to convince other physicists that there was, in fact, no singularity at the beginning of the universe."[30] The big bang theory requires the existence of a boundary between the expanding universe and the "stuff" that it is expanding into. So far, even with the far-reaching eye of the Hubble telescope, no boundary has been detected. According to Hawking, it might be the case that there is no boundary to the universe, and that the universe has always existed. Such thoughts led Hawking to write, "The universe would be completely self-contained and not affected by anything outside itself. It would neither be created nor destroyed. It would just BE."[31]

In searching for an alternative cosmogony, there is really only a single candidate. Creation occurs on a continual basis, its source being the world of wave functions described by quantum physics. According to physicist Julian Barbour, "Creation did not happen in a Big Bang."[32] Rather, it is an ongoing process in the here and now, a process that we can understand. Instead of having laws of nature, Barbour suggests that all that is necessary is a single law of the universe. He believes that such a law may exist in the Wheeler-DeWitt equation, an equation that does not require the concept of time. According to Barbour, the timelessness of constant creation emerging from the quantum mechanics is compatible with the structure of Einstein's relativity theories. For Barbour, time does not exist, and the perception of motion is illusion. What does exist is "now," as determined by the fusion of space and matter into a single configuration of the universe. There is no single now but an infinite number of them, all of which are instances of a common principle of formation. In Barbour's view, they are all different instants of time that we experience.

The importance of the Wheeler-DeWitt equation—a quantum-mechanical wave function that describes the universe—is that it holds true for the universe as a whole in the here and now as well as in its beginning, according to quantum gravity theory. For Barbour:

> The Wheeler-DeWitt equation is telling us, in its most direct interpretation, that the universe in its entirety is like some huge molecule in a stationary state and that the different possible configurations of

> this "monster molecule" are the instants of time. Quantum cosmology becomes the ultimate extension of the theory of atomic structure, and simultaneously subsumes time.[33]

In the Wheeler-DeWitt equation, there can be a variety of possible states of the universe, where the probability of each state can be determined. Which state actually occurs is dependent on us. Such a concept agrees with Walker's model of consciousness, in that the absolute state of existence is the energy wave, and that the nature of the universe itself determines what we are. Like Walker, who concluded that the brain is an interpretive device for state selection and operates as a result of state vector collapse, Barbour believes that "our existence is determined by the way we relate to (or resonate with) every thing else that can be."[34]

For Barbour, all that needs to exist to explain the cosmos is configuration space and a static, well-behaved wave function. Configuration space is not the empty space that we are accustomed to thinking of, such as the space beyond Earth's atmosphere, but an abstract arena, such as Schwaller's "quality," where everything exist as One. Such a model circumvents the reckless thinking that our existence is highly improbable, since it is fact that we are here, and states that our existence is inevitable, including the most fundamental elements of organic life. Existing within the Wheeler-DeWitt equation (wave function) is the method for the creation of complex molecules like proteins and DNA.[35]

The Anthropic and the Absolute

Our biological form is ingrained in the immediate, observable world. It seems ridiculous that the cosmos in general and life in particular should be the result of some unknown quality that inexplicably quantifies itself. Yet from a cosmogonical perspective, such a notion merits attention, particularly because of the anthropic principle. If a hypothesis or theory is put forth to explain the universe, that idea also has to explain why life in general, and mankind in particular, exists. In the Darwinian model, life is a grand accident brought about by the happenstance of Earth's particular position in the solar system, where the right conditions and events somehow enabled life to come forth. Yet there is no explanation for how inert chemicals combined to do so. There is also no explanation for the Cambrian explosion or the human explosion.

Life just seems to be, and perhaps it is only our madness that impels us to find out why. Such has been the condition of mankind from time immemorial. Through the efforts of the scientific disciplines, however, a massive body of knowledge exists that goes far in explaining the anthropic cosmos that we see as reality. After what we have covered, it is time to focus a little bit on the more tangible aspects of the mystery.

The human brain receives billions of bits of information each moment through our five senses; and the human mind is an extraordinary processor of the abstract occurrence of emotion, thought, concepts, and calculations. It is as if the human brain has been specially formed to translate the concrete into the abstract and the abstract back again into the concrete—the principles of nature, *of number.* Assuming physical reality is created through number, then we as humans have the ability to abstract that reality back into number, through mathematics/physics. Every conscious moment, we automatically perceive these principles that manifest as the concrete, immediate world. And in turn, the reality of the concrete is then perceivable. Our biologic consciousness is a circle moving between the concrete and the abstract, coalescing as a unity of mind and body.

Consciousness appears through form with the unique perspective of an observer, a fact that cannot be denied—a fact that has been the focus of study for a small but growing number of physicists. Although Schwaller's treatise on number is clearly a philosophical approach to the anthropic mystery, its rational exploration into the irrational foreshadows the subsequent efforts of the world's greatest physicists throughout the remainder of the twentieth century—Einstein, Bohr, Heisenberg, Schrödinger, Dirac, Wheeler, DeWitt, Walker, Penrose, Barbour, and of course Stephen Hawking.

Although conclusions as to what these great scientists have endeavored to discover as the fabric of reality are purely interpretational for each of us, the direction of their work is clear, in my opinion. By realizing and embracing that all of nature is a concretion of the abstract, the pure quality of consciousness and being, and that man is the pinnacle of this concretion, we come to a greater understanding of what life is. Schwaller's Abstract Man, Two created from One, is the creator, just as mankind is that which is created.

This principle of self-creation is true in the sense that the abstract quality of observation—that which some call God and the impetus of all that exists—is manifest in a stream of constant creation. The irra-

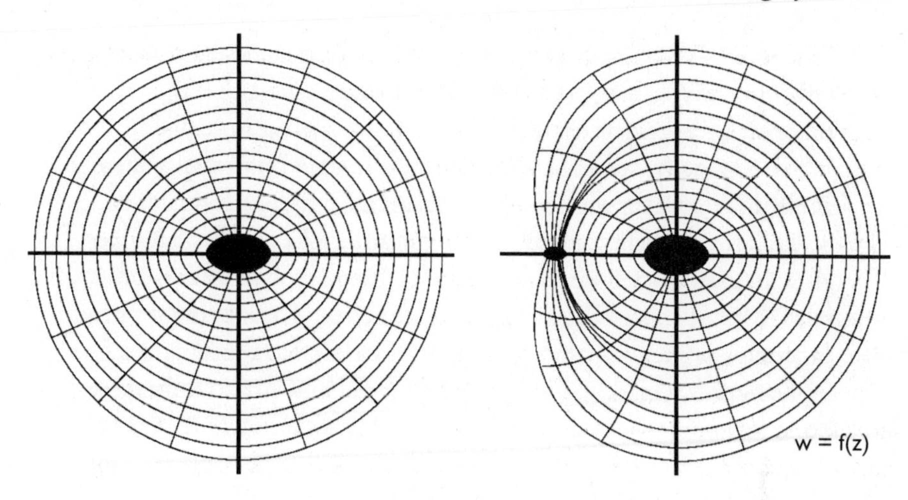

Fig. 4.11. The reflective principle

tional then becomes the rational, and mathematical and conceptual problems associated with infinity fade into meaningless mechanics of a linear world. The past and the future do not exist, nor will they ever, except as a concept of our imagination. There is only the "now" manifested from the absolute state, which Schwaller calls quality and Barbour refers to as configuration space. Exactly what this is, we may never know.

Since we cannot see it, the absolute is easily thought of as existing apart from the immediate observable world. It is unreasonable, however, to view the absolute apart from everything quantifiable, since it contradicts a widely accepted principle known as *reflection*.

In set theory, the reflection principle states that every conceivable property of the absolute (the class of all sets) is also part of any given set. In other words, every conceivable property of the absolute is shared by some lesser entity.

> Suppose that f is an analytic function that is defined in the upper half-disk $\{|z|^2 < 1, 3|z| > 0\}$. Assume that f extends to a continuous function on the real axis, and takes on real values on the real axis. Then f can be extended to an analytic function on the whole disk by the formula:
>
> $$f(\bar{z}) = \overline{f(z)}$$

The values for z, reflected across the real (horizontal) axis, are the reflections of $f(z)$ across the real axis. It is easy to check that the above function is complex differentiable in the interior of the lower half-disk. What is remarkable is that the resulting function must be analytic along the real axis as well, despite no assumptions of differentiability.

This is called the Schwarz reflection principle, and is sometimes also known as Schwarz's symmetric principle (Needham 2000, p. 257). The diagram above shows the reflection principle applied to a function f defined for [the] upper half-disk (left figure) and its image (right figure). The function is real on the real axis, so it is possible to extend the function to the reflected domain (left and right figures).

For the reflected function to be continuous, it is necessary for the values at the boundary to be continuous and to fall on the line being reflected. The reflection principle also applies in the generality of reflecting along any line, not just the real axis, in which case the function f has to take values along a line in the range. In fact, any arc that has a neighborhood biholomorphic to a straight line can be reflected across. The basic example is the boundary of the unit circle, which is mapped to the real axis by: $z \rightarrow (iz + 1) / (z + i)$.

The reflection principle can also be used to reflect a harmonic function (any real function $u(x, y)$ with continuous second partial derivatives which satisfies Laplace's equation), which extends continuously to the zero function on its boundary. In this case, for negative y, defining $v(x, y) = -v(x, -y)$ extends v to a harmonic function on the reflected domain. Again note that it is necessary for $v(x, 0) = 0$. This result provides a way of extending a harmonic function from a given open set to a larger open set (Krantz 1999, p. 95).[36]

According to this principle, assuming thought is a property of the absolute, the absolute's thoughts are also a property of some possible thought. So when an individual conceives some property called *X*, then the first thing that satisfies *X* will not be the absolute. It will be some rational thought that just so happens to reflect the absolute's feature that is expressed by saying it has property *X*. The point here is that if one accepts the existence of the infinite absolute, then one is, for the most part, committed to accepting the existence of infinite thoughts and sets. To deny the reflection principle is to practically assert that the absolute can be finitely described, which is most unreasonable.

Suppose that for every thought called *X*, the thought *X a possible thought* is also a thought of the absolute. By reflection, there must be some thought *Y*, such that for every thought *X* in *Y* the thought *X is a possible thought* and is also in *Y*. This *Y* then shares the property of the absolute. This *Y* must be infinite, however. So infinite thought exists. All attempts to analyze consciousness and self-awareness rationally appear to lead to infinite regresses, indicating that consciousness, essentially, is infinite.[37]

For society in general, however, skepticism abounds, since such ideas run contrary to everything we've been taught and everything we've been conditioned to, particularly in the Western world of the omnipotent, omniscient, omnipresent God. Such ideas also run contrary to belief in the scientific mainstream of "seeing is believing," where, ironically, seeing is the worst possible kind of evidence. Evidence is evidence, and data are data. Neither is an interpretation. Only the observer is offered that honor of interpreting data, which is a mystery in itself.

Anthropocosm: The Man Cosmos

Schwaller describes philosophically in *A Study of Numbers* what physicists allude to in their equations and essays. Who we are and why we are here, the ultimate mystery, will forever remain a mystery. Intuitively it can be understood, however, that the cosmos is the nature of Man, and form is the sole means of its expression. That is to say, without the perception of separation—without state vector collapse of the energy wave—there is no form to experience and, consequently, no expression. Without these principles that are a part of our nature, the cosmos would remain in an undefined state of absoluteness, the quality of all possibilities that *could* be the cosmos.

It is also the case that the cosmos is responsible for our form. All matter that makes up the cosmos is actually configured energy—movement—existing as a result of the polarized energy within hydrogen atoms undergoing the extreme process of nucleosynthesis. Carbon, nitrogen, oxygen, and other heavy elements—the building blocks of life—were created as a result of large stars collapsing under their own weight and then exploding with tremendous heat, spreading newly created elements into empty space to form interstellar clouds. New research even suggests that amino acids, important for protein synthesis, are also formed in interstellar clouds.[38]

Our existence as form can be traced to cosmic events, and the circumstances of our continued existence can be traced to a truly universal level. Our Earth is dependent on the sun and the solar system in which it is gravitationally trapped; which is dependent on the Milky Way galaxy, in which it is gravitationally trapped; which is also held in place by other forces including, but not limited to, other galaxies. Any interruption in this line of cosmic dependency would likely result in the cessation of our existence.

Although it seems as though we are insignificantly small compared to the rest of the universe, there is the reality of the observer. We observe and perceive an ordered, yet dynamic, arrangement of energy we naturally translate into sight, sound, smell, taste, and touch. It is interesting that the concrete can be experienced and appreciated only through the abstract. Would there be a cosmos without mankind to perceive it?

To take away the measures of this reality means reality's destruction, which suggests the universe was never concrete in the first place. We only perceive that it is. Therefore, we can conclude that the concreteness and form in which we live are really only the knowledge of such things. Why it is magical, as Perlmutter so honestly puts it, is that *we are the magicians*. We always have been and always will be. It is my opinion that this is true science in all its brilliance. René Schwaller de Lubicz, as much a scientist as he was a philosopher, called it sacred science.

PART II

The Roots of Ancient Wisdom

According to the anthropic principle, any theory attempting to describe the existence of the universe must also explain the presence of organic life, which includes mankind and man's essence—the conscious state of being self-aware. Along with the anthropic principle comes the question: Is conscious life cause or effect?

If biological life came into existence by accident, then the universe must be, somehow, self-creating. But such a concept is the same as believing that intelligence created the universe. If intelligence of some form or another created the universe, what could it create from except itself? Again, there is self-creating taking place.

Regardless of whether there was an intelligent cause—and no one really knows if there was or wasn't—the ultimate effect of this self-creation, as far as we know, is conscious life: the perception through which mankind exists in the physical universe. In applying the anthropic principle to cause and effect, the idea emerges that mankind is not so much form as abstraction: the perceiver of the cosmos. Between cause and effect (perception), both of which are abstractions, there stands the immediate, concrete world we experience. This is the mystery, which is truly unexplainable and must be solved by the individual.

We must embrace a functional mode of thinking, as opposed to rational thinking, to understand this mystery. To situate the abstract within a physical reality, a very different way of thinking about the world is required; the life of the symbol (its esotericism)—in its identification with the life we live (the symbol's exotericism) is what defines reality. The symbol is the experience of our consciousness, which is consciousness of the effect, summarized by the cause of which we are all a part. Taken to the limits of rational thought, this functional thinking determines cause and effect to be of a nature that results in consciousness of consciousness, which is another way of describing the unique human property of self-awareness.

The knowledge of cause and effect, and how it is expressed through the symbolic, is responsible for mankind's eagerness to continually construct approximations of this universal truth. It also explains why history is steeped in innumerable varieties of religious tradition. Behind the historical religious tradition and at the headwaters of religious thought, however, there was once a long-standing tradition simply referred to as the ancient mystery school. This school existed at the beginning of civilization, perhaps from time immemorial, although it is now hidden by the various perceptions and dogmas mankind has created along the way. Like today's hypothetical science mystery school, the foundation for the ancient mystery school was science, as the root of ancient wisdom. But the science of the ancient mystery school was based not on a consideration of ideas or objects that lead to rational thought. Rather, it was based on functional thinking and the action that creates objects or ideas. This is the point of demarcation between modern rational thought and the ancient world of symbol and its expression of truth through symbol and myth.

Whether functional or rational, science has always been mankind's birthright. Philosophy is its consequence—the question of why. With an understanding of ancient philosophical concepts, and the assumption of the unknown science from which they disseminated, the most ancient epochs of human history hint at a surviving knowledge that mankind has only recently achieved.

5

Changing Paradigms

Science and the Origin of Religion

> *In scientific investigations, it is permitted to invent any hypothesis and, if it explains various large and independent classes of facts, it rises to the rank of a well-grounded theory.*
>
> CHARLES DARWIN

Without some future scientific breakthrough, the questions of how and why life exists are clearly beyond the limits of science. Furthermore, given the unknowable disposition of events surrounding the universe's origins, there will likely never be a cosmogony established as fact. Nonetheless, inferences can be made from what scientific knowledge there is, and such inferences move science into the realm of the sacred. As a result, given the anthropic principle, an understanding of the origin and nature of the cosmos is possible. Such an understanding is the *Man Cosmos,* meaning that the Man is the cosmos. Although such a cosmogony is clearly a matter of interpretation, its conceptual foundation connects rationally with today's scientific evidence, particularly physics, as much as the irrational can be rationally explained.

At this stage of reading, you might wonder why the concept of God has not been addressed. The reason is that God is a conceptual matter, as opposed to being a scientific matter, and involves a faith-based system of belief. As a consequence, potentially every person may have a different idea of what or who God is—if he or she, in fact, believes that God exists. Although God may indeed exist in some manner unknown to us, there is no scientific evidence for an omnipotent, omniscient, and omnipresent entity. Science puts forth a very different view of God in the characteristics of the cosmos and the principles of nature.

Not quite two hundred years ago, however, Western civilization had a very different interpretation of Man and the Universe. From the fourth century through the middle of the nineteenth century, a biblically based Christian worldview was the social norm. Before the Enlightenment, which began during the latter part of the seventeenth century, there was little, if any, philosophical opposition to such a worldview. In some regimes, any opposition was thwarted by imprisonment, torture, and even death. The world was flat and believed to be a little over 5,500 years old, as calculated by the archbishop of Armagh, James Usher (1581–1656). Disease was believed to be the result of evil spirits, and magic was considered witchcraft. But all that was about to change.

During the Enlightenment, a string of scientific discoveries and inventions from Francis Bacon (the inductive and experimental scientific method), Galileo Galilei (the mathematics of mechanics, especially dynamics), Johannes Kepler (the laws of planetary motion), Nicolaus Copernicus (the sun-centered solar system), and Roger Bacon (the use of experimental methods in alchemy that led to chemistry) laid the foundation for the birth of science. With an understanding of his predecessor's works, the mathematician Sir Isaac Newton (1642–1727) developed differential and integral calculus, providing the methodology to study phenomena such as gravity and motion. Concepts brought forth by Newton in his widely popular book *Philosophiae naturalis principia mathematica* (Mathematical Principles of Natural Philosophy), first published in 1687 and then again in 1713 and 1826, propelled the Western world into a new age of understanding. Newton was just as much a Christian, however, as he was a scientist.

Newton also wrote about theological concepts such as biblical prophecy, believing its interpretation was essential to understanding the nature of God. His book *Observations upon the Prophecies of Daniel and the Apocalypse of St. John* addressed not only prophecy but also how Christianity went astray during the fourth century CE. For Newton, the first Council of Nicaea set forth false doctrines concerning the nature of Christ, particularly the concept of the Trinity. Despite this opposition to Trinitarian dogma and the Council of Nicaea, he was deeply spiritual and believed in divine creation. Science was going to need another boost from a few more observant and insightful individuals to further validate its precepts.

During the first part of the nineteenth century, strange fossilized bones, discovered accidentally, were making news. Although people had been finding these strange bones for thousands of years, in 1841 British scientist Richard Owen realized that the bones represented by such fossils were different from those of any living creature. He placed them into a new grouping he called Dinosauria, which means "terrible lizards." The remnants of a world existing before the known world soon became the focus of explorers and scientists alike. With this newfound knowledge, so would the questioning of mankind's origins.

In England in 1844, *Vestiges of the Natural History of Creation* was published, proposing that lower forms of animal life evolved into higher forms, then termed *transmutation*. Aware of the scandalous ramifications and how shocking it would be to Victorian readers, the author, Robert Chambers (1802–1871), chose to publish anonymously. Fifteen years later, his successor would not be so reserved.

In November 1859, the London publisher John Murray released *On the Origin of Species by Means of Natural Selection, or the Preservation of Favoured Races in the Struggle for Life,* written by Charles Robert Darwin. All 1,250 copies that were printed sold on the first day. By Darwin's death in 1882, over 22,000 copies had been sold. Now known simply as *Origin of Species,* it has become one of the best-selling books of all time. It has also started one of the biggest philosophical wars of all time, revolutionizing scientific, anthropological, religious, and social thought. Up until that time, any theory on the origin of man had to be squarely seated in the realm of biblical concepts. The scientific method spurred by the Renaissance now became the bane of traditional religious thought. Soon after *Origin of Species* first publication, the "evolutionary wars" began. Both scientists and theologians rose up against Darwin's ideas.

In the July 1860 *Quarterly Review,* Darwin was called a flighty man who had written an "utterly rotten fabric of guess and speculation." Even the great geologist Louis Agassiz of Harvard, who believed that each species was "a thought of God" by intelligent design, dismissed Darwin's theory as "a scientific mistake, untrue in its facts, unscientific in its method, and mischievous in its tendency." Religious leaders accused Darwin of destroying the foundations of faith, and even of demeaning the human race. In public response, they wrote that his work was "degrading" and "materialistic" and proclaimed that it insinuated "there is no God and the ape is our Adam." This "Darwinism," a term coined by Thomas Huxley in the *Westminster Journal,* was also likened to blasphemy and atheism, particularly in the United States. Yet there was very little in Darwin's book concerning the origin of man. For Darwin, natural selection was not incompatible with a belief in God, the architect and primary cause behind the physical universe. It was not theology, but instead a theory of scientific process to explain the diversity of life.

In March 1868, the radical thinking of Darwin and his supporters received another boost. French workers, laying a railway line near Les Eyzies in the Valley of Cro-Magnon, dug into an ancient rock shelter, exposing thousands of years of geologic history. Archaeologists Edward Lartet and Henry Christy soon discovered that it contained the skeletal remains of five people: three adult males, an adult female, and a child. Buried with them were stone tools, carved reindeer antlers, ivory pendants, and marine shells. Cro-Magnon man had been discovered. At the time, no one had seen or heard of anything like it. There were many more such discoveries to follow. As years of work continued and new sites were uncovered, it became clear that forty thousand years ago, a race of people had settled in the western regions of Europe in the modern-day countries of Spain and France.

Further discoveries provided evidence that prehistoric Cro-Magnon–style settlements stretched from South Africa to modern-day Israel and from western Europe to Siberia. But no other area was as densely populated as western Europe. These astonishing finds provided much information about prehistoric life.

Believed to be a nomadic hunter, prehistoric man thrived on plentiful herds of wild horses, deer, goat, bison, and mammoth, supplementing his diet with nuts, berries, and fish. Cro-Magnons hunted individually, as well as collectively, using various weapons that later included the bow and arrow and fishnets made from vine. Their tools were intricate and specialized for various types of prey. Harpoons for fishing were barbed to increase effectiveness. Cores of stone were used to mass-produce long, thin blades, further modified to create projectile points, knives, and scrapers. Spear-throwers aided hunters by increasing a spear's velocity. Bone and antler were expertly crafted into utensils, some artfully decorated. Needles of antler or bone were used to sew skins for clothing. It is also believed that some clans built canoes to catch larger fish farther from the shore. Intelligent and innovative, Cro-Magnons were well equipped to survive and thrived in their environment.

Then, in the first quarter of the twentieth century, just after the First World War, the invention of the mass spectrometer led to the discovery of more than two hundred isotopes. (An isotope is an element with one of two or more atoms having the same atomic number but different mass numbers.) All rocks that contain naturally occurring radioactive elements go through a process of decay. Most of these elements are found in igneous rocks, so the isotopes and dates for when a rock solidified indicate how long ago the molten rock cooled. Because these radioactive elements decay at a constant rate, their age can be estimated by measuring the amount of radioactivity that the parent element has by comparison with stable daughter elements. It is through these isotopes that rocks act as a geologic clock and can be used to estimate the age of Earth.

For example, uranium decays and produces subatomic particles, energy, and lead. As uranium-38 decays to lead, thirteen intermediate radioactive products are formed, called daughters. They include radon, polonium, and other isotopes of uranium, as well as eight alpha and six beta particles. Each radioactive isotope has its own unique half-life, which is the time it takes for half the parent's radioactive element to decay into a daughter product. The rate of decay is proportional to the number of parent atoms present. The proportion of parent to daughter products shows the number of half-lives, which is used to find the age in years. If there are equal amounts of parent and daughter, then one half-life has passed. If there is three times as much daughter as parent, then two half-lives have passed.

Using these radiometric-dating techniques, the scientific age of Earth was calculated to be nearly four billion years. This was (and is) at extreme odds with the prevailing notion at that time that Earth was only six thousand years old. Since the universe is governed by natural laws that scientists observe and mathematically describe, such a discrepancy was difficult to reconcile. After years of experimentation and testing, the relationship between time and radioactive decay was deemed to be correct. There was little evidence to defend a young Earth.

For the scientific community, the culmination of evidence was irrefutable. Darwin's revolutionary ideas of evolution and natural selection were to be taken seriously and soon won over the academic intelligentsia. The biblically based creationist doctrine was retired and would no longer influence the scientific method. It was a turning point for Western civilization and culture and a divorce from ancient mythical beliefs.

Without a creator, the mystery of life was actually a much bigger mystery than previously believed, and the world was certainly much older than previously thought. Scientists were free to develop a new model of mankind's history.

Science and the History of Man

Today, according to a multidisciplinary consensus that includes genetics as well as archaeology, mankind—*Homo sapiens sapiens*—has existed for 150,000, possibly 200,000 years; an archaic version *(Homo erectus)* may go as far back as two million years. The difficulty with investigating such a distant epoch is that recorded history stops at approximately 3000 BCE, which means that nearly all of mankind's history is unaccounted for. This raises one of the more fascinating prehistoric questions: What were our ancestors doing between 150,000 and 5,000 years ago?

The short answer is that they were explorers of unknown lands. Scientists who study mitochondrial DNA generally agree that all who are alive today are genetically the progeny of a single woman who lived in Africa a very long time ago. Given the multitude of prehistoric sites around the world, it seems we were homesteaders and explorers, searching for better lands and curious to see what lay beyond the next hill or valley. This raises another question: To accomplish a global occupation, what level of social sophistication had human cultures achieved early in prehistory?

For anatomically modern humans, who are naturally built to walk, 100,000 years is plenty of time for them to make their way by foot into all corners of the world. The case can also be made that early man was little more than a meanderer with no real need for sophistication. One of the older indigenous populations, however, sheds some light on just how sophisticated mankind was. According to geneticists, who are supported by the archaeological evidence, Australia was first settled 62,000 years ago. Here there is some of the earliest and clearest evidence for modern behaviors.[1] Although sea levels were likely several hundred feet lower at that time, the topography of Earth was similar to what it is today. Australia was an island continent, the same as it is today, so there was no way for ancient people to settle there without the aid of some type of watercraft. The obvious deduction is that the vessels they built had to be seaworthy. Not a trivial task in any age, this required cooperation and communication by teams of boatbuilders. Manning such a craft also required the skill of navigation. In essence, it required intelligence, language, and a methodology for guiding the craft where there were no landmarks.

Furthermore, it is not likely that humans first began to sail solely to reach Australia. They would not have known it existed. One possibility is that they marched out of Africa, then north through the Sinai Peninsula and east across the Indian subcontinent, then stopped on Asia's southeastern shores and contemplated how to continue out into the sea. But if they had never built a boat before, or contemplated it, for that matter, why would they build one simply to sail into oblivion? Moreover, walking was, and still is, a very inefficient means of travel. In addition, 80,000 to 65,000 years ago, the Middle East was occupied by groups of Neanderthals, possibly deterring any northerly migrations. Crossing the Red Sea by boat and continuing around the Persian Gulf along the coasts of Iran, Pakistan, and India would be a more efficient route.[2] Such a scenario suggests that humans were already navigators and had honed their skills long before reaching Southeast Asia, which would have given them the confidence to continue.

Not everyone chose to continue east, and in no way could the migration be construed as a grand campaign to find Australia. It was likely an undertaking of those with adventurous spirits. Camps were established along the way where families set up permanent homes, in much the same manner that North America was settled during the eighteenth and nineteenth centuries. As the population grew, camps turned into villages, and in ever-greater concentric rings, people fanned

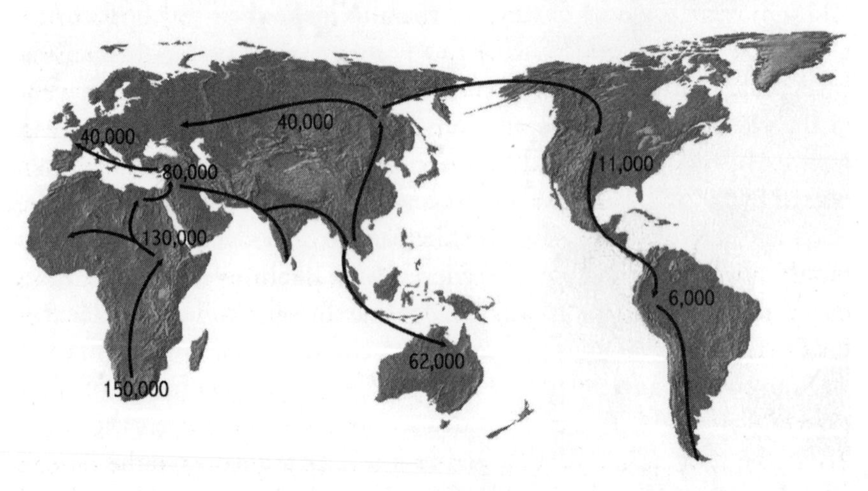

Fig. 5.1. Migrations based on genetic research

out to the north. By 40,000 years ago, these modern humans had made their way into eastern Asia, northern China, and Siberia. Also during that time, as one theory explains, another wave of travelers left Africa for Europe.

By 30,000 years ago, the Sangamon Interstadial, a relatively warm era in the geologic history of alternating cold and warm periods, was ending and a new age of ice was beginning. According to geologists, this new ice age locked up 5 percent of the Earth's water from the sea and gradually reduced the shallow areas of the world's continental shelves. By 25,000 years ago, lowered sea levels exposed the shelves beneath the Bering and Chukchi Seas, where Alaska and Siberia nearly meet. This expanse of new territory formed a land bridge, known as Beringia, over which humans and animals migrated from Siberia to North America.

Crossing this land bridge to the west likely posed no significant challenge to Asian nomads. But glaciers effectively blocked the passage south into the heart of North America. By 20,000 years ago, these ice sheets covered parts of southern Alaska and all of Canada. According to theory, at the end of the ice age, the glaciers may have receded enough to create an ice-free corridor through which people could have passed. A people referred to as the Clovis entered North America through Beringia, migrated south through the ice-free corridor, and burst onto the heart of the continent around 11,200 years ago, although recent investigations by James Adovasio at Meadowcroft suggest that the date is closer to 20,000 years ago.

There is archaeological evidence that shortly thereafter, humans reached areas in South America. Researchers are now seriously investigating the possibility that the first Americans made their way along the west coast of North and South America. Whichever way they traveled, by 6,000 years ago humans resided on every habitable continent. Between 130,000 and 6,000 years ago, anatomically modern man spread across the world to reside in all areas of the habitable continents.

This generally accepted view of human migration, however, commonly known as the Out of Africa model, contains a rather obvious curiosity. If recent human origins are African, then why don't all of us look more African?

A Genetic Bottleneck

To account for this curiosity, geneticists have argued that the human species recently passed through a "bottleneck" that drastically reduced the world's population. Yet the evidence to explain why such an event occurred has always been elusive.[3]

In 1998, paleoanthropologist Stanley Ambrose, of the University of Illinois, offered a solution to this dilemma. According to volcanologists Michael Rampino, Stephen Self, and Greg Zielinski, approximately 71,000 years ago, Mount Toba in Sumatra (near Malaysia) erupted, spewing an enormous amount of ash into the atmosphere. It was the largest volcanic eruption in the last two million years, nearly 10,000 times larger than the Mount St. Helens explosion in 1980, according to Rampino, associate professor of earth and environmental sciences at New York University. The resultant caldera formed a lake one hundred kilometers long and sixty kilometers wide.

The eruption had devastating and lasting climatic consequences. A six-year-long volcanic winter followed, and in its wake an ice age occurred that lasted for a thousand years. With its sulfuric haze, the volcanic winter lowered global temperatures, creating drought and famine and decimating the human population. According to geneticists' estimates, the population was reduced to somewhere between 15,000 and 40,000 individuals. Professor of human genetics at the University of Utah Lynn Jorde believes it may have been as low as 5,000.[4]

Ambrose's model, which he calls the "weak Garden of Eden/volcanic winter" model, is based on the "weak Garden of Eden" model proposed by Henry Harpending, Stephen Sherry, Alan Rogers, and Mark Stoneking. They propose an African origin for modern humans around 130,000 years ago, and credit the invention and spread of advanced stone-tool technology for population growth after the bottleneck, which occurred 50,000 to 40,000 years ago. Ambrose argues that volcanic winter caused the bottleneck and that populations expanded in response to climatic warming 10,000 years before the advent of Stone Age technology.

If this is true, modern human races differentiated abruptly only 71,000 years ago, mainly through the founder effect (the concept that a small group of people found a new population), genetic drift, and adaptation to local environments.[5] It seems doubtful that any human population, particularly explorers, could be so successful without typical human skills, such as the ability to organize, socialize, and even conduct business through some sort of cooperative trade network. Of course, these qualities require a spoken language.

The Development of Language

In his hunt for the origin and development of language, Ben Marwick, at the University of Western Australia's Centre for Archaeology, discovered that trade networks existed very early in prehistory. According to Marwick, who bases his model on University of Michigan professor Robert Whallon's idea that alliance networks and language evolved simultaneously, the distance raw materials or artifacts are transported by a population is related to the population's level of language sophistication. The more advanced the language, the greater the distance.

Marwick defines a transfer as one or more associated artifacts, made from a raw material that can be identified as originating from a specific location. For example, if obsidian artifacts appear at numerous excavations over a wide geographical area in the same stratigraphic context and can be traced to a single source, then that raw material has been transferred between various groups.

According to Marwick's studies, two million years ago African hominids began trading in raw materials with other hominids who lived close by, up to a range of thirteen kilometers. Sometime around a million years ago, the range expanded up to a hundred kilometers. According to Marwick, use of a protolanguage enabled African hominid groups to exploit much larger territories than before, although this proto-language would not have been recognizable as a modern language.

The language would have been closer to how primates communicate, typically via status-based negotiation. In trading for goods, bartering (negotiating) is often required, which necessitates payment in some way. So the groups engaging in trade had to have some planning ability. Marwick believes that through repeated exposure of one group to another over a long period, a language gradually developed. By 800,000 years ago, archaic humans were capable of a proto-language, a theory that is supported by anatomical as well as archaeological evidence.[6]

During the Middle Stone Age (250,000 to 40,000 years ago) in Africa, stylistic diversity in stone projectile points indicates that ethnic groups bonded together regionally and traded those points with other regional groups. By 130,000 years ago, from the changes in tools, weapons, and campsite organization, it is evident that a symbolic language, complete with syntax, had emerged in Africa. In Europe, the same conditions occurred 30,000 years later. Mankind had developed the ability to use a finite number of sounds to produce an infinite number of meanings. This included the use of words and sentences to arbitrarily represent objects, ideas, and emotional states as well as real or imaginary events beyond the present. According to Marwick, "Language is symbolic in itself, and the ability to express symbolic categorizations of social systems allows individuals to identify and interact with other unrelated individuals in terms of symbolic categories rather than as unique individuals."[7] A way of verbally communicating that we would recognize as a language was well under way. So were trade networks, which required the verbal and mental abilities typical of modern humans.

By 100,000 years ago, trade routes reached 130 kilometers, then 400 kilometers 28,000 years ago. The area of trade networks for modern hunter-gatherer populations is similar to that of anatomically modern humans between 60,000 and 30,000 years ago. Furthermore, during the Aurignacian period, between 35,000 and 28,000 years ago, goods being exchanged began to include marine and fossil shells as well as stone. For Marwick, the long-distance transfer of these items suggests that the open trade networks were becoming social and ritual as well as functional. With trade networks reaching a range of 800 kilometers by 9000 BCE, civilization was right around the corner.

Farming and the Dawn of Civilization

According to traditional teachings, civilization began in the Mesopotamia Valley at the beginning of the third millennium BCE. More than 11,000 years ago, however, the seeds of civilization were already sprouting. In the fertile fields of the Jordan Valley just north of the Dead Sea, people banded together to erect a rectangular platform, possibly a shrine, then surrounded it with stone walls. Although archaeologists refer to these early builders as Natufians—from the Wadi Al Natuf, where their culture was first identified—precisely who they were or where they came from is somewhat of a mystery. They chose that spot to build a shrine because of the natural irrigation afforded by the Jordan River four miles to the west and from underground tributaries from the Central Mountains. Even today, the area is supplied with freshwater. As time went on, the shrine grew into a gathering place. Within a few thousand years, with its resident population booming, mankind, or at least the Natufians as our representatives, crossed the threshold into urban life. Mankind's first known city came into existence.

Archaeologists have verified that the city was abandoned once or twice, but it was continuously inhabited from the fourth millennium BCE until its destruction at the hands of the Egyptians in 1580 BCE.

During this stage it was known as Jericho. Ten thousand years later, Jericho still holds the record for the oldest-known city.

On top of the twelve-foot-thick remains left by its first inhabitants lies a more telling story from later inhabitants. Referred to as a "pre-pottery layer," it displays the remnants of a second occupation between 8350 and 7370 BCE.[8] Those who lived there built circular brick homes and surrounded the entire village with a perimeter wall. A tower with an internal stairway reinforced the wall. Although Jericho's tower was originally believed to be for defense, it is more likely that it was to protect the population from floods and mudslides, since the tower faced the west and was inside the wall. Only towers built outside a wall provide the advantage of shooting or throwing objects at those attempting to scale the defensive barrier. In later occupations, 7220 to 5850 BCE, Jericho's inhabitants built rectangular houses with plaster floors and walls.

The Natufians, who first appeared nearly 13,000 years ago,[9,10] were responsible for the earliest settlements of the Levant (the modern states of Israel, Lebanon, Syria, and Jordan). They were a unique and inventive people. According to Ofer Bar-Yosef, professor of anthropology at Harvard University, no field studies outside the region have discovered any other prehistoric group whose culture resembles theirs.[11] Evidence of their society exists throughout the area. The caves in Mount Carmel, the Judaean hills, Nahal Oren, Hayonim Cave, Rosh Zin, Rosh Horesha, Wadi Hammeh, Wadi Judayid, and the lower layers at Beidha have provided archaeologists with a wealth of information about their culture.[12] Their villages, dwellings (pit houses), graves, stone and bone industries, jewelry, and art were superior to those of all foraging societies in the Near East. One of their most significant undertakings, while living in oak and pistachio woodlands, was to harvest wild cereals from the undergrowth and plant them as crops.[13]

The emergence of this culture was a major turning point in the history of the Near East. According to Bar-Yosef, the Natufians played a major role in the agricultural revolution.[14] Seeds of barley, wheat, legumes, and other plants have been found within the remains of their dwellings, but there is no agreement on how they acquired the seeds.[15] Other evidence suggests they achieved a relatively high degree of technology early on. Although rare, examples such as the sites at Hayonim Terrace and Ain Mallaha indicate a sophisticated domestic life. They dug pits and coated them with plaster for storage; they may have used baskets for aboveground storage, although the evidence for this is less direct.[16]

Who were these first farmers? According to British archaeologist James Mellaart in *The Neolithic of the Near East* (1975), they were descendants of the Cro-Magnon—people of rugged Euro-African descent from the Mediterranean area, Caucasians with dolichocephalic skulls[17]—an odd cranial shape (oblong) by today's standards, but typical of ice-age peoples. Bar-Yosef refers to them as proto-Mediterranean. Wherever they came from, they were experts in making and using tools.

Until the advent of mechanized farming, the sickle, a tool with a curved blade and a short handle, had been used worldwide for cutting grass and crops. Natufians likely invented it, as well as the pick, which is considered the forerunner of the ax and the adze. These tools have been found in abundance in the Natufian homeland, and according to scientific analysis, they were used to harvest cereals. The Natufians created stone utensils and used them in processing food, as well as for crushing burned limestone and red ocher. Some were decorated with patterns such as the net, zigzag, and meanderer. These types of designs often appear on spatulas, stone bowls, shaft-straighteners, and rare ostrich-egg containers; broken examples have been found at a site in the Negev. Whetstones, made of sandstone, were used for shaping bone into objects. Shaft-straighteners for making arrows indicate they were using the bow.

Natufian bone carving was far richer in quantity and elaboration than any earlier or later culture before the Neolithic period. Teeth and horns from gazelles, wolves, fallow deer, roe deer, and birds served as raw materials. Some bone tools were used for hide-working and basket-making. Others were barbed and used for hunting (spears or arrows) and fishing, as well as for making handles for the sickle blade. By grinding and drilling bone, they shaped beads and pendants, many of which bear specific decoration. In their propensity for carving, they also decorated tools. Sickle handles have been found with elegantly carved naturalistic images of humans and animals.

Other carvings, such as limestone figurines, depict young ungulates (possibly gazelles), owls, dogs, and, although rare, humans. One figurine was carved with an owl at one end and a dog's head at the other. Another, this one from horn, boasted a man's head with a bull's head at the other end. Other figurines include a tortoise, a kneeling gazelle, and possibly a baboon. A unique carving found at Ain Sakhri has been interpreted to represent a mating couple.[18]

More symbolic and perhaps religiously significant art were human female figurines molded from limestone or clay. Several depict a kneeling female, while others are of seated women. Some scholars believe this art indicates the emerging role of women in an agrarian society, an early development responsible for the later, major shift in religious thought that brought about the cult of the Mother Goddess.[19]

Natufians honored their deceased and likely believed in an afterlife. Typically, they buried their dead in deserted dwellings or outside their houses in a pit. Most graves were single and ordinary, rarely lined with stones or plaster, but a few special burials exist. Limestone slabs surrounded one grave at Nahal Oren, and at another grave deep mortars, called stone pipes, marked a sealed tomb.

During the late Natufian period, corpses were beheaded, and the severed skulls, less the lower jaw, were placed inside homes or in another special-purpose building. A later development of this practice involved the restoration of the skull's facial features in plaster, sometimes set with cowry shells in the eye orbits. A number of these plastered skulls have been found beneath floors and in open spaces, suggesting ancestor veneration. Other scholars believe this was a public ritual aimed at attaining equality among the inhabitants of the village.

Although most burials contained no grave goods, those that did often contained head decorations, necklaces, bracelets, belts, earrings, and pendants of marine shells, bone, teeth, and beads. A few graves held more-elaborate items, such as a bone dagger, the bone figurine of a young gazelle, and a small model of a human head carved from limestone.

One of the more unusual mortuary practices was discovered at Ain Mallaha and the Hayonim Terrace (northern Israel). At each of these sites the remains of a dog were found alongside those of a human, the latter dating to the ninth millennium BCE.[20] According to Eitan Tchernov, from the Hebrew University of Jerusalem, and François F. Valla, of the Laboratoire d'Ethnologie Préhistorique, a detailed analysis and a comparison with all known Natufian remains suggest that genuine (domesticated) dogs were already living with humans during that time. When compared to wolves of the region, the Natufian dogs were smaller and had shorter snouts—classic signs of early domestication. It has been argued that these early domesticated dogs were merely aberrant specimens, but Tchernov and Valla concluded that this was not the case.[21] The dogs were pets that mated with other village pets, creating a unique canine group.

In support of Marwick's research on prehistoric trade networks, archaeological evidence also suggests that the Natufians maintained contact with other villages over long distances for mutual benefit. Marine shells used for jewelry were collected from the Mediterranean shore and, less often, from the Red Sea. At Ain Mallaha, a tusk shell was discovered that came from the Atlantic Ocean, and a freshwater bivalve from the Nile River.[22] Obsidian from central Anatolia (a plateau east of the Carsamba River in the central region of modern-day Turkey) has also been found in Jericho and, in smaller quantities, in Netiv Hagdud, Nahal Oren, and Hatoula.[23] This connection to Anatolia seems to have been an important relationship.

In 2001, three Israeli scientists, Yuval Goren, A. Nigel Goring-Morris, and Irena Segal, found evidence in the plastered skull from Kfar HaHoresh (in the Nazareth Hills of Lower Galilee) that cinnabar (mercury sulfide), known as vermilion in ancient times, was used as pigment in the plaster. Further study found that antimony inclusions, which appear as crystals, and traces of lead were present in the pigment.[24] This specific mineral composition does not occur in the eastern Mediterranean. The closest sources are in the Transcaucasian region and western Turkey. According to Goren, Goring-Morris, and Segal, it is the first clear indication found thus far linking the later tradition of Anatolian skull decoration to the southern Levant—to the best of their knowledge. Although traces of cinnabar have also been detected from a skull at Abu Hureyra, on the Euphrates,[25] other evidence suggests a strong connection to Anatolia and the prehistoric city of Çatal Hüyük.

Urbanization and the Dawn of Religion

There are remnants in Anatolia of an ancient, but distinctly modern, city called Çatal Hüyük (which means "forked mound"). Discovered in 1951 by James Mellaart, of the British Institute of Archaeology at Ankara, the area was first excavated between 1961 and 1965. Since 1993, an international team of archaeologists led by Ian Hodder has continued research and excavation.

During the early years of excavation, Mellaart uncovered a striking network of buildings and shrines erected by a sophisticated people who lived more than 8,000 years ago. At its zenith, Çatal Hüyük appears to have housed a population of 5,000 to 10,000 people. The oldest layer has been dated to 6,500 BCE, although sterile soil has not yet been reached. It was occupied until 5,600 BCE and was then abandoned for unknown reasons. Çatal Hüyük may be the first large city ever by ancient standards, twice the size of Jericho at its peak.

What Mellaart discovered during the 1960s is that the mound contained a large and closely packed city, although streets were not a part of the design. Continuing excavations have revealed a plan of rectangular houses built in a labyrinth-like arrangement and centered on small courtyards. Twelve layers of this specific style of construction are piled one on top of the other. Each house has its own separate walls but is built alongside its neighbor. New layers were built by partially demolishing an existing layer, then building new houses atop the ruins. This method of construction created the mound topography discovered by Mellaart. An interesting feature of the city is that no walls for defense were ever found, suggesting that the city was never in need of fortification.

Houses consisted of mud brick, wooden beams, and plaster. Each unit was built according to a general plan: a main room, a kitchen, a storage room, and an area interpreted as a shrine. Wooden columns between the houses supported horizontal beams, which provided the frame for flat roofs. Ceilings were made of clay pressed into reeds. Window openings were placed in the topmost portion of the walls near the roof. Entry was provided by a hole in the roof, accessible only from a ladder—an ideal way to prevent predators of any kind from entering without the installation of a door. It also allowed the smoke from cooking and heating fires to escape.

Floors were created from lime-based plaster and covered with mats woven from reeds. Plaster walls were painted with designs in white, red, yellow, and black. Along the walls, benches and platforms accommodated those who wanted to sit. Small niches carved into the walls served as beds. Kitchens used nearly one-third of the available floor space, with small, hearth-style ovens set into the walls. Plaited baskets (for grains, tools, and other supplies) were found in the storage rooms near the kitchen.

One of the more bizarre aspects of Çatal Hüyük's culture was its cemetery. When a family member died, the body was excarnated (denuded of all flesh) and the bones placed beneath the sleeping quarters of the surviving generation. It appears that after the passing of a generation, the walls of the existing home were torn down and the area was filled in with soil in preparation for the next tenants. They literally lived on top of their cemeteries.

As in Jericho, early stages of farming and animal domestication are evident in Çatal Hüyük. In the summit area of excavation, a high percentage of sheep (goat) and cattle remains were found proportionally to pig (boar), horse, dog, fox, and hare.[26] Botanical remains of cereals (barley and wheat), pulses (peas, chickpeas, lentils, and wild legumes), and other seeds (pistachio, bulrush, pepperwort, and hackberry) have been found. Various fragments of other botanical remains are yet to be identified.[27]

It appears that Çatal Hüyük domestic life was fully developed. Obsidian from a nearby volcano was used for stone tools: blades, projectile points, and even blades with serrated edges. Eighty percent of the stone tools were made from obsidian, as were highly polished mirrors. The rest were made from flint and other stone. Wood was used to make boxes, bowls, platters, cups, and spatulas. Animal bones were carved into points for piercing, needles, tools for carving plaster, cups, spoons, spatulas, jewelry, fishhooks, hammers, and handles for blades. Baskets, woven in spiral fashion, were made of straw or other coarse plant fibers. Mats for floors were of the same material. Clay was used for making pots. Numerous clay balls, both large and small, were unearthed. Although something of a mystery, these may have been used as heat-transfer devices in cooking, which would be an ideal way to heat water without subjecting the container to an open fire.

The people of Çatal Hüyük made many different types of figurines of clay and stone. Some were crudely made animals. Other, more refined figurines include female statuettes (some holding animals) and seated male figures. The female figures are generally very plump. Some depict women giving birth to either humans or animals.

Articles of personal adornment, as well as utility, also suggest that the people who lived here were modern. Some items discovered in the ruins have been interpreted as bone toggles, belt buckles, bone rings, stone and clay beads, bone pendants, awls made of bone, beads for anklets, bracelets, and an exceptional flint dagger with a decorative bone handle.

There is also evidence of the manufacture of textiles, possibly from wool or flax. With Anatolian trade goods found throughout the Middle East, the city must have been a hub for commerce—probably trading in obsidian, textiles, skins, food, and even technology. Stamp seals were found of various designs, possibly used to decorate fabric or walls, but they may have also been used to stamp exported products ("Made in Çatal Hüyük").

The art of Çatal Hüyük is striking and offers great insight into the residents' belief system. Among the motifs used in their art are geometric designs, flowers, stars, circles, and, in some parts, depictions of life. There are also human hands, deities, human figures, hunting scenes, bulls, birds, vultures, leopards, wild deer and pigs, lions, and bears. A mural depicting the eruption of a volcano (most likely nearby Mount Hasan) is the oldest known landscape art, probably painted around 6200 BCE, when the city had reached its zenith. This painting features the block-style settlement in the foreground and a twin-peaked red volcano in the background with smoke billowing from its summit. Obsidian, the mainstay material of tools and other utilitarian items, may have been seen as a sacred material charged with the power of the gods, though the Mother Goddess seems a more likely candidate for a principal deity.

Figures of the Mother Goddess made of baked clay, as well as carved from stone, have been found consistently since 1961 throughout the continuing excavation. During the 1997 season, fifty-two figurines were found of various types (animals as well as humans), usually outside the living space.[28] The most common is the Mother Goddess.

She is often seen as a woman with large breasts, with her hands resting on her protruding belly. She is almost always depicted in the nude, lying down or crouching, and possibly giving birth. One of the more symbolically powerful figurines was found by Mellaart himself. It depicts a woman seated between two leopards, her hands resting on their necks. Mellaart believed that some structures were shrines to the goddess, but recent excavations have found that many of the religious ceremonies took place within individual homes.

Ian Hodder, current director of the Çatal Hüyük Archaeological Project, disagrees. He claims there is not enough evidence to suggest the Mother Goddess was worshipped, but admits it is clear that her figure was held in high esteem.

The bull was also revered in Çatal Hüyük. Bulls' heads and horns were plastered into walls and adorned interior rooms. Murals of bulls were painted on shrine walls. Female breasts, suggesting the room itself became the body of a goddess, often accompanied bulls' heads along interior walls. These bulls' heads were formed in high relief like statues. Some were genuine skulls, covered with clay and baked hard. Several shrines were filled with horns. Benches where worshippers would sit or lie were cradled by the huge sweeping horns of the now extinct Aurochs bull.

The practice of excarnation is portrayed in eerie frescoes, where the dead are placed in strange, open funeral houses. Griffin vultures strip away the soft tissue of the deceased. One painting displays a vulture with human legs, wings outspread over a small headless figure. Vulture skeletons have also been found in bull shrines, hidden in clay breasts, with the beak creating the tip of the nipple. Interestingly, one skeleton was found in an excavation precisely as the vulture murals portray, headless and in his grave with his left hand over his genital area. Although thought to be the god of some funerary cult, the vulture and practices of excarnation are still a mystery. One possible explanation is that the practice involved belief in sky burial. Since their flesh was removed by vultures, in essence, they were buried in the sky by way of the vulture.

According to Mary Settegast in *Plato Prehistorian,* Çatal Hüyük was the Rome of the sixth millennium BCE, where power, religion, and art—especially symbols of transformation—reached a zenith of expression.[29]

One such fresco, which was really two paintings (one superimposed over the other), shows the life cycle of the bee. In the older painting, on the left there is a sealed honeycomb and to the right bees emerge into a field of flowers. The later, superimposed painting depicts winged chrysalises (the insect just prior to becoming a butterfly) resting on the branches of a tree, and the bees being replaced by butterfly-type insects.

This scene, and other scenes like it that decorated the shrines, suggests that the principle of nature, notably regeneration, was the focus of Çatal Hüyük rituals.[30] It seems there was an understanding of the nature of life—the regenerative cycle of birth and death—in Çatal Hüyük, and that those who built the town believed in the soul, as well as its continuation after bodily death.

Thousands of years later, the Greeks also identified the human soul with the bee and the butterfly. The bee was a common symbol in Malta and Egypt. In Lower Egypt, the pharaoh's symbol was a bee named Bit. Malta's symbol was also a bee, with its hexagonal, honeycomb-style cells. Furthermore, Malta's ancient name is Melita, which is derived from the Latin word for honey.

Why was honey significant? Although we think of honey solely as a food source or sweetener, ancient peoples knew that honey was medicinal. In ancient Egypt, honey was not only used for the preservation of meat, but was also the most widely used medication. Of nine hundred remedies recorded in various papyri, over five hundred were based on honey.

Honey not only contains a series of nutritive elements, but is also tolerated even in very large doses. Honey stimulates the appetite and facilitates the digestion of other foods. It also has laxative, sedative, antitoxic, and antiseptic properties. Until recently, it was believed that honey's syrupy consistency kept air out of wounds, and that its high sugar content slowed bacterial growth. New evidence suggests, however, that honey has other properties that kill bacteria. Honey has been shown to stop bacteria from growing, even strains resistant to antibiotics. Compared to an artificial honey of the same thickness and sugar concentration, natural honey kills bacteria three times more effectively, according to Rose Cooper, a microbiologist at the University of Wales Institute.[31]

The Goddess as the Creatrix of Life

Numerous androgynous figurines and carved images were discovered at Çatal Hüyük, as well as others that were obviously females, since they were depicted with swollen abdomens, indicating pregnancy, or in a birthing posture. The excavators described some of the figurines as already being old and worn at the time of their burial, suggesting that they were much older than the cultural context in which they were found—possibly heirlooms from Çatal Hüyük's own prehistoric roots.[32]

Although most of the human-figure symbols throughout the complex were disfigured when the shrines were rebuilt during a later occupation, Mellaart was able to determine that most bore an extension on either side of the head, possibly a hairstyle to simulate a horn. In Greek mythology, the goddess Persephone (queen of the underworld) was often depicted as being horned; she wove shrouds for the dead and gave birth to Zagreus, a horned child.

According to ancient texts such as Proclus's commentary on Plato's *Dialog of Timaeus,* Athena's veil was the essence of her divinity. In the Egyptian version, it was said that no mortal had ever seen through the veil of the goddess Neith, who can be viewed as the always-changing world of nature that we live in and the principle of becoming. Yet the initiate could do so after gaining immortality through understanding the life force behind bodily forms.[33]

A shrine dedicated to a "twin" goddess was elaborately decorated with bull benches and pillars and contained a considerable number of weapons, important items for male burials: daggers, lance heads, and polished-stone mace heads. Some of the niches cut into the walls contained limestone concretions from caves, suggesting for Mellaart that the shrine was oriented to the underworld and that the Great Goddess was the mistress of the underworld.[34]

The origin of this goddess religion can be traced, in part, to the Upper Paleolithic cultures of Europe twenty-five thousand years ago. Although theories remain speculative as to the underlying theology of the goddess culture, there is evidence suggesting that the cultures were honoring the creatrix, the giver of life.

According to some anthropologists, these early matriarchal societies did not understand the connection between sexual intercourse and pregnancy, because of the nine-month delay between conception and birth.[35] In these societies, the mother would likely have been viewed as the parent of her family and the sole producer responsible for the next generation. The most tangible evidence comes from archaeological sites throughout Europe.

Stone, bone, and clay figurines of steatopygic (fat-buttocked) women have been discovered as far west as Spain and France and as far east as Russia, including numerous locations in between.[36] These figurines, commonly known as Venuses, are the earliest known symbolic representations. Although it is speculative what beliefs they held for Paleolithic peoples, some tribes, according to Johannes Maringer during the 1950s, still carved such figurines.[37] To a tribe in central Asia, these idols (called *dzuli*) represented the human origins of the entire tribe, suggesting that they may have been a form of ancestor worship.

Other researchers believe that the concepts behind the Venus figurines are much more complex, and that multiple roles are involved for the theological goddess. According to the research of Marija Gimbutas, many types of goddess figurines appeared in prehistory, but they did not form a pantheon. In essence, they represented different functions of the same goddess. The deity was nature itself—the nature that is life giving, life taking, and life regenerating. These were the three important functions of the goddess, which is the natural cycle of life. Perhaps it is the origin of the common term we use for naturally occurring phenomena, Mother Nature.

6

The Historical Basis of the Logos

Greek Philosophy's Influence on Religious Thinking

In the beginning was the Word [Logos], and the Word was with God, and the Word was God. He was in the beginning with God; all things were made through him, and without him was not anything made that was made.

JOHN 1:1–3

A human being is part of the whole, called by us "Universe," a part limited in time and space. He experiences himself, his thoughts and feelings as something separated from the rest—a kind of optical delusion of his consciousness. This delusion is a kind of prison for us, restricting us to our personal desires and to affection for a few persons nearest to us. Our task must be to free ourselves from this prison by widening our circle of compassion to embrace all living creatures and the whole nature in its beauty. Nobody is able to achieve this completely, but the striving for such achievement is in itself a part of the liberation, and a foundation for inner security.

ALBERT EINSTEIN

Whatever the origin of religious beliefs, from the oldest known sacred symbols of the ice age to ancestor worship, from animism (all natural objects have an individual spirit) to polytheism and on to monotheism, mankind has consistently developed the concept of God to explain the mystery of existence, of life and death. There is a tendency on the part of religion, as well as science, to historically view the concept of God in a linear fashion. The worship of nature gave rise to animism, which gave rise to polytheism, which gave rise to monotheism and today's dominant religions.

A linear development for the concept of God, however, is entirely a matter of interpretation. The significance of ancient symbols, as well as the underlying cultural theology and cosmology, may be more sophisticated than just a simple association of animal symbols to totemic beliefs. Approaching the beliefs of ancient peoples from the assumption that because they lived long ago their beliefs were primitive may not be the best approach.

Today primitive societies and sophisticated, highly technical societies exist side by side, as do (what some may call) primitive and sophisticated religious beliefs, the relative point of view being one's own belief system. Deciding if a certain culture's belief system is primitive (or not) is an interpretive matter resting in the knowledge of the culture's method of expression and choice of symbols. It is also a possibility that over a long period, the original philosophical precepts behind a given belief system have become blurred by the personal opinions of religious leaders who have the ability to mold and shape the opinions of their followers. Taken to the extreme, the original theology might become a completely new and different theology from what was intended by its founding principle (principles).

Unchanged for a thousand years, Western civilization emerged from the Dark Ages with Christianity as the dominant point of view. This is quite clear historically, but how the Christian concept of God developed during the first few centuries of the Common Era is much more elusive. It is a fascinating and complex story.

The belief that Jesus was born of a virgin as the Son of God—the incarnate God—is the foundation of Western civilization's Christian theology. This was determined at the request of the emperor Constantine, who ordered the council of bishops to set the story straight once and for all. These bishops convened at Nicaea in 325 CE and produced an edict:

> [We believe] in one Lord Jesus Christ, and the only-begotten Son of God, Begotten of the Father before all the ages, Light of Light, true God of true God, begotten not made, of one substance with the Father, through whom all things were made; who for us men and for our salvation came down from the heavens, and was made flesh of the Holy Spirit and the Virgin Mary, and became man, and was crucified for us under Pontius Pilate, and suffered and was buried, and rose again on the third day according to the Scriptures, and ascended into the heavens, and sits on the right hand of the Father, and comes again with glory to judge living and dead, of whose kingdom there shall be no end.[1]

By the fourth century, Christianity had spread throughout the Mediterranean world. This standardization of what people should believe was decided by the church leadership to achieve a unity of faith. For the social order, it was a pragmatic way to allow the governing body to keep up with the times. But three hundred years removed from the time of Christ, did the Nicene Creed truly reflect Jesus's teachings?

Division in the Early Christian Church

In Upper Egypt, in 1945, while digging for *sabakh* (a soft soil used for fertilizing crops), an Arab peasant named Muhammad Ali al-Samman made an astonishing discovery. He uncovered a red earthenware jar that contained thirteen papyrus books bound in leather. Unimpressed by the find, since he was hoping the jar contained gold, he dumped the papyri next to his stove to be used as kindling. At the time, Muhammad was under suspicion of murder in avenging his father's death. In fear of the police searching his house, he asked a local priest to keep a few of the books. A history teacher somehow obtained one of the books and, suspecting it had value, sent it to a friend in Cairo. Ultimately, the books attracted the interest of the Egyptian government. The authorities purchased one of the books, and the rest were eventually confiscated.

Deciphered, the books were found to be a set of fifty-two Gnostic treatises written in Coptic, including ten noncanonical gospels: the Gospel of Philip, Apocryphon [secret book] of John, Gospel of Truth, Gospel to the Egyptians, Testimony of Truth, Secret Book of James, Apocalypse of Paul, Letter of Peter to Paul, Apocalypse of Peter, and the now famous Gospel of Thomas, believed by some scholars to be the source of all the other gospels.

The discovery of the texts shed surprising light on the political and theological struggles of the early Christian Church, explaining events in the life of Jesus in a whole new way. A few of the books described mankind's origins very differently from the usual reading of Genesis.

One such text, the Testimony of Truth, tells the story of Genesis from the viewpoint of the serpent. Known to appear in Gnostic literature as the principle of divine wisdom, the serpent persuades Adam and Eve to experience knowledge. God tries to prevent them from doing so, threatening them with death. Finally they are successful, but God expels them from paradise.[2] Another text, entitled Thunder-Perfect Mind, contains a poem in the voice of a feminine divine power. Still others range from a quasi-philosophical thesis on the origin of the universe to secret gospels, myths, and mystical practices.

As Elaine Pagels points out in *The Gnostic Gospels,* these texts were virtually unknown for two thousand years because of their suppression during critical struggles in the formation of the early Christian Church. Until their discovery, the only available information on the various early Christian movements was written from the perspective of those whose theological stance ultimately won out. "Heretical" theology is, of course, labeled as such from the viewpoint of orthodoxy. Naturally, those who wrote and embraced the teachings deemed heretical did not consider themselves heretics.

Interestingly, orthodox as well as Gnostic authorities accepted the Gospel of John as a part of their scriptural reference base and used it as a primary source for teaching. A highly philosophical text with some Gnostic overtones, the Gospel of John is the only canonized gospel that attempts to describe the relationship among God, Jesus, the Holy Spirit, and Man. Because of its Gnostic tendencies, it was opposed by some factions of orthodoxy but was included in the official Bible because of verses such as 14:6, where Jesus says, "I am the way, and the truth, and the life; no one comes to the Father, but by me."

Emerging from the complex cultural and political climate of the first century were two different ideas of what Christianity was supposed to be: the orthodoxy, which was taught by Paul, and Gnosticism. Why two different versions of Christianity? What could be the cause for such a split among early Christians? As it is today, the concept of God was, in the first century, a highly interpretive matter. So were Jesus's teachings; and for the Gnostics, his teachings had little to do with the orthodox claim that Jesus was himself God.

The Gnostics

The Greek word *gnosis*—from which *gnostic* comes—has a very special meaning. Traditionally, *gnosis* has been interpreted to mean knowledge, but not an ordinary type of knowledge. The Greek language distinguishes between scientific and reflective knowledge. Gnosis refers to a knowledge gained through observation and experience, best described by the word *insight,* and not by learning from instruction. Gnosis, therefore, involves an intuitive process by which one comes to know one's self. To know oneself is to know human nature and destiny. According to the Gnostics, at its deepest level, to know oneself is to know God, clearly in violation of Judaism's principal tenet that God is wholly other. The Christian Gnostics subscribed to such a view. This was reason enough for excommunication.[3]

Although likely existing within Judaism before the birth of Christianity, at the beginning of the second century the Gnostics were known by various names. The Ophites (from *ophis,* Greek for serpent), a collective name for several Gnostic sects that regarded the serpent as a symbol of creative wisdom, were also known as the Brotherhood of the Serpent. Not all Ophites, however, agreed on a standard theology. Some held to Jewish principles, others to Christian, while still others were anti-Christian. Only the idea and symbolism of the serpent was a foundational belief.

At the end of the second century, Irenaeus (ca. 125–ca. 202, Greek theologian, bishop of Lyons, and church father) wrote a history of heresy, but he did not know the Gnostics under the name of Ophites. Clement did, however, and cited another group called the Cainists, whose name was taken from the object of their worship. In the fourth century, Philaster believed that the Ophites, Cainites, and Sethites were the source of all Christian heresy because of their belief that the serpent was the true origin of mankind and that the body (physical existence) was, in fact, evil.

These Gnostic groups declared the serpent of paradise to be wisdom itself, since wisdom came to Earth through the knowledge that the serpent brought. They exalted Cain and Seth as heroes of the human race, whom they felt were granted this knowledge. All Ophistic circles believed in seven spirits under the dominion of the serpent (a demonic hebdomad or primordial power). Last mentioned of the seven spirits is the son of fallen wisdom, *yalda bahut* (which means "son of chaos"), and from him, continuing in successive generations, were Jao, Sabaot,

Adoneus, Eloeus, Oreus (or light), and Astaphaeus. They are said to be expressions of the God of the Hebrew scriptures. The Ophites claimed that Moses himself had exalted Ophis by picking up the serpent, and that Jesus also had recognized it by saying,[4] "Just as Moses lifted up the snake in the desert, so the Son of Man must be lifted up, that everyone who believes in him may have eternal life."[5]

According to the Theosophists, the Gnostics were not a Christian sect in the common use of the term. Although they believed in a Christos principle, their Christos was the Eternal Initiate (the Pilgrim), typified by hundreds of ophidian symbols for several thousand years before the Christian era. Its name was Ophis, which was the same as Chnuphis or Kneph, the Logos, or the good serpent. A living serpent, representing the Christos principle, was displayed in their mysteries and revered as a symbol of wisdom. This Christos of pre-Christian thought (and the Gnosis—wisdom that is revealed esoterically from God) was not the god-man Jesus Christ but rather the divine ego, made one with Buddhi—the path to eternal knowledge. The Gnostics' androgynous iconography can be seen on the Belzoni tomb in Egypt as a winged serpent with three heads and four human legs. On the walls descending to the sepulchral chambers of Rameses V, divine ego is found as a snake with a vulture's wings (the vulture and hawk being solar symbols). According to the Encyclopedic Theosophical Glossary:

> "The heavens are scribbled over with interminable snakes," writes Herschel of the Egyptian chart of stars. "The Meissi (Messiah?), meaning the Sacred Word, was a good serpent," writes Bonwick in his *Egyptian Belief.*[6]

The crowned serpent of goodness mounted on a cross was a sacred standard of Egypt. The serpent—symbol for gnosis—was borrowed by the Hebrews in their "brazen serpent of Moses," the healer and savior. The Ophites were referring, therefore, not to Jesus or his words, but to Ophis, when they cited John 3:14. Tertullian, one of the early church fathers, knowingly or not, had confused the Ophites' gnosis with Jesus. This was instrumental in the formation of Church doctrine that still exists today. The four-winged serpent is the god Chnuphis, the good serpent, that bore the cross of life around its neck or suspended from its mouth, and became the Seraphim (angel) of the Hebrew tradition. In the eighty-seventh chapter of the Egyptian Book of the Dead, in a vignette of a serpent with human legs, the human soul is transformed into the serpent Seta, and the omniscient serpent says, "I am the serpent Seta, whose years are many. I lie down and I am born day by day. I am the serpent Seta, which dwelleth in the limits of the Earth."[7] Gnostics would say he is the ego.

The concept of the Logos—Ophis, the good serpent—which the author of the Gospel of John identifies as Jesus (being the Word), and that the Gnostics used as an expression of the Christ principle, suggests a much older origin for the philosophical concepts that helped define first-century Christianity.

Greek Philosophy and Logos

One of the early Greek philosophers, Heraclitus (535–475 BCE), believed that the world is composed of a unity of opposites and that these opposites succeed each other. Hot and cold, light and dark are present in the same object, but only a single property is exposed at a time. Day and night are one, as well as the living and dead. Such philosophical thinking led people to nickname Heraclitus "the Obscure."

Heraclitus also believed that all opposites (qualities) are simultaneously present in nature, and that harmony, which makes physical reality what it is, consists of opposing tension. To illustrate the point, Heraclitus used the example of a lyre (bow). When the bowstring is pulled one way (by one end of the bow) and the other way (by the other end), the tension between these opposing forces allows the bow to perform its function. So beneath the apparently motionless exterior of the bowstring is the constant tension between opposed forces. The bowstring appears static, but it is really dynamic. Thus, the tension in the bowstring is balanced by the outward tension exerted by the arm of the instrument, so that a coherent, unified, and stable composite is produced.

For Heraclitus, a play on the word *biós* also helped illuminate his point. The name of the bow is life, but its work is death. In Greek, the bow is *biós* and life is *bíos*. (The two words are spelled the same, but the accent is on different syllables.) The stressed bow represents the tension between opposites in conflict; this is expressed metaphorically in the name of the bow and, through wit, is the opposite of the bow's work.

In taking this idea to the extreme, if the balance between opposites were not maintained for the world, then the tension (unity) would cease to exist, and the world would also cease to exist. The essence of

Heraclitus's philosophy is found in his phrase "Nature loves to hide," meaning the forces that determine objective reality are invisible, even though the objects themselves are visible.

Heraclitus also believed in the Logos, the principle that something exists that governs all things. This Logos is something that is said and can be heard and understood. Things come to be in accordance with Logos. Therefore, man is wise to listen to it. Its nature is so deep, however, that it is unrecognizable and incomprehensible by the human mind. Its true essence can never be known. Nonetheless, the Logos is responsible for the underlying order and change in the universe, as well as the way in which we perceive it.

As it relates to the cosmos, Heraclitus's Logos is a natural process of ordered change in the universe, providing the diversity of nature not found in matter itself. It is the energy that balances opposing forces—what keeps the lyre functioning. The Logos is responsible for the harmony of the lyre, as much as the Logos is the continued existence of everything. (Everything depends on continual change and motion.) Logos, then, is the principle of constant creation.

Heraclitus explains constant creation with the example of a flowing river, a river that is always changing. Since the composition of the river changes from one moment to the next, it is not the same river for any length of time, so one can never step into the same river twice. All nature is like the river, and no object retains its identity for any length of time. There really are no persisting objects; all that exists is only our perception of persistence.

Heraclitus's student Aristotle (384–322 BCE), who was also Plato's student, carried the idea even further. Acknowledging that all nature is in motion, and that nothing is true of what is changing, he believed it was not possible to fully explain all aspects of this constantly changing world. In reference to Heraclitus's river, he believed that one couldn't step into the same river even once (an idea that requires an initial creation of matter). Although the river is still the river, it is never in the exact same configuration because the flowing water is always changing the bank. Furthermore, the water is always different because it keeps moving downstream. Although the river is the same object, its water is always different as a result of change. The river is continually being created through an orderly process of change. This change is the Logos of the river.

For Plato, the concept of constant creation, where no object retains its qualities or characteristics from one moment to the next, seemed to be a slight misunderstanding of Heraclitus's teachings. According to Plato, an object could persist despite continual change. Together, change and permanence coexist. One may step in different waters, but the river is the same river; only its composition changes, of which the Logos is the acting principle.

Around 300 BCE, a Phoenician named Zeno of Kition founded Stoicism,[8] a new philosophical school of the Hellenistic period that honored ancient teachings. In Stoicism, the Logos represented the intrinsic ordering principle of the universe. Nature and Logos were often treated as one and the same. The Logos was identified as the overall rational structure, however, and was synonymous with reason. Not all creatures have Logos (reason); the Logos was intended for mankind only.

Stoicism stressed the rule "Follow where Reason [Logos] leads." One must resist the influence of the passions—love, hate, fear, pain, and pleasure. Within the cosmos is the individual's source of potentiality, vitality, and growth. These were the seed of the Logos. By reason, all mankind shares in the divine reason.

Much later, Philo of Alexandria (20 BCE–50 CE) mixed Jewish tradition and Platonism, referring to the Logos as a divine mediator, the healer of the soul, comforter, and ambassador, identifying Logos as the invisible supreme cause (God) as it relates to the world. The Logos was God's model or plan in harmony with the traditional Jewish account of creation, and in the word of YHWH, pronounced Yahweh. As Logos relates to the plurality of being in creation, however, for Philo it becomes a more personal figure, as in Genesis, where God says, "Let Us make man in Our image."[9]

The Origins of Greek Philosophy

According to traditional historians, Western civilization began with classical Greece around 500 BCE with profound intellectual and social changes: the rise of philosophy, logic, axiomatic mathematics, and the beginnings of democracy and individualism. Although the Greeks did not invent the modern world, their work in government, philosophy, literature, and art was the basis of emerging ideologies that would later spread throughout Europe. Most of the credit for their influence lies in the literary and intellectual accomplishments of their philosophers, such as Socrates, Plato, and Aristotle. These learned figures, however, did not appear out of nowhere. At the birth of Greek philosophy centuries before, the great classical philosophers were the likes of Thales of Miletus, his student Anaximander, and Pythagoras of Samos.

After the fall of Mycenaean civilization and the Greek Dark Ages that lasted from 1100 to 800 BCE, history identifies Thales of Miletus (624–560 BCE) as the first Greek philosopher and the founder of the Ionian school of natural philosophy. Thales predicted the solar eclipse of May 28, 585 BCE, proved general geometric propositions on angles and triangles, and, through the "laws of prospectives," calculated the height of Egypt's pyramids. Anaximander (610–545 BCE) introduced *apeiron* (the concept of infinity) and composed a theory for the origin and evolution of life.

A second generation of Greek philosophers included Anaximenes (570–500 BCE), a pupil of Anaximander, and Pythagoras (569–500 BCE), born on the island of Samos off the coast of Asia Minor. According to some accounts, at the age of twenty, Pythagoras visited Thales in Miletus, where Thales encouraged him to pursue his interests in mathematics and astronomy. To do so, Thales advised him, he had to visit Egypt. Whether Pythagoras followed this advice has been a point of contention among historians. Some historians believe such travels are stereotypical features in the biographies of Greek sages, legend rather than fact.

But Egypt at the time of Pythagoras was already two thousand years old and a well-known center for learning during ancient times. Most accounts of Pythagoras's life agree that he traveled on numerous occasions to the East, where he acquired the basis of his philosophy and science. Schwaller observes:

> It is incontestable that he journeyed to Egypt, as all authors concur on this point. Iamblichos specifies that Pythagoras remained for twenty-two years in Memphis and in Thebes, where the priests are said to have taught him mathematics and astronomy.[10]

Whatever the case may be, it is clear, through the later writings of Greek philosophers, that Pythagoras was responsible for important developments in mathematics and astronomy, as well as the theory of music. Euclid of Alexandria (325–265 BCE), a student of Plato, gathered together in his mathematical text *The Elements* fundamental knowledge that was generally considered to be of Pythagorean origin: the geometry of plane figures in Books I through IV, theories of proportions applied to plane geometry in Book VI, geometry of planes and solids in Book XI, and the construction of the five regular polyhedrons. According to Philolaos of Croton (ca. 470–ca. 430 BCE), astronomy was also a Pythagorean discipline. "Spherics," as Pythagoreans referred to it, included the spherical shape of Earth (a radical belief at that time), the tilt of the terrestrial axis as an explanation of the seasons, the rotation of Earth resulting in day and night, and the movement of Earth around "the central fire" (the sun).[11]

According to these later Greek philosophers, it seems Pythagoras was already teaching a cosmological system comparable to that of Copernicus long before the sixteenth century. Although this type of interpretation of Renaissance science is vigorously refuted, it is possible that Copernicus's heliocentric system had a very distant origin, and that he was inspired to develop the proof for it. Furthermore, the cosmology of Philolaos was and still is considered to be the heliocentric system itself, which is attributable to Pythagoras. Therefore, if the knowledge of the heliocentric solar system goes back to Pythagoras, we have to assume that a very long scientific development preceded him. If true, then human history is likely to be quite different from what orthodoxy presents.[12]

Later in life, Pythagoras moved to Croton and founded a philosophical school that attracted numerous followers. For the initiate, learning astronomy was a duty. Pythagoras taught the heliocentric system, the spherical Earth, and that the moon is a "dead planet," reflecting light from the sun. More than a thousand years later, both Giordano Bruno (1548–1600) and Galileo Galilei (1564–1642) derived their own theories of astronomy from written fragments known to be Pythagorean. Bruno was burned alive in Rome for not recanting such theories.[13]

Pythagoras knew that any triangle whose sides were in the ratio 3:4:5 was a right-angled triangle—evidence that he did visit Egypt. One of his more important discoveries was that the diagonal of a square is not a rational multiple of its side, but a number that can be expressed as the ratio of two whole numbers. At the time, irrational numbers was a revolutionary idea for mathematicians. Although Pythagoras taught geometry, most of his work was dedicated to describing the universe through number, where the initiate learned the secrets of the universe.

Pythagoras also taught that the soul of man comes from the "World Soul" and is thereby immortal. According to Pythagoras, the World Soul accomplishes its evolution by means of numberless incarnations on Earth.

Pythagoras

Despite the fact that Pythagoras was one of the ancient world's most renowned individuals, little of what he wrote has survived except in the writing of later philosophers such as Plato, Aristotle, and Iamblichus, or in ancient essays from other Greek writers known as the Doxographists. As a result, it is difficult for scholars to know how much Pythagorean doctrine can be attributed to his personal beliefs and how much is from later development. Be that as it may, texts from other ancient philosophers that did survive represent him as a scientist; other texts portray him as a mystic. It is certain, however, that he discovered numerical ratios that determined the concordant intervals of the musical scale. It is also likely that his geometry can be attributed in part to the tradition of Thales, who worked with the 3:4:5 triangle. According to Dick Teresi in *Lost Discoveries: The Ancient Roots of Modern Science,* "To this day, the theorem of Pythagoras remains the most important single theorem in the whole of mathematics."[14] Pythagoras was most likely both a mathematical genius and a mystic.

Key to Pythagorean philosophy was the idea that there is unity in multiplicity, the concept of the One evolving and pervading the many. Pythagoreans simply refer to it as the Science of Numbers, in that "all things are numbers." It was not pure mathematics, as contemporary society would likely view it, however, but a description of principles according to the creative laws of nature. What Pythagoras established in this philosophy was a fundamental characterization and description of nature translated into numbers. Possibly for the first time, a description of the laws and principles of nature (the universe) was devised by Man.

In *The Life of Pythagoras,* Iamblichus (250–325), one of the more important Greek Neoplatonic philosophers, wrote that Plato believed the study of numbers awakens an organ in the brain, which the ancients described as the third eye or "eye of wisdom," now known to physiology as the pineal gland. In relation to number and mathematics, Plato himself wrote in *The Republic* (Book VII), "Every soul possesses an organ better worth saving than a thousand eyes, because it is our only means of seeing the truth."[15]

For Plato, the philosophical possibilities in the knowledge of numbers, and the study of arithmetic—meaning theory, since calculating or doing sums was considered logistic—should be required for those managing the affairs of state, "not in an amateur spirit, but perseveringly, until, by the aid of pure thought, they come to see the real nature of number."[16] Plato insisted that the science of numbers was not to be used for mere buying and selling, but "to help in the conversion of the soul itself from the world of becoming to truth and reality."[17] Plato's acceptance of numbers as science is a Greek philosophical tradition reaching back to Pythagoras and the origins of Greek philosophy.

At the heart of Pythagoras's science of numbers was the belief that all relationships could be reduced to number associations, and that all things are in fact numbers. His approach was that the world could be understood through mathematics, which later became central to the development of the various scientific disciplines.

The Pythagorean Science of Numbers

According to Aristotle (*De Caelo,* Book I), Pythagoreans developed and applied themselves to the sciences and believed that numbers were the first principles of all things.[18] In Pythagorean terms, number relates to all numerals and their combinations, except for the numerals one and two. Numbers are the extension and energy of causation contained in One (the monad) and express metaphysical concepts. They represent the functions and principles by which the universe is created and animated. As numbers increment, each successive number symbolizes a specific function and incorporates all combinations of previous numbers.

How Pythagoras discovered this science of number was through the art of music. He, or one of his followers, noticed that the differences in vibrations characterizing musical notes could be calculated. Although the actual number of vibrations could not be calculated, the string of a musical instrument could be measured: the shorter the string (or the greater the tension), the faster the vibration and the higher the tone when the string was plucked. Harmony is produced when the vibrations from two or more strings resonate together. When one string is exactly half the length of another string, a ratio of 1:2 is achieved when both are plucked, which is pleasing to the ear. Additional intervals were found in ratios of 3:2 and 4:3. So the basic intervals of the musical scale were expressed in four numbers: 1, 2, 3, and 4. Together, they add up to 10.

The significance of this was profound. Music could be composed and arranged through numbers. Numbers could be converted reliably into meaning, and there was a hidden order to the immediate world. Pythagoreans took this concept another step and applied it in a holistic way to describe the universe.[19]

One: The Monad

This section ties in to Schwaller's philosophical treatise on number. It is also important to the ancient point of view. Pythagoras was a very important philosopher (one of his math theorems is still in use today) and had bearing on the philosophical thought that helped Christianity to emerge as the dominant religion in the fourth century.

The number One is absolute and the unity of all things that exist. It is the unchanging All from which all other numbers are produced. Before One, there exists nothing, or naught, represented by a circle, for Pythagoreans the most fitting symbol for Deity. In modern terms, we have little difficulty perceiving One as omnipotence, or God. Cosmologically, One can be considered the nondimensional reality from which the entire physical universe emanates.

Since One never changes and is separate from all other numbers, it is referred to as the *monad*. Because the mind is stable and has preeminence over physical form, One represents the universal Mind. It contains all natural qualities of the universe, particularly male and female, odd and even. As such, One is the symbol of wisdom.

Pythagoreans have referred to the monad by many names: chaos, obscurity, chasm, Tartarus, Styx, abyss, Lethe, Atlas, Axis, Morpho (Venus), Tower, and the Throne of Jupiter, because of the great power of the universe that controls the motion of the planets. The monad was also known as Apollo and Prometheus—Apollo because of his relationship to the sun, and Prometheus because he brought light to mankind. Whatever name was applied to One or the All in ancient times, and there were others, it was always in reference to the primordial One.[20]

One represents the beginning as well as the end for everything, yet has no beginning or end in itself. It is also the receptacle of matter, and creates from itself Two, the duad. Schwaller refers to the movement (or separation) from One to Two to as the Primordial Scission.

Two: The Duad

Two represents duality, or polarity. The duad is the dual expression of unity and the most fundamental aspect of all natural phenomena. Two represents not the sum of one and one, but a state of primordial tension. It is the metaphysical concept of unreconciled opposites that invokes such principles as evil, darkness, inequality, and instability. In a world where there are only One and Two, existence would be static. Nothing would ever happen.

With movement away from the Divine One, in Two there exists separation, which in a sense is the creation of ignorance. Where One symbolizes wisdom, so Two symbolizes ignorance. But through natural ignorance, wisdom is born. So Two also represents the mother of wisdom. Pythagoreans have referred to Two as Juno, since she is wife and sister of Jupiter, and Maia, the mother of Mercury, as well as other names such as Diana and Venus.[21]

Through the power of Two, the metaphysical "deep" was created in opposition to heaven, serving as a symbol of illusion. The deep, or that which is below (the physical world), is a reflection of the above (the spiritual world).

Three: The Triad

Three, the triad, is the first odd number. Symbolized by the triangle, the triad is sacred in that it is composed of the monad and the duad—the Divine Father and the Great Mother. Androgynous, it exists as a result of One becoming Two and is the symbol of divine creativity. Its principal function is to create equilibrium between One and Two.

For Pythagoreans, Three was the number of knowledge, music, astronomy, and geometry. Its properties are friendship, peace, and justice—the principle of reconciling opposing forces.

Four: The Tetrad

Four, the tetrad, is the primogenial number: the first number generating meaning that is born from the combined principles of One, Two, and Three. As such, the tetrad serves as the basis for all nature. Symbolically, the tetrad represents God, in that the sum of 1, 2, 3, and 4 is 10, the decad. As a result, Four is the most perfect number because it connects all beings, elements, numbers, and seasons. It is also the first geometric solid.

Four, as the number representing God, should not be confused with One. Whereas One is abstract and unknowable, Four is intelligible, representing such human qualities as mind, science, opinion, and sense—the four qualities of the human soul, according to the Pythagoreans. It is also representative of the four elements, the cause and maker of all things.

It is important to note that the elements of the ancient world—earth, fire, air, and water—were not physical elements, as science today defines elements, but rather principles in which the physical universe operates. The ancients used these four common, natural phenomena to describe the functional roles for the principle of matter. Earth, fire, air, and water represented the abstract principles of reception, activation, mediation, and material. Earth is the receptive and formative principle; fire the active, coagulating principle; air the subtle, mediating principle; and water the material principle.

Everything that physically exists operates by at least one of these principles, and most involve a combination. For example, without exception, everything in the physical universe is active, in motion. Our planet spins and moves in orbit around the sun. Our solar system also moves, as does the galaxy. Movement pervades the quantum level as well. In fact, action or motion defines existence.

An atom, the most basic unit of matter, is nothing more than energy (vibration) with a particular charge. For example, a uranium atom is made of 92 electrons orbiting a nucleus of 146 neutrons and 92 protons. (Breaking apart this atom releases a tremendous amount of energy.) On a slightly larger scale, everything physical, except air, is also formative, which means it can be formed into other compounds. Everything we see has been formed from some process that is linked to the creation of heavier elements through the stellar life cycle.

Air, the mediating aspect, separates all physical objects. Physical objects are composed of various elements and require the formative, active, and mediating principles to exist. Water is the only substance that can serve as the combined representative of these three principles. It serves as a mediating factor in the same way air does; yet it is also formative. It combines into a form and is active because it flows.

The tetrad, then, is the concept of matter. It is insubstantial, yet contains the four elements to describe physical reality. The principles manifested are life. As a cosmic symbol, it represents the Pythagorean term "Key-bearer of Nature," and as such represents the universe as chaotic matter before becoming form through the Spirit. The cross, created by the intersection of the Spirit's vertical line and matter's horizontal line, represents spiritual man "crucified" into the flesh, consciousness being manifested into the immediate world.

Five: The Pentad

Five, the pentad, is a union between Two and Three (the first even and odd numbers). Unaffected by disturbances of the other four elements, the pentad was symbolic of a fifth element known as the *ether,* the space above the sphere of the moon inhabited by the planets and stars. Five is also the monad (One) plus the tetrad (Four), the elements manifested by the unity of the absolute. Symbolizing nature, this concept of manifest reality is seen in the pentagram, the five-pointed star. Four points of the star represent the animal principle; the fifth represents *manas,* the life force believed to dwell in a person or sacred object. The ancient Greeks viewed the pentagram as the sacred symbol of light, health, and vitality.

In essence, Five was symbolic of the universe manifest, the concept of naturally occurring phenomena being dual in nature and triple in principle. All harmonic proportions and relationships are derived from Two, Three, and Five, in which matter is formed and the process of growth occurs. Male (odd) numbers represent functions that are initiative, active, creative, positive, aggressive, and rational. Female (even) numbers are correspondingly receptive, passive, created, sensitive, and nurturing, representing a state that is acted upon. Four accounts only for the concept of matter. Five is its creation.

Six: The Hexad

The hexad, Six (see fig. 6.1), represents the formation of matter into what we recognize as the cosmos, the physical universe including space and time. It is the creation of all form, and the ordering of the form that brings perfection to all that exists.

The hexad depicts the six directions of extension that all solid bodies comprise: up/down, left/right, forward/backward. Sacred to Venus, Six was the union of the two sexes, and the composition of matter by triads, necessary to develop the generative force inherent in all bodies. The six-pointed star embraces the spiritual and physical consciousness,

Fig. 6.1. Interlaced triangles

viewed by Pythagoreans as the symbol of creation. Interlaced triangles, also known as King Solomon's seal, portray the union of spirit and matter, male and female.

The white triangle's apex represents the divine monad, and the dark one, the manifestation of the world. The upward-pointing triangle evokes spirit, consciousness, and hidden wisdom, whereas the downward-pointing one represents matter, receptive space, manifestation, or wisdom revealed. Together they represent the universe manifest from a central point within the circle of time and space, symbolizing the descent of spirit into matter and its reemergence from the limitation of form.

Seven: The Heptad

Seven, the heptad, represents the mystical nature of Man. The cube symbolizes the heptad with six surfaces emanating from a seventh, hidden point within. From within the cube's center radiate six pyramids—One actualized by Six.

The heptad, however, is also a marriage of Three and Four. The threefold creative quality of mankind (spirit, mind, and soul) and the concept of matter (Four) together represent man as abstract, as well as physical, beings conscious of themselves and of their surroundings. Together, the marriage of Three and Four is the septenary (sevenfold) man.

On the noumenal plane—where an object is independent of the mind—the triangle is Father/Mother/Son, or Spirit. The quaternary

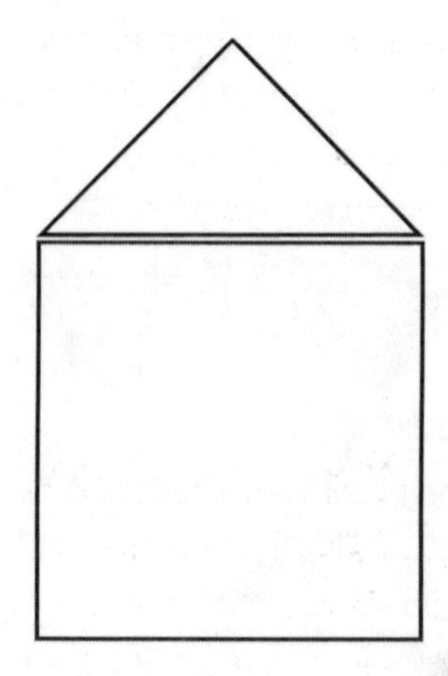

Fig. 6.2. The triangle and quaternary, the symbol of the septenary man

represents the ideal root of all material things. The triangle represents the three higher principles, immortal and changeless, while the quaternary refers to the four lower principles, which are mortal and always changing. As a compound of Three and Four, Seven not only governs the cycle of life on the physical plane, but also dominates the series of chemical elements, as well as the world of sound and color.

Eight: The Ogdoad

Eight, the ogdoad, is derived from the cube, since a cube has eight corners. The ogdoad reestablishes the monad in that Eight is divided into two Fours, each of which is divided into two Twos, each of which is divided into two Ones. Therefore, Eight is a new unity analogous to the first unity, representing renewal or self-replication. In ancient Greece's Eleusinian mysteries, eight was a mysterious number derived partly from the twisted snakes of the caduceus and the serpentine motion of the celestial bodies.

Nine: The Ennead

Nine, the ennead, is the first square of the first odd number, Three, and represents spiritual and mental achievement. It is also the limit of all numbers, since all other numbers that exist come from the first nine. All numbers from One to Nine create an infinite amount of numbers. In ancient Greece's Eleusinian mysteries, nine was the number of the spheres through which man's consciousness passed on its way to birth.

Ten: The Decad

Ten, the decad, comprises all arithmetic and harmonic proportions. Ten perfects all numbers, and within it is the nature of all that exists. Its essence is a return to unity. In Ten exists the perfect triangle, the Sacred Tetractys. Pythagoreans believed that its marvelous properties were the source and root of eternal nature. In essence, it is an expression of metaphysical reality, the "ideal world" of Plato.

The Pythagorean Tetractys

The relationships and principles of these ten numbers formed the basis of philosophical understanding for Pythagoreans, visually represented in the form of a triangle, called the tetractys. Composed of ten dots, the Pythagorean tetractys is arranged in four rows, where the three higher numbers represent the invisible, metaphysical world, and the lower seven refer to physical phenomena. It was the creed by which Pythagoreans were sworn:

> I swear by him who the Tetractys found,
> Whence all our wisdom springs and which contains
> Perennial Nature's fountain, cause and root.

Arranged from most to least dense, a gradation from more collective to more individual experiences, the Four Elements (the fourth row of the tetractys) refer to the physical world and the way man experiences it. Earth, which is dry and cold, refers to material aspects of nature and form. Coldness is a uniting power, and dryness represents the power of rigidity. Water represents emotions and relationships, such as love, hate, friendship, fear, and dependency. Thus water, being cold and wet, represents the uniting power of relationships and the flexible power of emotional reaction. Air represents reason and ideas, particularly those that are creative or require thinking. Air, being wet and hot, represents our adaptability and ability to draw distinctions as well as to innovate. Fire, referring to ideals, inspiration, and spiritual matters, represents intuition and aspiration. Fire, being hot and dry, represents the power of separation and the discriminating force of intuition.

The Three Principles (row three of the tetractys) refer to time (change) and how we experience it. Salt, which is closely related to earth and water, represents memories and the unchangeable. Our mem-

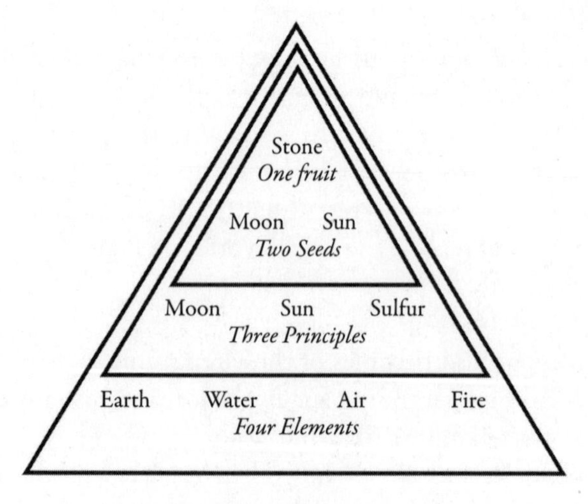

Fig. 6.3. The Pythagorean tetractys

ories comprise the objective past as well as our emotional appraisal of it. Quicksilver (mercury), which mediates between salt (concrete past) and sulfur (abstract future), represents the variability of the present moment. Quicksilver's position at the center also represents our control over the situation and the ability to change between the inner and outer orientations of existence. It is the present choice, based on emotional appraisal and rational analysis of the situation. Sulfur represents the future and our intentions, particularly aspirations and goals.

The Two Seeds represent mind, its conscious (the sun) as well as unconscious (the moon) aspects. The moon, closely related to salt and quicksilver, represents obscure or hidden aspects of a problem. The sun represents the visible aspects of a given situation, particularly what is distinct in conscious awareness.

The One Fruit refers to the *Unus Mundus,* the Self. It is the integrated mind of man Pythagoreans refer to as the World Soul. The Stone represents the totality of the universe, the abstract as well as the concrete, and a unitary event that manifests simultaneously in both worlds.

When all ten dots of the tetractys are connected with a line, nine triangles are formed. When the three dots that form the triangle are excluded, along with the center dot, the remaining six dots form a cube. When lines are properly drawn, a six-sided star is revealed between them. When the center dot is included in the formation of the cube and star, the three unused corner dots represent the threefold invisible cause of the universe. It is symbolic in number of the abstract becoming form and of conscious life manifesting as the physical universe.

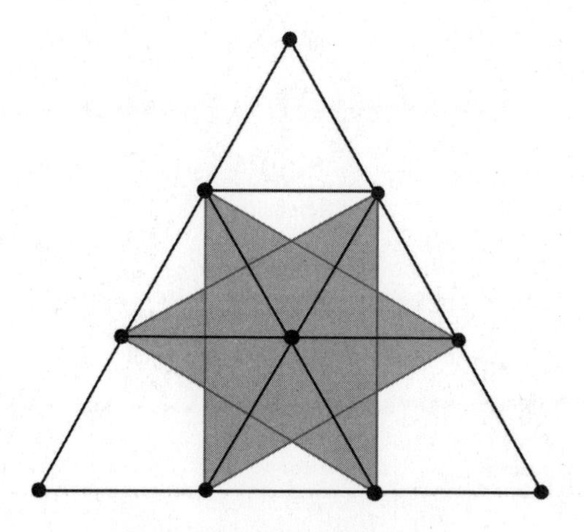

Fig. 6.4. The star and the cube

Relating Number to Form

According to Plato, Pythagoras believed that "Deity geometrizes," meaning that physical forms in the universe are created in an outward motion from within. From a point, the force of nature radiates equally in all directions, creating a sphere. The point then extends into a line and becomes a diameter, dividing the sphere into two equal parts, where one part is negative and the other positive. This polarity within the sphere serves as the basis for all action and reaction. Likewise, the point extends vertically, creating a cross within the sphere. This cross, according to Plato, is representative of the divine form that serves as the basis for the structure of the universe. From it, an infinite number of lesser forms are created.

According to Pythagoreans, in demonstrating that all physical forms are related to number, Pythagoras discovered that there are only five symmetrical solids. In a symmetrical solid, each and every face must be identical. These five symmetrical solids are the tetrahedron (four equilateral triangles as faces), cube (six squares as faces), octahedron (eight equilateral triangles as faces), icosahedron (twenty equilateral triangles as faces), and dodecahedron (twelve regular pentagons as faces).

A face requires at least three sides, so the first Pythagorean number in form is Three, representing the triangle. The second number is Four, representing the square. One and Two, not being numbers, symbolize the two abstract spheres of existence, the Supreme World and the Supe-

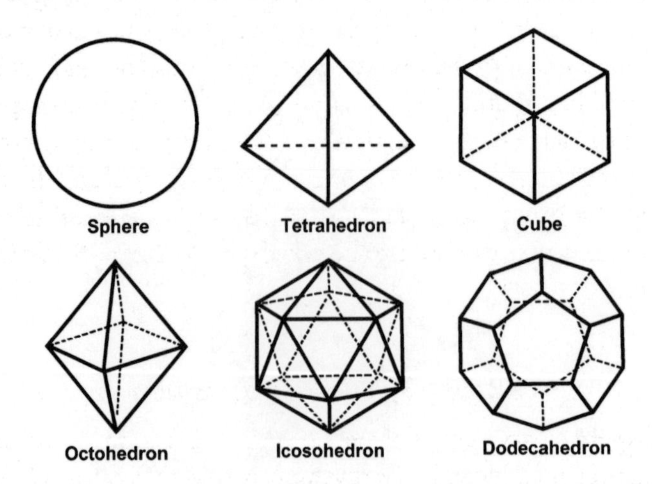

Fig. 6.5. The Pythagorean solids

rior World. Accordingly, it is these numbers—One through Four—that produce Ten, and thereby create the archetypal universe.

When numbers represent digits, they are symbolic of a quantity. In Pythagorean philosophy, they represent entities that are symbolic of the nature and origin of the universe. For example, the point (or dot) symbolizes the power of One, what today would be viewed as a singularity. The power of Two is the line, the power of Three the surface, and the power of Four the solid. The essence of these numbers is that they represent the manifestation of the three-dimensional world emanating from a nondimensional space.

Following this logic of two worlds, the Supreme and the Superior, and then a world associated with form called the Inferior World, in Pythagorean cosmology there are three levels in which physical reality exists. The first two are abstract and hold the intangible. Referred to as receptacles, the Supreme and Superior Worlds are reserved for principles and intelligences, respectively. Only in the third receptacle (the Inferior World) does quantity exist.

The Supreme World is the arena of the divine. In it is the essence of the universe and all that the universe comprises. The Superior and Inferior Worlds exist within its omniscience. Within the Superior World, the home of the immortals, exist all archetypes that provide the symbolic imagery for the Inferior (material) World. The Inferior World is, of course, the immediate, observable world.

All matter (the four elements—earth, air, fire, and water) was believed to be composed of these symmetrical solids. Being the most stable of forms, earth particles were cubical. Fire was made from particles in the shape of the tetrahedron, the simplest and, therefore, the lightest solid. Water particles were in an icosahedral shape, since they were the heaviest. Air particles, as a result of being intermediate between water and fire, were octahedral-shaped particles. The dodecahedron, on the other hand, was mysterious and the most difficult to construct, and was considered to be the divine pattern in planning for the universe.

Number Symbolism Summary

The Pythagorean science of number is exemplary in its attempt to describe nature in a logical, nonmythical way—a way not all that different from any modern mathematical attempt. In a sense, this ancient science of number is a foreshadowing of modern physics, in that the atoms of different elements are defined by their atomic numbers. For the various elements, the number of protons determines form—number defines form. Saying that there are only six essential forms of which everything is made is not all that different from saying there are only ninety-two naturally occurring elements. (In fact, there are only eighty-eight.) Where modern science ceases in its quest to understand the principles of nature, however, Pythagoreans conceptually exerted numerical logic in a cosmogonic sense. Where did everything visible—or invisible, for that matter—come from? And why do things of nature seem to be interdependent on everything else in nature?

This cosmic interconnectedness of naturally occurring phenomena, whether quantum or classical, has been the basis of scientific investigations for quite some time. Most notably, conscious life on Earth depends on stellar factors such as a large moon and the radiant energy from the sun needed for water to exist in a liquid state. Certainly, there is a natural unity and order to the universe, which is why it is referred to as the cosmos. Pythagoreans noted as much, and asserted that being and unity are the nature of objects. Number was a way to explain this and to define the most fundamental level of existence.

For Pythagoreans, numbers were real and had quantity. They were not just man's conceptualizations needed for record keeping. Whether the origin or the finality of all things, there was One before anything else. Two was a result of One dividing, thereby creating duality—in principle, the essential nature of everything. It is only through Three, however, that One and Two have the ability for a dynamic relationship in which action and reaction occur. In essence, Three is the principle of that interaction and the source of all understanding. Four is the first generated number (generated by One, Two, and Three), creating the principles of physical existence. Five is the dynamic manifestation of Four into the physical universe—the creation of space. Six is the creation of all form as we perceive it and relate to it—the consciousness of man. Seven is man relating to the One and a realization of man's mystical nature. Eight is a renewal, in that man reaches a new unity analogous to the first unity (a spiritual death and resurrection). Nine is the spiritual and mental achievement that a new unity allows. Ten is its perfection and a return to unity, comprising all arithmetic and harmonic proportion, the origin and aspiration of phenomenal nature.

The Triangle and the Trinity

According to this Pythagorean cosmogony, Three is the magic number for identifying the nature of deity as it relates to creation. Three contains all qualities that describe physical existence. One, the monad, is a point and the source of all numbers. In Two, the duad, is the creation of polarity or opposites. It is only in Three that unity or harmony is restored. Aristotle, in relating this concept, wrote:

> For, as the Pythagoreans say, the world and all that is in it is determined by the number three, since beginning and middle and end give the number of an "all" and the number they give is the triad. And so, having taken these three from nature as (so to speak) laws of it, we make further use of the number three in the worship of the Gods.[22]

This concept of the Divine being composed of three terms provided the basis for the triangle as the most profound of all geometrical symbols. For the ancient Greeks, the letter *D* (Δ the triangular delta) became the "vehicle of the Unknown Deity."[23] While in Athens, the Christian missionary Paul referred to this concept of deity as a means of explaining the message of Christ: "For as I was walking along I saw your many altars. And one of them had this inscription on it—'To an Unknown God.' You have been worshiping him without knowing who he is, and now I wish to tell you about him."[24]

The idea of relating Christ to the Unknown God of Greece would later have great theological ramifications for the entire church. In essence, it was the birth of the Trinitarian concept. So the trinity

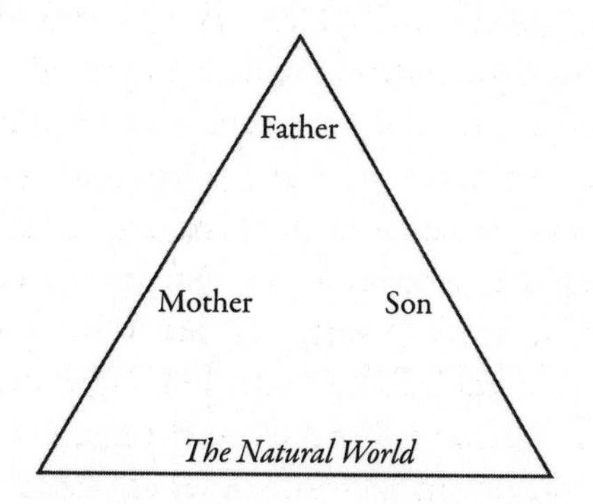

Fig. 6.6. The higher trinity of the universe

was not an invention of the Christian movement but rather a synthesis of already existing (Pythagorean) beliefs into a new framework of thought. In fact, the idea of the Trinity and the Son as part of its triune nature was already accepted among ancient intellectuals, as we have seen.

In this higher trinity of the universe, the first number, One—the monad, an indivisible, impenetrable unit of substance—appears on the plane below where the form (geometry) of the universe emerges. It is the first and, therefore, the divine number. It is also the apex of the manifested equilateral triangle, the Father of the immediate world. The triangle's left line is the duad or Mother. The triangle's right line represents the Son, described in ancient cosmogony as being one with the Father, the apex. The triangle's base represents the universal plane of nature where Father, Mother, and Son are unified.

This philosophical understanding gives us the concept that man and the universe were made in the image of God. Furthermore, since both were made in the same image, the understanding of one predicates knowledge of the other; so there is a constant interplay between the grand or archetypal Man (the universe) and the individual man (the little universe). Paul's theology, which he spread throughout the ancient world, later became the foundation of the Nicene Creed and was little more than the anthropomorphizing of an already existing concept of God.

With the Hellenistic conquest of the Mediterranean region during the third century BCE, Greek culture and language *(koine)* proliferated throughout the known world and effectively became the common international language. Greek concepts of humanity and divinity were philosophical and esoteric and were a part of the cultural context that existed at the time the Gospels were written. Not only is the preamble to John's Gospel highly suggestive of this philosophy, with the identification of Jesus as the Logos, but the main thesis of the text is as well. Jesus as the son of man is not only a recurrent theme in John's Gospel; it appears to be the main theme in all the canonical Gospels. Still, these themes appear to be the philosophical legacy of an even older culture—pharaonic Egypt.

7

Egypt

The Ancient Source of Knowledge

In accordance with the method of concealment, the truly sacred Word, truly divine and most necessary for us, deposited in the shrine of truth, was by the Egyptians.

Clement of Alexandria, *The Stromata*

Almost a thousand years before Heraclitus's great work *On Nature,* the oldest of all Greek prose, the ancient Egyptians already had a concept of the Logos, believing that the world came into being through the word of Thoth. One of the more famous early-twentieth-century Egyptologists, E. A. Wallis Budge (1857–1934), in describing Egyptian thought, noted:

> Inanimate nature likewise obeyed such words of power, and even the world itself came into existence through the utterance of a word by Thoth; by their means the Earth could be rent asunder, and the waters forsaking their nature could be piled up in a heap, and even the sun's course in the heavens could be stayed by a word.[1]

For Budge, however, the concept of the power of the word was not indicative of the philosophical character of Egyptian life, but rather an implication of magic and childish superstition. Budge found it difficult to conceive why the Egyptians carefully preserved, in their writings and ceremonies, so much gross and childish superstition. Nonetheless, he concludes that they did "believe in One God who was almighty, and eternal, and invisible."[2]

He added:

> Although they believed all these things and proclaimed their belief with almost passionate earnestness, they seem never to have freed themselves from a hankering after amulets and talismans, and magical names, and words of power, and seem to have trusted in these to save their souls and bodies, both living and dead, with something of the same confidence which they placed in the death and resurrection of Osiris.[3]

Such has been the view of Egyptian civilization for most of the twentieth century. Even so, ancient historians, such as Proclus (411–485 BCE), recorded that the earliest of the Greek philosophers, Thales of Miletus, traveled to Egypt in his quest for knowledge and was introduced to geometry. Thales became well known for this theorem: If A, B, and C are points on a circle, and the line AC is a diameter of the circle, then the angle ABC is a right angle. A number of scholars argue that Thales' theorem was, in fact, Egyptian.

Although there is little doubt that classical Greek society contributed immensely to the advancement of civilization, particularly in art, architecture, and philosophy, the idea that the Greeks were never influenced by other cultures (notably Egypt) is largely unsustainable, since it is clear from ancient texts that those cultures did not live in isolation. The notion that Egypt had no influence on the ancient Greek state does not fit well with the historical evidence. Egyptian civilization flourished for two thousand years prior to the establishment of the Greek state.

The main difficulty in understanding the relationship between Greek concepts and those of Egypt is that the Egyptian language of hieroglyphics was completely lost by the onset of the European Dark Ages. It wasn't until 1822, when Jean-François Chompollion deciphered the Rosetta stone, that the Egyptians' symbolic style of writing again became readable. But their concepts and stories appeared primitive, largely because of their reliance on what has been perceived by modern scholars as a menagerie of gods.

For Wallis Budge, how Egyptian civilization continued to progress toward a high level of intellectual development from the earliest to the latest period of its history was very difficult to understand. It is no different today. Yet continued research into the cultural milieu and intellectual reasoning of the ancient Egyptians has led to some fascinating insights into Egyptian thought, such as that of the Memphite theology.

The Shabaka Stela

There is a remarkable stela (an upright stone on which text is inscribed) housed in the British Museum that is attributed to the Twenty-fifth-Dynasty Egyptian pharaoh Shabaka (ca. 760–ca. 695 BCE). Although its most recent use was as a millstone, originally this basalt carving of sixty-one columns and three rows described what scholars refer to as ancient Egypt's Memphite theology. Its cosmological focus is the creator's heart and mind, where physical reality is born of thought and the spoken word. According to most scholars, the original papyri from which the inscriptions were copied come from the Nineteenth or Twentieth Dynasty (1295–1075 BCE), although a few believe the source is even older and dates to the Old Kingdom, during the middle portion of the second millennium BCE.

While there is still much scholarly debate about whether the ancient Egyptians had the capacity to be truly philosophical, the Shabaka stela does appear to be the conceptual forerunner of the Greek

Fig. 7.1. The Shabaka stela (photo by the author)

Logos. According to the inscription, the creator (Ptah) created the cosmos with his heart and tongue—a thought and an utterance:

> There comes into being in the heart; there comes into being by the tongue as the image of Atum! Ptah is the very great, who gives life to all the gods and their kas. It is all in this heart and by this tongue.[4]

> His [Ptah's] Ennead is before him as heart, authoritative utterance, teeth, semen, lips, and hands of Atum. This Ennead of Atum came into being through his semen and through his fingers. Surely, this Ennead (of Ptah) is the teeth and the lips in the mouth, proclaiming the names of all things, from which Shu and Tefnut came forth as him, and which gave birth to the Ennead [of Ptah]. The sight of the eyes, the hearing of the ears, and the breathing of air through the nose, they transmit to the heart, which brings forth every decision. Indeed, the tongue thence repeats what is in front of the heart. Thus was given birth to all the gods. His [Ptah's] Ennead was completed. Lo, every word of the god came into being through the thoughts in the heart and the command by the tongue. Thus all the faculties were made and all the qualities determined, they that make all foods and all provisions, through this word. [Life] to him who does what is loved, [death] to him who does what is hated. Thus life is given to the peaceful and death is given to the criminal. Thus all labor, all crafts, the action of the arms, the motion of the legs, the movements of all the limbs, according to this command which is devised by the heart and comes forth by the tongue and creates the performance of everything.[5]

Ptah is the principle of the Supreme Being, and through his words "Atum's image" was used to create the world. In ancient Egypt, the heart referred to the intellect or mind, whereas the tongue was the realization of thought. Accordingly, in the Shabaka inscription, through an authoritative utterance, Ptah creates the world and all that it contains. For anthropologist Henri Frankfort (1897–1954), these passages are the "true Egyptian equivalent of John's, 'In the beginning was the Word, and the Word was with God, and the Word was God.'"[6]

Likewise, James Henry Breasted (1865–1835), founding director of the University of Chicago's Oriental Institute, believed that this Egyptian concept of creation through the Word was critical to a revised understanding of ancient history. When Egyptian texts were first being deciphered, little sense could be made of them. They appeared to be a hodgepodge of myths, spells, and stories. It was also believed that the Greeks were fully responsible for the development of the Logos concept, and that it was introduced later into Egypt from abroad.

The discovery of the Shabaka stone clearly shows that the concept already existed in Egypt at an early period, well before the "Greek Miracle" of civilization and science. Breasted believed that "the Greek tradition of the origin of their philosophy in Egypt undoubtedly contains more of the truth than has in recent years been conceded," and that a philosophical interpretation for the functions and relations of Egyptian gods had "already begun in Egypt before the earliest Greek philosophers were born; and it is not impossible that the Greek practice of the interpretations of their own gods received its first impulse from Egypt."[7]

Contemporary scholars agree. Stanford associate professor of comparative literature Tom Hare notes, "For this reason the theology has been cited as an antecedent to the first verses of both the Book of Genesis and the Gospel of John."[8] According to Cornell professor Martin Bernal, in *Black Athena: Afroasiatic Roots of Classical Civilization,* volume 1, "Proof that Egyptians could think in terms of abstract religion, which was published eighty years ago, has received so little attention. The proof comes from a text generally called Memphite Theology, which dates back to the 2nd and 3rd millennium."[9] Bernal also finds that it "looks remarkably like the Platonic and Christian Logos, the Word which 'already was, The Word dwelt with God, and what God was, The Word was, The Word then was with God at the beginning, and through him all things came to be.'"[10]

The Memphite Theology

The Memphite theology inscribed into the Shabaka stela states that Ptah was preeminent in every body, in every mouth of all gods, all people, all flocks, and all crawling things that live. Ptah was the creator or first cause. He created Atum's image from within the heart and through the tongue. Atum was the primal creative force, the father of all gods, as well as the pharaoh. As the creative principle, both Ptah and his creation, Atum, were the embodiment of life energy.

Through Ptah and from Atum, the Ennead—the nine essential gods—developed. Shu and Tefnut emerged from Atum, and from them the other gods of the Ennead. And from the Ennead, through the will of the heart, there were eyes for seeing, ears for hearing, and the nose for breathing the air. Thus, Bas (the embodiment of soul) was made of the male and female. As a result, the ability to construct existed—all craft, the doing of the hands, and the going of the feet.

The Memphite theology describes a mystical process: Through the uniting of the heart and tongue of Ptah, creation occurs. It is a description of the eternal mystery that every person encounters when he or she deliberates on his or her origins. For the ancient Egyptians, this mystery was embodied and symbolized by the heart, tongue, and hand; the heart was the eternal seat of the soul, the tongue the seat of the mind, and the hand the seat of creation.

"Becoming" and the Ennead (the Number Nine)

For the ancient Egyptians, the biggest mystery of all was the "becoming" of the creator from the unseen into the seen, the One who manifests as many. This becoming was revealed through four successive stages: Atum (or Re) at Heliopolis, Ptah at Memphis, Thoth at Hermopolis, and Amun at Thebes. According to the Leyden Papyrus of Qenna, written during the Twenty-eighth Dynasty:

> All the gods are three: Amun, Re, and Ptah, who have no equals. He whose nature [literally, "whose name"] is mysterious, being Amun; Ra is the head, Ptah the body. Their cities on Earth, established forever are: Thebes, Heliopolis, and Memphis [stable] for eternity. When a message comes from heaven, it is heard at Heliopolis, it is repeated at Memphis to Ptah, and it is made into a letter written in the letters of Thoth [at Hermopolis] for the city of Amun [Thebes].[11]

This idea of a message represents the progress of "the becoming" from heaven to earth. Because Heliopolis was considered to be "the ear of the heart," it is there that the message was heard. In esoteric texts, the sun was deemed to be the heart of the solar system, so Heliopolis was the heart of Egypt, the city of the sun.[12] The name Heliopolis, as it is used in the funerary texts also means "the absolute origin of things,"[13] not to be strictly confined to the physical city of that name. When it is stated in Egyptian texts "I have come from Heliopolis" or "I am going to Heliopolis," the meaning is "I have existed from the beginning" or "I am returning to the source."

According to the teachings at Heliopolis, the One who began the "becoming" is Atum, whose name means "all" and "nothing" and represents the unmanifest potential of creation. Atum is one with Nun, which is the indefinable cosmic ocean. Atum's first act was to distinguish himself from Nun, which is described in Egyptian mythology. As Atum (the All or Absolute) realizes himself, he emerges from Nun as the primordial hill and creates Shu, the principle of space and air, and Tefnut, the principle of fire, which, according to the Sakkara Pyramid texts, he spits into existence. (The Sakkara Pyramid texts are a set of hieroglyphics dating to the Old Kingdom's Fifth, Sixth, and Eighth Dynasties, 2705 to 2213 BCE, that were inscribed on the walls of ten pyramids, although they are believed to have actually been composed much earlier, around 3000 BCE.)

In another version, Atum gives birth to himself by masturbating and creates Shu and Tefnut. In a third version, he creates himself by projecting his heart,[14] bringing forth the eight primary principles known as the Great Ennead of Heliopolis. The Great Ennead was the nine great Osirian gods: Atum, Shu, Tefnut, Geb, Nut, Osiris, Isis, Seth, and Nephthys. The term is also used to describe the great council of the gods, and as a collective term for all the gods. Osiris, Isis, Seth, and Nephthys represent the cyclical nature of life, death and rebirth, none of which is apart from Atum, according to the pyramid texts.

Atum represents the unknowable Cause. It can be thought of as the modern Western concept of God. From him everything is created. He is at the top of the Ennead. From him, all other principles of the universe emanate. From Atum is born Shu (air/wind) and Tefnut (water/moisture), the most import elements for life, representing the establishment of social order. Shu puts forth the principle of life and Tefnut, the principle of order. From Shu and Tefnu, Geb and Nut, Earth and sky, were created. From Geb the sun is born. When Nut and Geb meet, darkness occurs. Nut and Geb give birth to Osiris, Isis, Seth, and Nephthys.

In applying the four principles (unity, duality, reconciliation, and the concept of matter), Osiris represents incarnation and reincarnation, life and death, which is renewal. Isis is the feminine aspect of Osiris. Seth is the principle of opposition, or antagonism, and Nephthys, the feminine aspect of Seth.

These creation events are taking place outside of the limits of terrestrial time, beyond the realm of the temporal. They occur in heaven, not on Earth. According to Schwaller, such mysteries are not to be understood by the reasoning process of the mind's intelligence.[15] It is an esoteric mystery, which is not comprehended by the rational mind and can be perceived only by what symbolists call the "intelligence of the heart." What is really being dealt with here is the primordial mystery of God and his creation, Atum, who becomes One and then Two, and so on up to Eight.

> *I am One that transforms into Two*
> *I am Two that transforms into Four*
> *I am Four that transforms into Eight*
> *After this I am One*
>
> Coffin of Petamon, Cairo Museum, [artifact] no. 116023

The manifestation or proliferation of one into many that occurred at Heliopolis is the abstract principle of creation. At Memphis, Ptah carries this abstraction further and brings down fire from heaven. At Hermopolis the divine fire begins to interact with the other elements (water, earth, air) within the terrestrial world. At Thebes, a reiteration of these three processes is combined into one, represented by the triad of Amun.

According to John Anthony West in *Serpent in the Sky,* the Great Ennead emanates from the absolute or "central fire." The nine *neter*s (principles) are bounded by the One (the absolute), which becomes both one and ten, and is the symbolic likeness of the original unity. The Great Ennead is repetition and a return to the source, which is seen in Egyptian mythology as Horus, the divine son who avenges the murder and dismemberment of his father, Osiris.

The Egyptians espoused a natural, holistic philosophy that described the creation of man not as its being thrust into a dangerous and violent world, but as the incarnation of the divine. Man was the cosmos, and the role of the individual was to realize this in order to achieve eternity. Pythagoras understood the Egyptian philosophy and described it quite coherently in his writings and teachings. The Egyptians spoke of it as myth that held an esoteric truth.

René Schwaller de Lubicz's Pharaonic Egypt

While living in Egypt during the late 1930s, René Schwaller de Lubicz realized that a certain picture at the tomb of Rameses depicted the

~~A rep~~ ~~Image~~ Image of things unseen = Symbolism ⊗

from the Unseen to the seen and
the ONE who manifests a Many.
The Becoming was revealed through
4 successive stages
{ Atum (or Re) at Heliopolis
Ptah at Memphis
Thoth at Hermopolis
Amun at Thebes
Pg 184

(All - Nothing) (? son of Nun - Ephraim)

Atum is at one with Nun
cosmic ocean – —— ⊕

pharaoh as a human figure carved in the representation of a right-angle triangle. Proportionally the triangle's sides were 3:4:5, and his upraised arm added another unit. For Schwaller, this demonstrated that the Egyptians knew about the Pythagorean theorem centuries before Pythagoras was born. It also suggested that the roots of medieval masonry construction were in ancient Egypt. From 1936 until 1949, Schwaller de Lubicz and his family stayed in a hotel in Luxor investigating the hieroglyphic and architectural evidence, particularly at the Temple of Amun-Mut-Khonsu, looking for an ancient system of psychological, cosmological, and spiritual knowledge.

Luxor's Temple of Amun-Mut-Khonsu, originally constructed during New Kingdom dynasties (1570–1085 BCE), is an enormous asymmetric complex built on three separate axes over a fifteen-hundred-year period. Additions to the temple were also made by the last of the Egyptian-born pharaohs, Alexander the Great, and even the Romans. All additions were aligned according to the original axes, suggesting that the architectural guidelines ordering the temple were handed down from one generation to the next. Construction of such a temple seems to defy modern logic because it was built on three axes. But Schwaller saw within it a deliberate expression of harmony, proportion, and symbolism, realizing that the ancient Egyptians embraced a form of communication forgotten, to a large extent, by the modern world.

The symbol and symbolism in the Egyptians' art and architecture address thought in a way different from rational, exoteric styles of communication. Although the brain's left and right hemispheres are not as polarized as was once thought, and their relationship is surely more complicated than can be explained, research shows that the hemispheres do engage in different modes of thinking. (The human brain is a paired organ, composed of two equal sections called cerebral hemispheres.) The left hemisphere of the brain centers on the logical, sequential, rational, analytical, and objective modes of thinking; the right side engages in random, intuitive, holistic, and subjective thought. Modern thought tends to emphasize the left-brain mode of thinking, with its rational and scientific approach to life. According to Schwaller, the symbol is a holistic style of communication appealing to the right brain, what he refers to as "intelligence of the heart." We do not read a symbol as we would read a magazine or newspaper, but understand it implicitly, as we grasp the music of a symphony, except that the symbol is perceived visually—assuming we know the concepts behind the imagery.

After more than a decade of research in Egypt, Schwaller published his work in a series of books describing the nature of symbolism and how the Egyptians used it in their sacred writings (hieroglyphics). In some respects, his conclusions were startling and not well received by traditional Egyptology, since they profoundly confronted the prevailing archaeological theories concerning the development, sophistication, religion, and culture of the ancient Egyptians. Schwaller proved categorically, however, that dynastic Egypt possessed mathematics far superior to that of the Pythagorean Greeks, whom they preceded by more than fifteen hundred years, as well as that of the Europeans, an additional fifteen hundred years later. He demonstrated that Egyptian culture represented a magnificent worldview in which science, religion, philosophy, and art were all part of a single discipline based on man's innate and intuitive knowledge of nature and creation. Despite the fact that traditional Egyptology generally rejects Schwaller's ideas, they have never been refuted since their publication more than fifty years ago.

In inscriptions and the architecture of the temple itself, Schwaller believed, the Egyptians inscribed their knowledge of cosmic laws, creation, and how spirit (energy) manifests as matter. Their insight into God, man, and nature was progressive, positively brilliant, and anything but primitive. What appears at first glance as a pantheon of animal gods was really a way of expressing cosmic principles, describing how consciousness is manifest as the universe. Symbolically, each animal represented a certain principle of nature and, as such, evoked that principle in man when its head, in portrait, was placed on a human body. Each was associated with number symbolism and commanded the geometry of the temples erected to commemorate it, evoking in the beholder the concept of that specific principle.

The Neters

These specific principles of nature were referred to in the ancient Egyptian language as neters and can be categorized as metaphysical, cosmic, and natural. We can think of the metaphysical neters as the triune principle of creation embodied in the characters of Atum, Ra, and Ptah. These three vital principles represent the universal harmonic directive, or the laws of the universe. For the Egyptians, they were created directly rather than procreated and were not engendered by nature. As such, they were represented in art as having no navel. They were also incapable of exercising their own judgment and were therefore portrayed in art without a skullcap. Depicted by a crown, symbol, or animal head, the neters acted in a set way and could not modify the orientation of their activity.

Ra, for example, was not the sun itself. Rather, Ra represented the solar energy that, during the course of its daily cycle, animates all the organic functions of the human body, one after another, at each hour of the day and night; as such, all human beings are subject to the sun.[16]

The cosmic, seasonal neters, in general, such as ram-headed Amon—Aries—obey the laws of creation, as do those associated with the five extra-calendar days (the Egyptian year was 360 days, with five extra days between the end and beginning of the new year): Osiris, Seth, Isis, Nephthys, and Horus. In essence, they are the response to the harmonic summons, the harmonic vibrations of the ecosystem that makes our planet sustain life where summons means to call forth, and depend on astronomical, as well as astrological, synchronization.

Natural neters are the functional life of natural objects and events, such as Renenuret for harvests, Hapi for the Nile, Selkit for childbearing, Apet for gestation, and the vulture Nekhebit for incubation. From mineral to man, they preside over all reproduction and regeneration. The aggregate of natural neters can be thought of as the essence of Mother Nature.[17]

According to Schwaller, every living organism is in contact with all the rhythms and harmonies of all the energies of the universe. It is not possible to separate one's energy from the surrounding energy, except, perhaps, through cerebral presence (mediation, focusing within).[18] We often speak of "instinctive" actions but fail to consider their origin, which is the true source of natural "magic." For one who can consciously regain the instinctive state, there is real power. Although it is rare to achieve such a state—through prayer, for example—because of the concentration and dedication required, gaining real power incites us to strike a harmony with the energy of nature. In ancient Egypt, the functional name of such an act (prayer) was "to summon the Neter." Foolishly, Schwaller contends, "we treat this as superstition."[19]

In the case of plant life, a visible and tangible product exemplifies such "magic." Sunshine is a visible and perceptible cause. The water lily, with its roots in submerged soil, stretches up to the water's surface to blossom in the sunlight, where it opens with the rising sun and closes with the setting sun. The primal cause of the water lily is not understood. We rationalize such "magic" and call it nature, leaving it at that. Nonetheless, there is a relationship between the lily and the sun vital to its propagation. This relationship, or harmony, between organic objects in nature and the cosmic forces (the sun or, in the case of the tide, the moon) is the essence of the word *neter* and indicative of "magic."[20]

For the ancient Egyptians, according to Schwaller:

> Every essential moment can actually be designated by a name: the name of a Neter. In the history of all of sacred science, the name has always played a preponderant role, and knowledge of this name (says the Egyptian Book of the Dead) is indispensable for crossing the gates of the Dwat, the world of the transposed sky, the netherworld. To know the name of the Neter means to know its particular activity, because the Neter is a functional principle and not a god, as popular custom would have it, be it Jewish, Greek, or Christian.[21]

This same idea can be understood in the biblical Genesis, where Adam walks with God naming each creature. Implied in this act, Adam (Man), who is created in the image of God, knows all functions and harmonies of the universe.[22]

Mankind, with its tendency to humanize everything through myth, has done so to these principles. The deeper significance is that these images symbolize certain natural functions and influences. The various neters must be given a human form so that the ordinary person is able to grasp, as well as transmit, essential truths that are metaphysical in character. Also in this way, a person's instinct, says Schwaller, "grasps it and this leads [him] to imagine the Neter of a mountain, a valley, a river, or of any other object—or phenomenon—which strikes [his] emotional nature."[23] Yet the true name of the neter remains the function that it incarnates.

The Ancient Egyptian Meaning of Logos

For Schwaller, the introduction to John's Gospel is in the style of Euclidean geometry, where the point is the intersection of two lines and the line is a point in motion. That seemingly irrational moment at the origin of things, a moment that John defines as the Logos, identifies the Word with its source of emanation. The Word represents the fragmentation of a homogeneous state of oneness (everything is of a single nature) into a state of heterogeneous objects (things are differentiated).[24]

Although the Logos of the Gospel is a term that has many meanings, it generally means "word" (Greek *verbum*). To be more precise, it needs to be related to the term *weaving* in its traditional, symbolic significance, where it represents the intersection of complementary notions. In the craft of weaving, two threads, by themselves, cannot be situated. Yet when woven together, they become an intersection, which gives rise to form. With this understanding, the term *Logos* contains the significance of "the manifestation or divine work of creation as well as the immanence of the creator of this work," and can be likened to a "circuit phrase." (One of Schwaller's analogies to express his philosophy was the electrical circuit—e.g., three conductors carry a voltage in a three-phase circuit.) The Word itself is not magical, but the action it implies is. The utterance of certain letters or words, which have no semantic meaning in themselves, such as the *oooommmm* sound in the Eastern meditative tradition, excites certain nervous centers.

The pharaonic texts of ancient Egypt contain many examples of litanies that play a "magical" role through the repetition of sounds. Although translation of such words into English is impossible, since their pronunciation is unknown, hieroglyphic writings confirm their properties. The spoken language of ancient Egypt was still in use up until the sixth century. Asklepios and others[25] affirm the existence of this magical effect of the pharaohs' words. In a letter to King Ammonopens, Asklepios insists that a formal recommendation is so important that it must not be translated into Greek:

> For fear that the arrogant elocution of the Greeks, with its lack of nerve and what can be called its false graces, might pale and eradicate the profundity, the solidity, and the efficacious virtue of the vocable of our [Egyptian] language. Because the Greeks, O King, have but empty discourses good for producing demonstrations, and therein lies all the philosophy of the Greeks, a clamor of words. As for us, we do not use simple words, but sounds all filled with power.[26]

This use of sacred language explains the mysterious character of Egypt's pharaonic teachings (myths). According to Schwaller, the conditions for comprehension are "purity, selflessness, and the mastery of instincts."[27] It must be a revelation, as opposed to a mental understanding.

In these teachings, "evil" stands for all that reduces the spiritual to the material, which must die in order to return to its spiritual source. The "fall" of the divine into bodily form, however, creates a blemish that must be eliminated for the return to be complete. This requires the Passion of the individual to understand the mystery of life.

The Logos, the divine Word of the beginning, is the All, or the undifferentiated state of existence that is the nature of the Self. It is what we call spiritual, and at the same time what caused the Self (or consciousness) to become mortal. This is the cause and struggle to attain final deliverance from being mortal. It is also an absolution of the irrational. Sin, in Schwaller's view, is simply the fact of our immortal essence being mortal. It is a justification of the irrational for the irrational does in fact exist. If it did not exist, there would be no mystery to Original Cause.

According to Andre Vandenbroeck, who befriended Scwhaller during the late 1950s:

> The irrational is eliminated by absolution in the beginning. The absolute, that is the only irrational, and it is put first, in the beginning, the same beginning where the Greek John hears the Logos. Therefore it is absolved from any particular participation; the whole without detail must be in the beginning, the homogeneous whole: oneness, the ultimate irrationality. The beginning and the end are the same; between them lies the passage from one to two, with two being a new one. In the absolute, in principle, nothing can be told, not even its own irrationality. From the beginning absolute to the ultimate irrationality: consciousness experiences this as a path. But it is an identity, an identity by absolution.[28]

Schwaller also refers to it as the Pharaonic Opus Magnum, which ends in the Christian revelation of divine incarnation and revealing of the ultimate phase. In the pharaonic mysteries, this phase was called the reconciliation of Seth and Horus, but it was never realized in Egypt's history. It explains why the pharaonic sages considered the precessional transition from the age of Aries to Pisces to be a natural progression toward Christianity as foreseen by the Egyptian temple, and why the first Christians chose the sign of the fish as their standard. *Life is spiritual.*

It is not just spiritual but also organic, however, encompassing the physical realities of fertilization, gestation, birth, growth, maturity, aging, and death. It is also cosmic, encompassing the principles of polarity, process, relationship, substantiality, and potentiality. Logos is the principle of divine origin (which is esoteric) and the manifestation of the principle in the physical (which is exoteric). The principle whose source is undefined and the manifestation of that principle together form life, as we know it. Understanding the universe will always contain these two points of view, the esoteric and the exoteric.

For example, according to the second law of thermodynamics, also known as the law of entropy, energy tends to flow spontaneously from a concentrated state to a diffused state. This law states that in its most general form, the world acts spontaneously to minimize potentials and maximize entropy. With this active end-directedness (time-asymmetry), our physical cosmos is given a universal physical basis. In other words, entropy, since it governs all celestial movement, is the basis for our sense of time. It is the most powerful, most general idea in all science, and is why paper, trees, coal, and gas burn; why sand and dry ice (even in a pure oxygen atmosphere) cannot ever burn; why there are hurricanes or any weather at all; and why the sun will eventually cool down. It is also why everything living has to die.

Entropy, being the tendency for all matter and energy in the universe to move toward a state of inert uniformity, is exoteric based on the observation of science; but it is also esoteric. Without entropy, nothing that exists now could ever have been created. In fact, from the disintegration and destruction of whole solar systems, new stars and star systems are created. The exoteric fact of destruction is the esoteric principle of creation. Together, they are the cosmic reality of constant creation.

Greece and the Ancient Model of History

With the emergence of the Greek educational system after the fifth century BCE, Pythagorean concepts helped inspire a new generation of philosophers. With the works of Socrates, Plato, and Aristotle, a whole new world was born, a world that eventually became Western civilization as we know it. Thus, for many scholars, the "Greek Miracle" is the starting point, the beginning of philosophy and science. But much as it is today, cultures of the ancient world were not isolated from each other. Through trade and commerce, the Mediterranean Sea served as a thoroughfare for not only goods, but ideas as well. The ancient historians, such as Herodotus and Clement of Alexandria, are clear on this point. This is known as the ancient model of history. But the Ethiopians and Egyptians developed all the elements necessary for civilization *long before* such elements became evident in Europe and Asia. The conditions in North Africa not only required that mankind adapt to survive, but also allowed the invention of sciences complemented by art and religion.[29] As a society, in Egypt, there was little difference among these three disciplines, as their monuments still testify today.

The fundamental nature of Egyptian culture, despite academic enthusiasm and meticulous research, has always been an enigma. Most scholars believe that, philosophically, they were primitive and polytheistic, even though their construction techniques were highly advanced. Nowhere else in the ancient world exist such magnificent structures built on such a massive scale.

A number of scholars believe that the world, and specifically the Hellenistic world, owes a debt of gratitude to the ancient Egyptians. For African native Cheikh Anta Diop (1923–1986), born in Senegal, Egyptologist, historian, and director of the Radiocarbon Laboratory at the Fundamental Institute of Black Africa, "The Greeks merely continued and developed, sometimes partially, what the Egyptians had invented. By virtue of their materialistic tendencies, the Greeks stripped those inventions of the religious, idealistic shell in which the Egyptians had enveloped them."[30]

Of course, there is great opposition to this interpretation of Greek and Egyptian history. According to Mary Lefkowitz in *Not Out of Africa,* "Aristotle did not steal books from the library of Alexandria and try to pass them off as his own. Nor did any of the other Greek philosophers learn their ideas in Egypt, because even if they went there (and not all of them did), they would not have been able to study with priests in the Egyptian Mystery System."[31] For Lefkowitz, the Greeks were a very capable people and created their own literature, literature very different from that of the Egyptians. For example, she argues that the Memphite theology is of a completely different nature than Aristotle's works. It describes the creation of the world as the Egyptians viewed it; it refers to the uniting of Upper and Lower Egypt.[32] Aristotle's argument was that the universe did not need a creator or divinity, and that there exists an "unmoved mover" responsible for the state of the universe: the force that keeps the heavens in eternal motion—gravity.[33] For

Lefkowitz, "All that the two texts have in common is a concern with the creation of the universe."[34]

Of course Aristotle did not steal books from the library of Alexandria. Greek civilization and culture were different from those of the Egyptians and were renowned and inventive in their own right. The key concepts of this argument favoring Greece or Egypt as the progenitor of civilization are influence and inspiration, as well as approach—different perspectives on the same idea. The Greek approach was materialistic, public, and secular, whereas the Egyptian approach was mystical and esoteric and held secret by the temple priests. Diop argues that the Greeks did not view the gods in the same light as the Egyptians, and that they "reduced them to the level of man."[35] For Diop, this anthropomorphizing of the Egyptian gods was nothing more than acute materialism, characteristic of the way the Greeks thought. Its full effect was felt a few hundred years later with an emphasis on the material sciences that Aristotle so clearly displayed in his works.[36]

A modern example of differing perspectives on the same concept can be seen in the efforts of Carl Sagan. Not metaphysically oriented, Sagan delved deep into science to explain the cosmos and the biological life that arose from it. He did so in a marvelous manner. Yet in no way are his views profoundly different from those of the Greeks and Egyptians, except that he describes the details and chooses not to anthropomorphize the mystery of creation and existence. It's not difficult to read between the lines of Sagan's scientific oratory and wonder if he held back a little just to stay within the bounds of scientific objectivity—"the cosmos is all there is, ever was, and ever will be" suggests that we are, in some way, the cosmos.

Despite all the wonderful Sagan books and documentaries, the mystery of why life exists continues. Science is a perspective, but it is not the only perspective. Nor does it corner the market on truth.

Sagan's "cosmos" philosophy is a product of contemporary science, but how far back must its roots be traced in history to arrive at the true, original source of inspiration? It is the fundamental nature of science (and humans) to continue to build on the efforts of previous generations. Ironically, in the famed documentary series *The Cosmos: A Personal Journey,* Sagan explains the categorization of atoms by atomic number and casually mentions how Pythagoras might have admired the numbering system of the periodic table of elements.

Despite Pythagorean contributions to the origins of philosophy, there is a dark side to the sixth-century BCE Pythagorean movement in southern Italy.

How heavily involved Pythagoras himself was in politics is unknown, but the Pythagoreans recruited aristocratic members of society and attempted to gain the advantage in its government. Unfortunately, their political ideology was decisively undemocratic, which did not fare well with the existing government and the population in general. Furthermore, their secrecy and rites of initiation, which were unduly harsh, requiring five years of silence, led the rest of society to view the Pythagoreans as a cult. As a result, society—its culture and government—was threatened. War broke out when the Pythagoreans subdued the city of Sybaris, known for its luxurious way of life, but their military success was short-lived. By the fourth century BCE, the people of southern Italy had had enough. Revolts ensued, and the Pythagoreans all but disappeared from social and political life.

Since no records exist from Pythagoras and the Pythagoreans themselves, any cause proposed for the events in southern Italy during the fifth century BCE remains speculative. Nonetheless, by all accounts, the Pythagoreans attempted to install a theocratic system of government ruled, of course, by them. These ruling elite would likely have been the *esoterics,* those who had passed all initiation rites and were allowed to see and speak with Pythagoras himself "behind the veil." (Those initiates who had not completed their five years of silence were allowed only to *hear* the words of Pythagoras—that is, in front of the veil.)[37]

Why the Pythagoreans wanted society to operate in this manner, in my opinion, is suggestive that Pythagoras did, in fact, visit Egypt, and studied within the temple mystery school. After his arrival in southern Italy, I believe, he and his followers attempted to re-create the success of Egyptian civilization in the tradition of the pharaonic theocracy. The Greeks, however, would not have it. The greater Greek civilization, its government and culture, remained definitively Greek. But this is not to say that knowledge did not pass from the Egyptians to the Greeks, in much the same manner as knowledge today is often shared across many cultures through the university system. How could a civilization that had already existed for two thousand years, built an immense infrastructure, and served as a refuge in times of famine not influence neighboring cultures?

Language and Number

In *Stolen Legacy: Greek Philosophy Is Stolen Egyptian Philosophy,* George James argues that not only is the Memphite philosophy the source of Greek philosophy, but that it is also the basis of modern scientific belief.[38] For James, there is no conceptual difference between the role of the Egyptian gods of order (the Ennead) in creating and arranging the cosmos and Laplace's solar system/nebular hypothesis of 1796. As the hot stellar core spins off nine planets in Laplace's cosmology, so it does in Egyptian cosmology, except that the Egyptian names represent what we think of as gods, strictly in the Western sense of personified divine beings. As Schwaller so brilliantly concluded, their gods (neters) were never viewed as (Western) gods but rather as scientific principles—that is, laws of nature. *The difference is language.*

Our modern languages seem to lack the means of communicating the Egyptian concept of origins. Any attempt to reconcile the ancient Egyptian ideas of cosmogony requires a study of symbols and numbers, and how those symbols and numbers relate to the physical universe. Why numbers? Number is the only nonmythical way to understand how the Egyptians defined reality, its cause and effects. According to Schwaller's analysis of Luxor's Temple of Amun-Mut-Khonsu, it seems that Pythagoras understood what the Egyptian temple priests were teaching, and why he and his followers were so enthralled with numbers and geometry.

There are great minds today that share the same point of view. Roger Penrose, Oxford mathematician and colleague of Stephen Hawking, believes that mathematical truth goes beyond mere formalism:

> There often does appear to be some profound reality about these mathematical concepts, going quite beyond the deliberations of any particular mathematician. It is as though human thought is, instead, being guided towards some eternal external truth—a truth which has a reality of its own, and which is revealed only partially to any one of us.[39]

It had been previously believed that mathematics was a concept securely tied to the human mind. In 1931, however, mathematician Kurt Gödel proved fundamental theorems about axiomatic systems, showing that in any axiomatic mathematical system there are propositions that cannot be proved or disproved within the axioms of the system. Gödel's work ended years of attempts to establish axioms that would put the whole of mathematics on an axiomatic basis. Gödel's results were a landmark in twentieth-century mathematics, showing that mathematics is not a finished object, as had once been thought; they also implied that a computer can never be programmed to answer all mathematical questions. Gödel's work proved that the abstract world is already "out there," preexisting in some nonphysical state.

Although modern science has taken mathematics to the limits of the solar system in a quest to understand the universe, cosmology (the quintessential science of numbers) began long ago. For the Egyptians, cosmology was embraced in a holistic approach. For the Greeks, it was exoteric, particularly in the works of Aristotle. So the separation of (materialistic) science and religion began not with the modern world, but with the ancient. It is the legacy we live with to this day, and a force for great theological misunderstanding, particularly in the message of Christ.

8

Moses and the Mystery School

Egyptian Esoterism and the Birth of Religion

So Pharaoh said to Joseph, "I hereby put you in charge of the whole land of Egypt." Then Pharaoh took his signet ring from his finger and put it on Joseph's finger. He dressed him in robes of fine linen and put a gold chain around his neck. He had him ride in a chariot as his second-in-command, and men shouted before him, "Make way!" Thus he put him in charge of the whole land of Egypt.

GENESIS 41:41–43

In his youth, the biblical Joseph lived as a shepherd tending his father's flock. Out of jealousy, as the story goes, Joseph's brothers sold him to Midianite merchants for twenty shekels. These merchants then took him to Egypt and sold him to Potiphar, one of Pharaoh's officials or possibly the captain of the guard.[1] Years later, a famine swept across the region. Egypt, with its stores of grain, became a refuge for her neighbors. Joseph, who had seen the famine coming and advised the pharaoh to store grain to avoid times of famine, rose to nobility. Joseph's family (seventy in all) were among those who immigrated to Egypt. Thus begins the story of "the immigrants" in the land of the Nile.

Some time later, Joseph's father died, and was buried with all the trappings of a royal funeral: "All Pharaoh's officials accompanied him—the dignitaries of his court and all the dignitaries of Egypt—besides all the members of Joseph's household and his brothers and those belonging to his father's household. Chariots and horsemen also went up with him. It was a very large company."[2] Clearly, if the biblical story of Joseph is true, then he and his family had become an important part of Egyptian nobility.

In the eastern part of the Nile Delta, at Tel ed-Daba, archaeologists unearthed what could be Joseph's house and his tomb near the ancient city of Avaris. Egyptologist and historian David Rohl, a scholar at University College in London, believes this large Egyptian-style palace was the retirement home of the biblical Joseph that had been constructed in the geographical center of the Hebrew (immigrant) population. According to David Rohl, in the Discovery Channel documentary *Pharaohs and Kings* (based on his book *A Test of Time: The Bible—from Myth to History*), Joseph's Egyptian name was Zaphenat-Pa'aneah and he lived during Egypt's Twelfth Dynasty, when Amenemhat III was pharaoh. To calculate the period when Joseph lived, Rohl relied on Nile flood records to identify the years of abundance and famine in the Joseph story. He also believes that the Bahr Yussef canal was one of Joseph's projects on behalf of the pharaoh.[3] Rohl's evidence, and other evidence such as the Beni Hasan tomb (nineteenth century BCE) in the eastern delta, suggests that Semitic people came to Egypt not as slaves but as traders, herdsmen, and farmers over a long period, between 3000 and 1150 BCE.

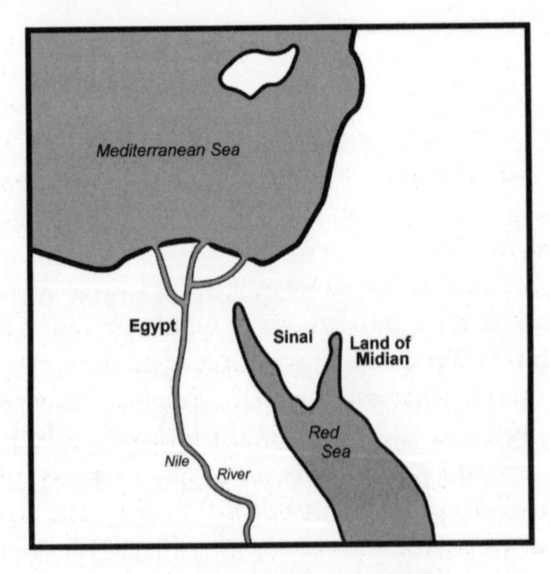

Fig. 8.1. Map of Egypt and Midian

There is a way to determine Egyptian beliefs from the time of Joseph, or at least the legacy of those beliefs. During the time of Joseph, a written history of the Hebrews, whom the ancient Egyptians refer to as "the immigrants," did not exist. Generations later, however, with the Hebrews growing to a purported population of six hundred thousand, the most famous Egyptian of all documented the aforementioned events and created a theology for the people he chose to lead. His name was Moses.

Though the existence of Moses has not been proved as fact, someone did write the books of Moses, and his influence and position in Hebrew culture have been immense. Nearly everyone knows the story of Moses, how he was born in Egypt to a Hebrew woman who placed him in a papyrus basket among the reeds along the Nile. Fortunately for Moses, Pharaoh's daughter, Tharmuth, and her attendants found him. A part of the story that is typically overlooked is that Pharaoh's daughter returned the infant Moses to his mother, although she may have been a wet nurse and not his true biological mother. Whatever the case, Moses was adopted by Princess Tharmuth a few years later.

While only four paragraphs of biblical text are dedicated to Moses's childhood, those years certainly were highly influential. The formative years of a child's life, the first ten years or so, are the most important. The consequences of his Egyptian upbringing are often overlooked. In fact, the name Moses is Egyptian, not Hebrew. *Moses* means "born of," just as *Tutmoses* means "born of Thoth."

Moses grew up in the house of Pharaoh, which means that he lived a life of luxury, attended the best schools, and then matured into a position of nobility within the ruling elite. Since he was adopted at a very young age, he would have little, if any, memory of his biological parents. There can be little doubt, therefore, that Egyptian thought and philosophy heavily influenced how he viewed the world, particularly religion and government. For all intents and purposes, Moses was culturally Egyptian, just as the daughters of Reuel referred to him as they explained to their father that an Egyptian rescued them.[4]

Thus far, exactly what the Egyptians taught in their educational system—what Moses or any other royal youth would have learned—has remained a mystery. No textbook or any evidence suggesting a curriculum, or what may be construed as a curriculum, has ever been found. Much of what is known concerning the Egyptian educational system comes from the earliest of Greek philosophers, particularly the Pythagorean movement in southern Italy during the fifth century BCE. At that time, southern Italy was a part of greater Greece.

According to the ancient model of history, Pythagoras traveled the world in search of knowledge, and when in Egypt managed to gain enrollment as a student in the temple at Thebes. Precisely what he learned is unknown. What is known is that when he arrived back in Croton, he started a university and began to recruit the society's upper class. Much of what Pythagoras and the Pythagoreans taught was geometry, a cosmogony based on number, and a cosmology of harmony—specifically the "music of the spheres."

As previously mentioned, the Pythagoreans were as much a political movement as they were philosophical. Unfortunately, their brand of politics was decidedly nondemocratic and foreign to the people of southern Italy.

From recorded history it is apparent that the Pythagoreans attempted to gain control of society and install a theocratic style of government run by nobility who had to endure an initiation process that took many years to complete, an initiation that can best be described as a rule of passage into the "mystery school" where the mysteries of life are revealed.

My opinion is that the style of government that the Pythagoreans prescribed and labored for was likely based on what Pythagoras learned while studying in the Thebes temple. In other words, the Pythagorean movement was attempting to create a system of government based on the Egyptian model. In the end, the movement failed miserably in a grassroots-style, anti-Pythagorean revolt.

Most modern scholars believe that Egyptian theology was unstructured and lacked a central doctrine, and that various cults held prominence in different areas or cities over the span of Egyptian civilization. One of the more common themes was sun worship, such as was practiced by the cult of Amun-Ra that flourished around the Eighteenth Dynasty; another was the strictly monotheistic ways of Akhenaten (1369–1332 BCE), who moved Egypt's capital into the desert and built a new city called Amarna, to the dismay of the traditional temple priests.

For the common folk, theology was expressed primarily through mythology in the tales of Osiris and Isis, Horus and Seth, and the flight of the phoenix. Within such myths, however, there likely lies a deeper truth that only the association of temple priests expressed in nonmythical ways—a truth to which Moses, as royalty, was probably privy.

After becoming an adult and settling into a position in Egyptian society, Moses's great weakness, we discover, was anger. After witnessing some injustice inflicted on a Hebrew man, he lost his temper and beat a man to death, then buried his body in the sand. Needless to say, he became a wanted man. He soon fled to Midian, in the region of the Sinai, where he started a new life by marrying the daughter of a priest named Jethro. He and his wife started a family and named their first son Gershom, which means "I have become an alien in a foreign land." For a long time Moses lived in Midian and tended his father-in-law's flock, before taking on the mission from God to lead his people out of Egypt.

Although the first book of Moses, Genesis, was not contemporary to Moses, the last four (Exodus through Deuteronomy) were; they tell the story of the flight from Egypt to the Promised Land. In those books he lays the foundation for Hebrew law and the philosophical precepts for their religious beliefs and practices. Genesis, in contrast, is a history of the Hebrew people, from the first man and woman through the life and times of Joseph, another wealthy and powerful Egyptian of Hebrew descent.

According to Moses, Hebrew history begins in a land "beyond the Euphrates River,"[5] clearly a reference to Mesopotamian origins, possibly the Semitic Akkadians. Before that, Hebrew history is Sumerian, since the earliest of stories relate to events that occurred around 4000 BCE, based on the patriarch's lineages as well as supporting archaeological evidence. The biblical text confirms this. According to Genesis, the founder of the Hebrew culture, Abraham, grew up in the city of Ur along the banks of the Euphrates River. The Book of Jubilees, a second-century BCE text telling the history of the world from creation to Moses, further clarifies during what period he lived by attaching a year to his birth, 1884 BCE.[6] Its historical context is the Third Dynasty of Ur, when Sumerian civilization was already more than a thousand years old.

How did Moses, who was raised Egyptian, learn of this history? A likely scenario is that while living in Midian, he heard the oral traditions from his priestly father-in-law, Jethro, who was a mysterious character in his own right. Jethro not only gave permission for Moses to return to Egypt, but also joined him in the Sinai Desert after the Exodus. Could it be that Jethro was the true inspiration behind the Exodus, or did Moses join an already existing migration out of Egypt?

As for the Exodus itself, the archaeological evidence paints a slightly different picture than does the biblical text; the Exodus according to the archaeological evidence is much less sensational and less orchestrated, although the essential facts appear to be correct. According to the documentary *Who Was Moses?* the Exodus was the result of increasing tensions between Egyptian society and working-class immigrants known as the Hebrews, although the Bible speaks of the Hebrews as slaves.[7] Whatever the case may be, this group of immigrants, the Hebrews, did in fact leave Egypt and migrate into an area just south of the Mediterranean Sea at the modern border of Egypt and Israel.

In their northward trek they did not cross the Red Sea, however. *Yam suph,* the term used by the biblical writer, was mistranslated. They crossed the "sea of reeds," not the Red Sea, which can mean any of the saltwater lakes between the Gulf of Suez and the Mediterranean. Even in biblical times, canals and seasonal tides fed these seas. But what could part these waters? There is a logical and fascinating explanation.

According to both geological and archaeological sciences, the island volcano of Thera (modern-day Santorini) exploded in 1628 BCE in a blast a hundred times larger than the Mount St. Helens eruption of 1980, causing a tsunami (as the core of the island sank into the sea). The pending tsunami pulled the water from the shallow marshes, only to send it crashing in again as a wall of water. Interestingly, the Egyptians noted this event in an inscription at the temple of Karnak dated to 1538 BCE:

> When I pharaoh allowed the abomination of the gods, the immigrants, to depart the Earth swallowed their footsteps. This was the directive of the god Nun, the primeval water who came one day unexpectedly.[8]

They also recorded the blast and the cloud of ash that darkened the sky:

> The gods caused the sky to come in a tempest of rain with darkness, unleashed without pause, louder than the cries of the masses.[9]

Core samples taken from the Mediterranean and the Nile Delta indicate that the ash cloud from the Thera eruption did in fact reach Egypt. According to professor of Egyptology Hans Goedicke, of Johns Hopkins University, volcanic glass shards are evident in the samples.

Such an event as the Thera explosion also seems a likely candidate as the cause of the ten plagues.

Further research into the story of the Exodus by Emmanuel Anati, of Italy's University of Lecci, confirms that the geographical elements in the story are also correct, although Moses's Mountain of God has been incorrectly identified in the past as Mount Sinai. According to Anati, the most likely candidate is Mount Karkom, near the Egyptian/Israeli border. There, over the past two decades, he and teams of archaeologists have discovered a wealth of evidence for not only human occupation, but religious worship as well. At the foot of the mountain, which is actually a plateau, there stands an altar made of stone, as well as twelve pillars. The area is also rife with the remains of circular dwellings and engraved stones, of which one appears to be representative of Moses's Ten Commandments, tablets sectioned into ten parts. (The engraved stones look like tablets although there is no writing on them.)

According to Anati, however, the structures around Mount Karkom date to approximately 2000 BCE, well before the traditional chronology of the Exodus. From his research, he has also determined that Mount Karkom, which is abundant with flintstone, has been a site of human activity since the Paleolithic. Anati's conclusion is that the Exodus story is likely based on archetypes, and that there are migration stories in the mythologies of various tribes on almost every continent:

> According to what we know, the entire human race, descended from the earliest *Homo sapiens,* acquired his skills and creativity, his capacity to develop philosophy and religion, when the common ancestors of today's humanity left their primordial territory, which transformed into the myth of the Garden of Eden. Waves of primordial migration left the African "paradise" and crossed the Sinai Peninsula, making it an age-old passage for men in search of their own "Promised Land."[10]

Mount Karkom shows evidence of prehistoric migrations, as well as the more specific Exodus of the Hebrews that gave birth to the kingdom of Israel. Whatever the case may be concerning the Exodus, history portrays Moses as the spiritual and political leader of the Hebrew people. If Moses was an Egyptian nobleman with a princely education, as the ancient texts claim, he was certainly qualified to be that leader. The important part, as it relates to the thesis of this book, is How much of his theology and cosmology was based on Egyptian philosophy and how much was truly the traditions of Abraham?

The Hebrew name for God was Jehovah-Nissi, which, according to the Greek historian Diodorus Siculus (first century BCE), was derived from the name of the Greek god Dionysus. Furthermore, only the Egyptian god Osiris was allotted a place of birth and burial. In Egypt, his birthplace was called Nissa, the Hebrew equivalent of Mount Sinai. In other words, Osiris was born on Mount Sinai—Moses went to Mount Sinai to receive the Ten Commandments. Moses went to Mount Sinai because he knew from his Egyptian upbringing that it was the Mountain of God. So, there is an Egyptian tradition that becomes Hebrew. Moses's burning bush that spoke the word of God has an Egyptian foundation also. Since Moses was in fact Egyptian, culturally speaking, he had to get the idea of a burning bush from somewhere. Although he could have made it up, since the tradition already exists and he was in fact Egyptian, it is more likely he grew up hearing the tales of a burning bush. Possibly acacia, this tree was purported to bow its leaves in hospitality to the traveler who sat under it. Sometimes the goddess Neith sat within its branches, inspiring the king/pharaoh with life and power. According to the story of Apollonius, when he was visiting Thebes, the tree spoke to him and announced that he was a teacher sent from heaven. In other stories this tree appears more like the Tree of Life and the Tree of Knowledge in the story of the Garden of Eden.[11]

In Egyptian processions, men carried a variety of standards upon which were mounted sacred images. One of the more common emblems was the image of the serpent. In a similar manner, Moses raised a standard of the bronze serpent in the desert to inspire the belief that whoever was sick would be healed if he or she gazed upon it.[12] The same is true, that it is an Egyptian idea, for the magical snake Moses used to cajole Pharaoh. Picking up a snake and applying pressure to the nape of its neck makes it become stiff as a rod. If it is then thrown to the ground, it soon recovers from its cataleptic state.[13] This was a common trick used to gain power over the minds of the people.

These outward, exoteric examples establish that Moses was relying on his Egyptian knowledge to promote his leadership and provide a system of belief for his followers. But if Moses was truly patterning Hebrew beliefs after Egyptian, then why was he so adamant that graven images were forbidden? Egypt was a land of graven images! The answer lies in the tendency for his followers to view the world in an exoteric manner. Idols were esoteric and *representative* of the principles of nature. Not understanding that the images were purely symbolic, they believed the images were, in fact, gods. By forbidding idolatry, it seems likely Moses was trying to instill in his followers a belief system based on esoterism. The story of the Garden of Eden provides the cosmological underpinnings for the rest of his works and is vital to the esoteric (Egyptian) tradition of Moses and his literary works.

The Garden of Eden

According to traditional Christian belief, the Garden of Eden was a perfect place both spiritually and physically. After man was exiled from its confines, it provided the conceptual foundation for a fallen world in need of deliverance, and for why God would incarnate on Earth as Jesus, making the sacrifice for all to restore mankind to its original, perfect state.

Is this concept of a fallen world the story's original meaning? The story's characters and elements hint that it is not. Snakes don't talk. Nor can a fruit, after eaten, provide knowledge of good and evil, or another fruit provide eternal life. Many theologians and scholars, past and present, believe the story was symbolic, or esoteric. But the question remains, Symbolic of what? To understand the significance of the Garden of Eden story, a historical approach is needed.

Locating the Garden of Eden

Despite the fact that Moses provides some intriguing clues that the garden was a physical place, its geographical location has remained a mystery. Nonetheless, in the Bible the Garden of Eden is described not as a mythical place but as an actual physical location where the headwaters of the rivers Tigris, Euphrates, Pishon, and Gihon meet. Furthermore, Moses describes Eden as the greater area, with the garden occupying its eastern end. Genesis also states that the river Gihon winds through the land of Cush, and the Pishon through the entire land of Havilah, where there is gold.

> Now the LORD God had planted a garden in the east, in Eden; and there he put the man he had formed. And the LORD God made all kinds of trees grow out of the ground—trees that were pleasing to

> the eye and good for food. In the middle of the garden were the tree of life and the tree of the knowledge of good and evil.
>
> A river watering the garden flowed from Eden; from there it was separated into four headwaters. The name of the first is the Pishon; it winds through the entire land of Havilah, where there is gold. (The gold of that land is good; aromatic resin and onyx are also there.) The name of the second river is the Gihon; it winds through the entire land of Cush. The name of the third river is the Tigris; it runs along the east side of Asshur. And the fourth river is the Euphrates.[14]

Except for the rivers Tigris and Euphrates, these geographical features have been lost to history. The Tigris and Euphrates run though the Mesopotamian Plain with their headwaters located somewhere in the Zagros Mountains, a range that separates modern-day Iran and Iraq. So if the Garden of Eden really existed, it would likely have been somewhere within the area of the Zagros Mountains. The identification of the rivers Gihon and Pishon would narrow the possible geographical area where Eden was located. Rivers do change their courses over time. And, given climatic changes, some river systems disappear altogether, which could help explain why these two geographic features are not apparent in current cartography. Their names may also have changed over the course of history.

David Rohl thinks both are true. According to his book *Legend: The Genesis of Civilization* and the corresponding television documentary *In Search of Eden,* the names of these two rivers have been masked by translations and changing vocabulary.[15]

During the seventh-century Islamic invasion of Persia, Arabic geographers referred to the Aras (Araxes) River in Iran as the Gaihun—close to the spelling of the biblical Gihon. Furthermore, the ancient name of this region through which the Gaihun flows is called Cush. Today, a certain mountain in the region is still called Kusheh Dagh, the mountain of Cush. So, according to Rohl, it is likely that the river Aras is the Gaihun, the second river in Eden. Rohl also pinpoints the first river, the Pishon, as the Kezzel Uizon. According to language experts, the Iranian letter *U* was changed to a Semitic *P* by the biblical author. As the Minian city of Uishteri is known by the Arabic name of Pishdeli, the Uizon translates into Hebrew as Pishon. *Kezzel* means "gold." Today, gold is still found in the Kezzel Uizon (Uizhun), just as the biblical story describes.

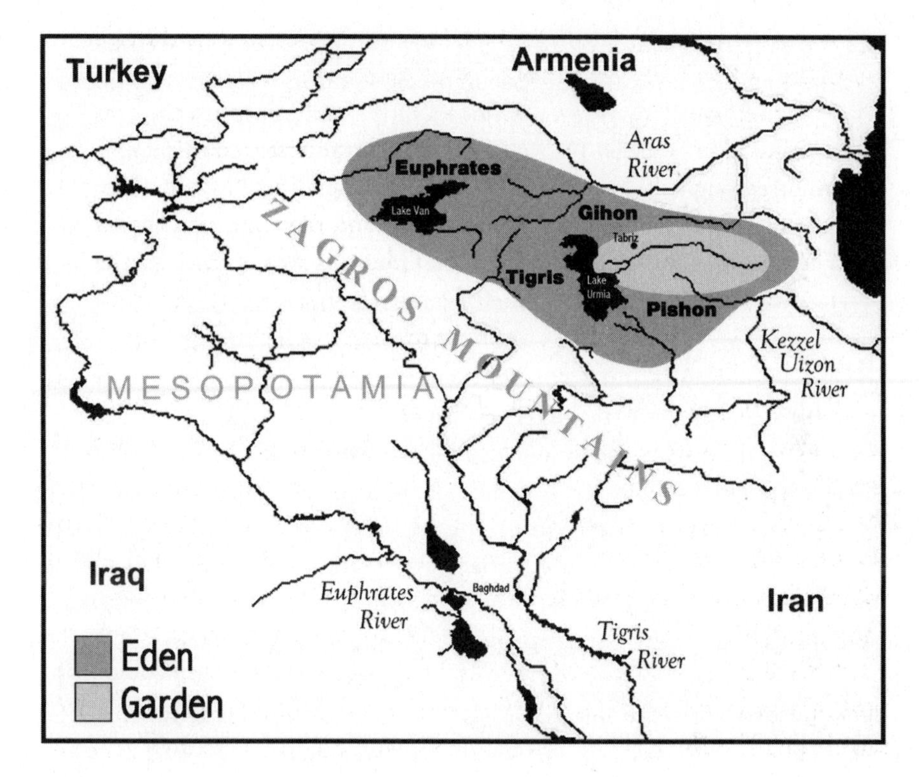

Fig. 8.2. The Plain of Eden and the Garden

With all these pieces fitting into the puzzle of the garden's location, Rohl concludes that Eden was located in the Zagros Mountains where the headwaters of the rivers Tigris, Euphrates, Aras, and Kezzel Uizon are found. The name Eden, meaning plain or uncultivated land, is derived from the ancient Sumerian word *edin,* which refers to an area of land outside a settlement. This description of Eden (or Edin) fits the description of the land around the Iranian lakes Van and Urmia.

Of the two locations the Genesis author describes, the greater one is a plain, the area of land called Eden, which is geographically defined as northwestern Iran and northeastern Iraq between the Caspian Sea and the north plain of Mesopotamia. The garden, the smaller locale, is in the eastern part of the plain. Garden (Hebrew *gan*) refers to an enclosed area, a walled garden.

The garden itself is between Lake Urmia and the Caspian Sea. It is known today as the Valley of Tabriz. Lake Urmia provides its western boundary, and the mountain ranges to the north and south provide the walls. At its eastern end is a gate—in more modern terms a pass—leading to a plain next to the Caspian Sea. Genesis refers to this area as "the land of Nod," and, according to Rohl, this land was referred to in ancient times as the Upper and Lower Nochdi, which means belonging to Nochd. He believes this was the land of Nod.

Everything else in the Eden story also fits. The mountain of God, Mount Sahand, overlooks the valley and the river of the garden. The red earth from which man was formed covers the foothills in the valley. (In Hebrew, *adam* means "red earth.") The eastern gate, where God placed the cherub with a flaming sword, is still known for its thunderstorms. It is an apt metaphor for the angel and his fiery weapon. At the entrance of the mountain pass is the village of Helabad, formerly known as Kheru-abad, which means "settlement of the Kheru people." Rohl believes that this may be a transformation of the Hebrew word *keruvim,* which is translated as "cherubs." The Kheru people were a clan of fierce warriors whose standard was a bird of prey, an eagle or falcon. In ancient times, perhaps they were the true guardians of the pass, Rohl muses.

The English word *paradise* means "a perfect place to live" and is synonymous with the Garden of Eden. Interestingly, it is taken from the Persian word *pare-da-sa,* which means "enclosed or walled park-land." The fourfold Persian garden, a pool at its center with four channels of water leading away, is called "paradise" and is patterned after the Garden of Eden.

Besides biblical text, other ancient texts refer to Eden (or Edin) and the garden. In ancient Sumerian texts, the word *edin* links Moses's story of creation with the land of Aratta, located in the Zagros Mountains. In Sumerian history, there are four other Eden-like stories that tell the tale of the Sumerian king Enmerkar and the lord of Aratta.

Sumerian Stories of Eden: The Land of Aratta

During the third millennium BCE, Aratta was a settlement, and possibly a city with which the Sumerian kingdom had religious ties and commercial interests. According to the Sumerian story "Enmerkar and the Lord of Aratta," Aratta was under the protection of Inanna, the sun god's daughter, who was the goddess of love and war. Enmerkar, the ruler of Uruk, took the goddess (possibly her statue) and moved her to his city, then beckoned the lord of Aratta to send goods to transform Uruk's shrines into places of spectacular beauty. Precious metals, precious stones, and building materials were required of Aratta as well as the craftsmen to perform the work. The lord of Aratta was willing to fulfill the request if Enmerkar would send him large amounts of barley in trade. Emmerkar agreed, but when the barley arrived, the lord of Aratta refused to fulfill his part of the agreement.

Years later, Enmerkar again sent his envoy to Aratta. This time, the lord of Aratta challenged Enmerkar to select a champion to fight with the Aratta champion. Enmerkar accepted, but because his response was so lengthy, he ordered it to be inscribed on clay tablets. (The author of the tablet implies that this was the beginning of writing.) At that time, the famine that had plagued Aratta lifted, so Aratta's ruler took courage that Inanna had not abandoned him. Although the ending to the story is fragmentary, Aratta eventually provides the materials and craftsmen for the refurbishing of Uruk's shrines.

In another Sumerian myth, "Enmerkar and Ensuhkeshdana," the lord of Aratta demands the submission of the king of Uruk, Enmerkar, and the return of the goddess Inanna to her home in Aratta. Enmerkar refuses and, likewise, demands Aratta's submission. After consulting with his advisers, who urge him to capitulate, in anger the lord of Aratta refuses. His priest comes forward, however, and boasts that through magic he will subdue Uruk. So he rewards the priest and sends him to Uruk, where he dies at the hands of an assassin. In the end, the lord of Aratta submits to Uruk.

Aratta is mentioned again in a third story, known as "Lugulbanda and Enmerkar." Uruk is under attack from the Martu people. In desperation, Enmerkar sends his messenger, Lugulbanda, to Aratta to the goddess Inanna, who is called his sister. Although Inanna's response is unclear, it appears that Aratta again supplies Enmerkar with metals, precious stones, and craftsmen. The story suggests that the materials were transported to Uruk by river.

Aratta appears in a fourth myth, "Lugulbanda and Mount Hurum." Enmerkar and his army are traveling to Aratta to make it a vassal state. When they stop at Mount Hurum in the neighborhood of Lake Van, Lugulbanda becomes ill and dies. His friends place his body on Mount Hurum with the intent to retrieve it after the war in Aratta. Lugulbanda, however, who is not really dead, prays to the sun, moon, and the planet Venus, then emerges from his trance and wanders the highlands. The rest of this story is lost.

These Sumerian myths portray Aratta as a wealthy and powerful state with which Sumer had a relationship during very early times.

According to the stories, it was some distance from the Mesopotamian Plain and protected by mountain peaks, yet close enough that the two engaged in trade. Aratta had building materials, precious stones, metals, and skilled craftsmen. It also appears that Aratta held primacy with regard to the religion of the Mother Goddess, Inanna, who resided in Aratta. Inanna was Aratta's patron goddess, and she was forcibly taken or lured south to Sumerian cities. Uruk and Aratta were also in contest for military superiority, each demanding the submission of the other. The method of transporting the "stones of the mountain" from Aratta to Uruk, and of transporting grain from Uruk to Aratta, seems consistent with such historical trade between the Armenian highlands and areas to its south—by boat from Aratta heading south and by pack animal from Uruk heading north.

If Aratta was located near the Mesopotamian Plain, the general implication is that these myths played an important role in the development of Sumerian religion. They may also be responsible for the construction of Sumerian cult structures, as well as trade and diplomatic ties between these two states. Trade relations were of such importance that writing was developed specifically for Aratta and Sumer.

Locating the Land of Aratta

Four general areas have been suggested for the location of Aratta. Two are in eastern Asia Minor: the Lake Van/Lake Urmia area and the Ayrarat district of historical Armenia. The Anshan-Hamadan area of western Iran was the choice of Sol Cohen, who translated "Enmerkar and the Lord of Aratta." Although the story does not specifically state where Aratta is, it leaves enough clues to construct a likely locale.

According to the story inscribed on the tablet, the Sumerian envoy's journey began in Susa, an ancient city on the broad plain in southwestern Iran. It was the capital of Elam, the biblical Shushan, which, from the fourth millennium BCE, was under the cultural influence of Mesopotamia. The emissary was instructed by his king to journey by the stars by night, and by day to travel with the sun. From Susa to the Anshan mountain-land, he and his company passed through the mountains. After traversing the seventh gate (a mountain pass), he lifted up his eyes as he approached Aratta—possibly a connection to the seven levels of heaven. Even today, we use the phrase "seventh heaven" to distinguish a state of extreme bliss.[16]

Another clay tablet, in Paris's Louvre Museum, describes a military campaign against the kingdom of Aratta by the powerful Assyrian king Sargon II. Just as Enmerkar's envoy crossed seven mountain ranges, so did Sargon's army. The description of his journey is in the form of a letter to the god Assur and relates the eighth military campaign against the kingdom of Urartu, a land that encompassed Armenia and Kurdistan. The 430-line text tells how the king led the operation and captured the holy town of Mushashir.[17]

According to the tablet, the king of Assyria and his army crossed a river in its flood stage and marched between high mountains that were covered with all types of trees, parts of which never saw the light of day. He traversed seven mountains with great difficulty. Against Surikash, a southern province of the biblical Mannai, he crossed the rivers of Aratta as though they were irrigation ditches.

Sargon was heading for ancient Surikash, which lies under the Kurdish town of Sakkez on the Marivan Plain. He traveled north through seven mountain passes across the river Aratta, and arrived on the plain (the *edin*) of Aratu, the biblical Ararat. Enmerkar's envoy also traveled north through seven passes and arrived on the plain of Aratta. The descriptions of the journeys are so similar that it is reasonable to assume they were headed to the same place: the place that Moses geographically describes as Eden, a land of wealth and riches tucked into the peaks of the Zagros Mountains.[18]

For the ancients, the Zagros Mountains provided a near-perfect place to live, fertile mountain plains that, with abundant rain, supplied an ideal climate for agriculture. It was also a safe haven from marauding enemies, extending from southern Azerbaijan to the Fars province near the Persian Gulf. The rough terrain is nearly impassable when traveling east or west. Many of the peaks crest at ten thousand feet or more, such as the Zard Kuh, the highest at nearly fifteen thousand feet.

Today, northern valleys of the Zagros support agriculture and are heavily populated. In the central range, shepherds graze their flocks in upland pastures. In southern parts, dates and cereals are grown in oases. It is home to some of the most ancient peoples of Iran. Kurdish groups settled there more than three thousand years ago and have cultivated its fertile plains ever since. Farther south, the Lurs, a very old culture that dates to the beginning of the Persian Empire, are also distinguished residents where life is based on herding and seasonal farming. There are also the famous Bakhtiari tribes, who migrate to cold climates in the summer and move south during the winter months.

Eden's Historical Context

With the geographic location of Eden and the garden now known, we may conclude that the Genesis story of Adam and Eve is literally true, insofar as Eden was a real place. There is no mention of creation elements—the serpent, the fruit, or Adam and Eve—in Sumerian stories, however. How could the events pertaining to Sumer and Aratta be interpreted as a creation story?

Based on the lineages of the biblical patriarchs, the events inspiring Adam and Eve's story occurred around 4000 BCE, according to Bible scholars such as the archbishop of Armagh, James Usher. Interestingly, this date fits well with the archaeological evidence of the period, the time of the Neolithic revolution when mankind began to organize into farming communities. According to anthropologists, human events leading up to this agricultural revolution played a significant part in the formation of Sumerian civilization, and likely served as a basis for Sumerian creation stories.

Origins of Sumerian Civilization

During the sixth millennium BCE, two cultures lived side by side in the Mesopotamian Valley. One culture occupied the north; the other, the valley's southernmost areas between the Tigris and Euphrates Rivers. The earliest settlements arose in the north within the Kurdistan Mountains, what we have discovered to be the mountain plain or *edin,* and then moved onto the Mesopotamian Plain. Over time, their culture spread farther east, west, and south, along the southern slopes of the Zagros Mountains.

Using various sources of clay, this culture created several types of ceramic art, including a fine and distinctive style of pottery that has been found from southeastern Turkey to Iran. To scholars, this early society was known as the Halaf culture, so named because it was first identified at Tell Halaf, in northeastern Syria. Its people were farmers and artists and relied on natural rainfall for their crops. Emmer wheat, barley, and flax were their primary crops, but they also herded cattle, sheep, and goats. Excavations of their stone houses have revealed exceptional pottery, jewelry, sculpture, and obsidian tools. From the evidence, it is clear that the Halaf culture was a complex prehistoric society whose trade contacts within its communities allowed its members to amass considerable wealth.[19]

Besides finely painted pottery, the Halafians made small baked-clay female figurines in a cultural and religious tradition known as the Great Goddess or Mother Goddess tradition, an indication that their social structure was possibly matriarchal. These distinctive clay figurines with large thighs, buttocks, breasts, and a long braid over the top of the head have been found throughout Halafian shrines. Two figurines found in the upper Tigris basin at Arpachiya and Chagar Bazar date to 5000 BCE. Figurines displaying exaggerated female characteristics have been found at other prehistoric sites in Turkey and areas of Europe.[20]

The site of Tell Sabi Abyad, in the upper Balikh Valley of northern Syria, has recently been a focal point of investigations aimed at clarifying the chronology, settlement, and organization of these late Neolithic societies, with an emphasis on the late sixth and early fifth millennia BCE. In 1988, one of the largest Halaf settlements in the region was discovered, and, according to archaeologists, what was found suggests a continuous and uninterrupted sequence of occupation.

One excavated Halaf settlement, measuring over 2,400 square feet, revealed multiroomed and possibly multistoried rectangular dwellings. Subsidiary annexes and other circular structures called *tholoi* surrounded these buildings. They were made from rectangular blocks of loam and pisé (a mixture of earth, chaff, and other things that give body to the mud) with walls that were occasionally built on a stone foundation. The outer wall contained a facade with regular, buttresslike supports protruding in all sections. Such supports most likely provided strength for an upper story, but they may have been purely ornamental or symbolic.[21]

The evidence suggests that the Halaf culture originated in the region of the river Khabur in modern Syria, where there are no known sites before 14,000 BCE and only two between 14,000 and 10,000 BCE. During the eighth millennium BCE, however, farming and herding were introduced into the area. Later (after 6000 BCE), the region experienced a remarkable increase in agricultural communities, an expansion of trade routes, and the development of a complex social organization. Plant and animal remains indicate that these Halaf communities relied heavily on domestic crops and livestock, although a small amount of wild plants and animals were also gathered and hunted.

According to demographic studies, the Taurus and Zagros Mountains were the site of an expanding population sometime between twelve thousand to five thousand years ago because of successive technological advances in plant and animal domestication.[22]

The Ubaid Culture and Sumerian Civilization

In 1966 at Nippur (in modern Iraq, a hundred miles south of Baghdad), potsherds scattered over a wide area provide evidence of the Mesopotamian Plain's earliest inhabitants, the other of the two Mesopotamian cultures. According to the finds, a large ancient settlement once existed, reaching more than a hundred hectares by 5000 BCE.[23] In 1905, while researching Parthian quarries, Clarence S. Fisher discovered occupation by this culture more than ten feet below the present surface.[24] This culture became known as the Hajji Muhammad.

According to archaeologists, the Hajji Muhammad provided the foundations of the Ubaid culture, which was fully developed by 4350 BCE and occupied most of southern Mesopotamia. British archaeologist James Mellaart maintains that cultural evidence (mainly ceramic) suggests that the Hajji Muhammad descended from the Zagros Mountains of Iran onto the plain.[25]

By 4500 BCE, Hajji Muhammad had expanded northward, and its people were living in small settlements, typically along rivers. Their temples were platformlike, built from mud bricks, and were located at the village center. At about the same time that the Ubaid culture appears in the archaeological record, the Halaf society seemingly disappears. Whether it was assimilated into or conquered by the Ubaid remains unknown. Either way, the Halaf culture faded into obscurity, and its characteristics were never seen again in the region.

Archaeologists interpret a shift in social organization around 4000 BCE as the beginnings of the Uruk culture and the seeds of Sumerian civilization. During this transition, society became distinctly stratified; ranks or classes were created, each group of people having a different function and level of social power. It also appears that the elite of the community demanded tribute from their subjects, not only to meet their own needs, but also to support craftsmen and laborers engaged in constructing temples and fortifications. By the late fourth millennium BCE, Uruk had become the largest city in lower Mesopotamia, covering nearly 250 acres.

By 3000 BCE, Uruk had entered its early dynastic period, with a majority of the population living within the city. A household-based economy developed in which people belonged to large extended families; Gu'abba, in Lagash, for example, housed as many as four thousand adults. Temples themselves constituted a household, although many others were dominated by leading families who actually lived in the temples. No longer self-sufficient, the common people contributed their labor in exchange for the necessities of life. As a result of this structured society, the elite now controlled the production, as well as the distribution, of goods. The first Mesopotamian kingdom was born, and according to Sumerian cosmology, its authority was handed down from the heavens above.

By 2500 BCE, Uruk's expanse covered a thousand acres and grew to a population of forty thousand people. The first ziggurats (step pyramids) were built so that city rulers could communicate with their gods. Artisans supplied the elite with luxury goods, while the common people used crude, mass-produced earthenware. A diversity of grave goods epitomized the growing chasm between the privileged few and the masses. Judging from the forty-four hundred-year-old neighboring Royal Tombs of Ur, those privileges included the right of kings to have their servants sacrificed and buried with them.

Although it is believed the Halaf culture was eventually absorbed into the Ubaid culture, the Halafians left no legacy in any other culture. How the Ubaid made the transition into a stratified society and became the kingdom of Sumer remains a mystery. Nevertheless, it is within this transition, from the Ubaid culture to Sumerian civilization around 4000 BCE that the story of Adam and Eve was supposed to have taken place.

The Real Adam

Some scholars believe the historical Adam may have been King Allum, the first in the Sumerian line of kings, or perhaps the first king of Eridu. He has also been equated with Adapa, the first sage of Eridu. Regardless of precisely who he was, there was something special about him. We know from Genesis that he was not created (born) in Eden, but migrated there, most likely as part of an adventurous band of travelers. Although there is no positive proof that Adam was indeed an individual, there are other Sumerian texts from which the Eden story may have evolved.

In 1985, former exploration geologist Christian O'Brien, along with his wife, Joy, published a translation of the Kharsag tablets from the library at Nippur.[26] According to the story, it is here in Kharsag, in the Zagros Mountains and another name for Aratta, that Enlil, the supreme god of the Sumerian pantheon, created mankind.

Written in cuneiform, these Sumerian-Akkadian texts tell the story of a thriving settlement of creator gods, the Anannage (or Anunnaki). Interestingly, the name of their settlement was Edinu, the same root word used to describe Adam and Eve's homeland in Genesis. O'Brien connected this tale of living gods, who were often described as having serpent eyes and shining faces (referred to as the Shining Ones), with Enoch's story of the Watchers. He concluded that they and the Anannage were one and the same.

O'Brien's book, *The Genius of the Few,* received little notice. A few scholars, however, including the British Museum Sumerian expert Irving Finkel, praised it. In 1996, British author Andrew Collins expounded on O'Brien's work in *From the Ashes of Angels,* an intriguing, well-researched book that examines the birth of civilization and Enoch's tale of the Watchers. According to Collins, the Anunnaki were the princely offspring of heaven and Earth who had arrived in the mountains and set up camp in a fertile valley. They called their land Kharsag or Edin. It describes either a central fenced enclosure or a lofty fenced enclosure.[27]

These mythical sons of God developed a sophisticated agricultural community that included irrigation as well as plant and animal domestication. Their homes, as the story goes, were made from cedar. Larger projects included a reservoir, a granary, and other buildings. One structure was dubbed the Great House of the Lord Enlil; it stood in grand fashion above the settlement. In the surrounding valleys there were tree plantations, other cedar enclosures, and orchards planted with trees bearing a threefold fruit. According to O'Brien, the community thrived for a very long time. Harvests were plentiful, with excess grain stored in the granary. Apparently, the residents allowed outsiders into the community as partners or helpers to share the labor as well as the bounty. In all respects, it was a Garden of Eden.

There were fifty founders of the community. The primary leader was Enlil, the Lord of Cultivation, and his wife, Ninharsag, the Lady of Kharsag, also known as Ninlil. She was the Shining Lady, as well as the Serpent Lady. The latter name led O'Brien to believe that the snake goddess was worshipped at Nippur. Also among the leadership were Enki, Lord of the Land, and Utu (or Ugmash), a sun god. The Anannage were said to have had a democratic leadership, with a council of seven that came together for major decisions. Occasionally, the Supreme Being Anu (meaning "heaven" or "highlands") would join the council as an adviser.

According to O'Brien's translation and interpretation of the Kharsag tablets, Eden was an agricultural settlement managed by a group of people known as the Anunnaki. On occasion, they received migrants and drifters (outsiders) to help with the work. It is not a large leap in reasoning to suspect that Adam was one of these workers taken in by the Anannage. Genesis 2:15 states:

> The Lord God took the man [Adam] and put him in the Garden of Eden to work it and take care of it.

From biblical text, we learn that Adam was said to be the first patriarch and part of a patriarchal culture. According to some researchers, this type of culture was a new development; earlier cultures, such as the Halaf, were more matriarchal, according to Marija Gimbutas, who spent a lifetime studying the goddess cultures of the ice age.[28] Her story of prehistoric European peoples is also a tale about a clash of cultures, and specifically the vanquished culture of the Mother Goddess.

Today, Europe is composed of many different ethnic groups with an assortment of languages. It is a widely held theory, however, that all these ethnic groups were once a single group called Indo-Europeans with a single language. Gimbutas's research provides evidence that before these Indo-Europeans, another culture dominated the lands during prehistory. It was a completely different culture that Gimbutas refers to as the Great Goddess or Mother Goddess culture. It was egalitarian, yet focused on the maternal as a cosmological foundation.

According to Gimbutas, the matristic society of the Great Goddess began to develop into an urban culture, especially in the Cucuteni civilization of central Europe in modern-day Romania and the western part of the Ukraine. Cities in the population range of ten to fifteen thousand were evident by 4000 BCE, but they did not survive because of a new, invading culture that Gimbutas defines as Indo-Europeans (also known as Kurgans or Aryans). They became the forefathers of western Europeans. Today, nearly all the world's languages are linguistic descendants of this ancient Indo-European language.

Where did these people come from? Around 7000 BCE, in south

Russia, east of the Black Sea and north of the Caucasus Mountains, a new culture developed that was based more on animal husbandry than agriculture. By 6000 BCE, this Caucasus Mountains culture had domesticated the horse. Men grew to importance in a patriarchal society that subjugated women. These men had to control large herds of horses and cattle. Horses provided mobility as well as food. From the sixth millennium BCE on, weapons of war, such as long daggers, began to appear in the archaeological record. Ownership concepts developed. Cattle stealing began, as did wars with neighboring groups.

By 4000 BCE, European art and sculpture had changed drastically. New gods were introduced, and a new style in social administration developed. Settlements were built on hilltops. Burial rites were typical of those from the Russian steppe, and were limited to males with weapons and symbols of power, such as the horse-headed scepter. Around that time, the spear became the Indo-Europeans' first god, which, according to Gimbutas, eventually evolved into the thunderbolt of Zeus.[29] Culturally, across Europe, everything changed.

It was a clash of two different cultures, religions, and ideologies. The new Indo-European gods were male. The three primary gods were the god of the shining sky, the thunder god, and the god of the underworld. The dagger symbolized the god of the shining sky, the ax the thunder god, and the spear the god of the underworld. Later, these ideas of god being a warrior in the sky provided the theological basis for the warlike and wrathful God of the Hebrew scriptures.

The ideology of the new culture was reflected in its patriarchal social structure, which had a focus on the psyche of the warrior. Every god was a warrior, whereas the goddesses became brides, wives, or maidens without any power or creativity. Dominance was an intrinsic part of this culture of seminomadic people who bore weapons and rode horses. Being more mobile, they began to dominate.

When this invading culture arrived, its people established themselves in the hills and in places that were difficult to access. But if and how much the people of the older culture defended themselves, according to Gimbutas, is difficult to tell. What is clear is that the culture of Old Europe was replaced by the new Indo-European culture.[30]

After 4000 BCE, the new patriarchal culture of the Caucasus was well on its way to being the established culture. Europe later became a hybrid of these two diverse cultures, which is evident in their mythology. Their ideas reached the Mesopotamian plain and were seen in

Sumerian and Semite stories. According to Gimbutas, Gilgamesh dethroned Lilith (Adam's mythical first wife, who did not like to be sexually dominated so she ran away to her home). Eve took the blame for paradise lost, and the Goddess was dethroned. Athena of Greek mythology was hybridized and became militarized, but she still kept some of her former qualities. Behind her was an owl and on her shield stretched a snake, in the tradition of Old Europe.

The ideology of separation between body and soul developed from this patriarchal culture through the concept of the sky god. The idea of an afterlife became more important in relation to the here-and-now. Life became transcendent. The idea was that one does not have to be concerned for this physical life because the "good life" will be in heaven after death, as opposed to the immanence it had in the Goddess culture.

Because Eden was possibly an annexed part of the native Indo-European lands, it is reasonable to speculate that the patriarchal Adam's original home and culture were from north of the Caucasus Mountains. He might have been part of a group that migrated south into the high plains of the Zagros Mountains and encountered existing agricultural societies. It is also possible that an encounter with a specific settlement in the Valley of Tabriz inspired the historical relevance of the Garden of Eden. The stories of the Garden of Eden (or Aratta) were very important to the Sumerian culture and to their descendants through the Akkadian. It was a place of significance, historically speaking, just as people in America stay attached to their parents' or grandparents' country of origin. After leaving Eden, Adam made his way onto the Mesopotamian Plain and continued his life according to his own culturally defined ideals. The rest of the story, and the story of his descendants, is continued in Sumerian and biblical history. They went on to create a kingdom that included the cities of Babylon, Erech, Akkad, Calneh, and Nineveh.

Of course, the story of Adam was written long after his death, so it is told in mythical fashion. As with most history, it was written by the victor of the cultural struggle between the Indo-Europeans and the existing culture they invaded and from a particular point of view, that of a patriarch. In the transition from a matristic to a kingdom-based culture, Adam may have been the first father figure on the Mesopotamian Plain to organize socially. From the perspective of his descendants, he really would have been the first man.

Adam marked a beginning point for his descendants and the growing patriarchal culture that he helped create. For posterity, religious conviction, and unity under this new culture's cosmology, he needed to be identified with, and created by, a male god. Many generations later, historical fact faded into myth, resulting in the idea that he was simply created by God. According to Sumerian mythology, the god Ea (the ancient name of the Hebrew God Yahweh) requested that man worship only one god. This, of course, was not Adam's doing. But, as with many cultural heroes even in contemporary times, fact becomes blurred in myth, and the memory of a person often takes on a persona bigger than life.

Eden: A Story of Becoming

The original stories of Eden—Aratta and Kharsag of the Eden/Zagros civilization—were traditional histories for those whose heritage was linked to the beginning of Sumerian civilization (which I will explain at greater length in chapter 12). Through the art of the story, this heritage was passed down the generations to Abraham and then to the Midian priest Jethro.

These Sumerian Aratta and Kharsag stories convey little in the ways of theology, however. They were more or less tales of conflict and order among neighboring lands, quite different from the biblical version of Eden. Societies struggle against each other (war), then a new order emerges at the end of the war. It is a cycle that is repeated in history over and over. The esoteric elements of the biblical Eden came later, when Moses, in his quest to provide spiritual significance and an identity for the fledgling nation of Israel, superimposed a symbolic meaning on these stories. So, with Moses's reinterpretation, the Garden of Eden took on an esoteric meaning, a philosophical understanding of life that he held from his days as a prince in Pharaoh's schools—an ideologically sound and necessary move for a man creating the philosophical and historical foundations for a new culture and nation.

In essence, the Eden story became a symbolic expression of the nature and identity of man—a story of becoming in the ancient Egyptian tradition of the mystery school. The serpent, as the principal character in the story, serves as the representation of what Schwaller de Lubicz refers to as the Primordial Scission, or the act of an absolute quality (God) manifesting the cosmos with mankind as the ultimate expression of that cosmos. In the Pythagorean science of number, the One becoming Two represents this Primordial Scission.

As a consequence of this ancient cosmology, particularly in Egypt, the serpent was revered as a symbol of intellect and wisdom. For example, the pharaoh's diadem displayed the serpent and the falcon, a way of expressing the divine act as the essence of man (the serpent) and man's spirit (the falcon) in its return to the source. In ancient Greece, the serpent represented the wisdom of the medical doctor as seen in the caduceus—the icon of the medical profession to this day.

Why the ancient Egyptians chose the serpent as a symbol of authority is that power itself is dual in expression. The serpent, a unitary creature, is actually dual in expression, both verbal and sexual in nature. It is simultaneously creative and destructive, in that multiplicity comes from unity, and creation represents destruction of an absolute state. The serpent bears both a forked tongue and a double penis, so the snake is an apt symbol. As such, Neheb Kau—the provider of forms and attributes—was the Egyptian name given to the two-headed snake representing the primordial serpent.

Duality and intellect, however, are not only human functions but cosmic functions as well. There is a higher intellect and a lower. The serpent represents the lower as well as the higher intellect. The higher intellect allows man to know God and is the heavenly serpent, the Serpent in the Sky—often depicted by the Egyptians as a man riding the back of a serpent to the stars.

How the serpent came to be the icon of evil is that duality unchecked is chaos (destruction). Knowledge, without its beneficial application, is a parody of God and representative of the destructive forces inherent in nature.[31] Through a simplification of Egypt's natural philosophy, the concept of duality unchecked, or chaos, came to be known simply as evil.

With Adam and Eve's eating of the fruit (the becoming), the Eden story is separated into the abstract and the concrete. Before they ate from the Tree of Knowledge, they were unaware of their nakedness. Afterward, they were ashamed and covered themselves. In other words, they became self-aware. Before they ate the fruit, they "walked with God" (existed as an abstraction, a quality). Afterward, realizing they were naked, they were aware of their flesh (existed as physical objects, a quantity).

In Moses's version of the Garden of Eden, he conveys the mystery of perception based on physical awareness; the difference before and after exemplifies a perception of awareness, a state of being, and a matter of knowledge and consciousness—an aspect firmly rooted in his Egyptian upbringing and indicative of Memphite theology.

Genesis as Esoteric Truth

> In the beginning, God created the heavens and the Earth. The Earth was formless and empty, darkness was over the surface of the deep, and the Spirit of God hovered over the waters. And God said, "Let there be light," and there was light. God saw that the light was good, and He separated the light from the darkness calling the light "day," and the darkness "night." And there was evening, and there was morning of the first day.[32]

It must not be overlooked that when God created the heavens and Earth, Earth was *formless* and *empty*. Furthermore, the sun and moon were created *after* light was created. According to Genesis, this happened on the fourth day. To account for the inconsistency of when "light" was actually created, some theologians have concluded that God provided the first daylight source (and that it was on one side of Earth), and that was the reason for the day–night sequence on a rotating Earth. Such reconciliation is not necessary, however.

The Genesis story of creation is intended not to be a literal account of creation, but an allegory of Man and Cosmos. The *idea* of a universe, its abstraction, is created first, and is why Earth was formless and empty. Everything that is concrete has form, so *formless* means that it has no material existence. This is further illustrated by the correct interpretation of "the beginning." The Hebrew word translated "in the beginning" is *bereshith*. Since it occurs without the article, it is a proper noun; it means the "absolute beginning." Moreover, the phrase "the heavens and the Earth" is a figure of speech, a Hebrew *merism,* where two opposites are combined into a single concept. There is no Hebrew word for *universe,* so the text is referring to the totality of physical creation, not just Earth or the solar system.

"Light" and the "waters" are principles of creation, and are also an allegory. They symbolize the abstract quality in its movement to quantity, to form, and ultimately to the concrete world. Light is the active principle, more commonly known as fire in the ancient cultures. Water is the composite, receptive principle, that which is acted upon. Together, these two principles form the Creative Unity and constant creation through harmony and disharmony. Light as the active source (the creative spirit) is not surprising, since the biblical text states that "God is light" in 1 John 1:5 and John 8:12, and also that God is "a consuming fire" in Deuteronomy 4:24 and Hebrews 12:29.

Light also symbolizes the Presence that is absolute Cause-Effect, for in it is that which acts and also that in which it acts. It is the discontinuous within the continuous, the concretion of the abstract. It is also interesting to note that light, the photon, is energy, and the basic form of movement we call energy. Polarized energy, the proton and electron, is the fundamental building block of all matter.

Time is the distance between cause and effect, but for the mental consciousness, as opposed to innate or knowledge a priori, it is a material reality, which is an illusion. Time exists only because of mental consciousness, with its ability to remember the past and anticipate the future. In creation (which is really constant), cause cannot be cause without first producing an effect, which will, in turn, be a new cause for its activity. So there will be the appearance of a sequence of effects. But there will never be a sequence from cause to effect. As a result of constant creation, time (as genesis) moves in the irreversible direction of the effect becoming the new cause. Since we are complete in form (the human body), no longer acting as a cause, we see the effects and not the cause. In this way, the body of man is symbolic of the cause; the body is symbolic of consciousness, as seen in the story of Adam and Eve.

When Adam and Eve lived in paradise, they were naked and had not yet acquired the knowledge of good and evil. This represents the state of consciousness prior to becoming form. Unmanifest consciousness is absolute quality without quantity, so the opposites of good and evil do not exist. First, Eve gains the knowledge represented by eating the forbidden fruit. Then Adam partakes. Although Man was created as male and female, the true order of creation, if you want to think of it that way, is that woman was "created" first and man second—not as an afterthought, but as a means for procreation. Nonetheless, knowing good and evil is possible only in a state where opposites exist.

Original sin, told as a story of man's disobedience to God's commands, in a sense is real and does result in death. More than two millennia of religious dogma, however, have altered its original intent. Disobedience is nothing more than identifying with the material world, the dominance of egotistical perception. Since, by nature, man's intuition is linked to an innate quality of existence, disregarding it is, in fact, disobedience to the true nature of the individual, the Self.

The manifestation of the abstract into the concrete creates a separation of complementary aspects that together are harmony, or unity. Original sin, therefore, is the mental state of the ego without the influence of innate consciousness. In a state of nonunity, or disharmony, the individual is left to his or her own physical wants and desires, resulting in "sin." In unity, the mental state of ego is no longer the prime interpreter of reality. Although the ego must continue for the individual to exist in the immediate world, the intuition of the Self provides an understanding that the present moment is eternal.

More than a thousand years after Moses, this esoteric prose on the nature of man would find a new champion, and a new audience, in the message of Christ. When recognized by the individual, the role of harmony and the Self would find new esoteric ways of expression.

9

The Son of Man

Esoterism and the Message of Christ

You diligently study the Scriptures because you think that by them you possess eternal life. These are the Scriptures that testify about me, yet you refuse to come to me to have life. I do not accept praise from men, but I know you. I know that you do not have the love of God in your hearts. I have come in my Father's name, and you do not accept me; but if someone else comes in his own name, you will accept him. How can you believe if you accept praise from one another, yet make no effort to obtain the praise that comes from the only God [or One]? But do not think I will accuse you before the Father. Your accuser is Moses, on whom your hopes are set. If you believed Moses, you would believe me, for he wrote about me. But since you do not believe what he wrote, how are you going to believe what I say?

JOHN 5:39–47

For first-century Judaea, Jesus's claim that Moses specifically wrote concerning one's self must have been a bold statement. Moses was not only the Jews' most famous political leader, but also a spiritual leader who defined the kingdom of Israel, its laws, and its culture. Such a claim was at least audacious and definitely offensive.

What did Moses write concerning Jesus? According to traditional interpretation, the passage from the Gospel of John quoted above is referring to prophecies concerning the messiah in the Hebrew scriptures, such as Genesis 12:3, where Moses writes that God says, "I will bless those who bless you, and whoever curses you I will curse; and all peoples on Earth will be blessed through you"; and especially Deuteronomy 18:15: "The Lord your God will raise up for you a prophet like me [Moses] from among your own brothers. You must listen to him."

The basis for this interpretation comes from the evangelist Paul, who, in addressing King Agrippa, said that the gospel of Christ is "nothing beyond what the prophets and Moses said would happen—that the Christ would suffer and, as the first to rise from the dead, would proclaim light to his own people and to the Gentiles."[1] This proclamation was the heart of the message Paul was spreading throughout the Mediterranean world, and it has been the heart of Christian doctrine ever since. But was Moses really referring to a messiah?

According to Zondervan's King James Reference Bible, Moses wrote of the coming messiah in the following passages:

> And I will put enmity between you and the woman, and between your offspring and hers; he will crush your head, and you will strike his heel. (Genesis 3:15)

> I will bless those who bless you, and whoever curses you I will curse; and all peoples on Earth will be blessed through you. (Genesis 12:3)

> Abraham will surely become a great and powerful nation, and all nations on Earth will be blessed through him. (Genesis 18:18)

> And through your offspring all nations on Earth will be blessed, because you have obeyed me. (Genesis 22:18)

> The scepter will not depart from Judah, nor the ruler's staff from between his feet, until he comes to whom it belongs and the obedience of the nations is his. (Genesis 49:10)

> The Lord your God will raise up for you a prophet like me from among your own brothers. You must listen to him. (Deuteronomy 18:15)

> I will raise up for them a prophet like you from among their brothers; I will put my words in his mouth, and he will tell them everything I command him. (Deuteronomy 18:18)

The difficulty in applying any of these passages specifically to the life of Jesus is that they are too vague for an accurate comparison. Nor do they mention anything concerning a messiah. (In personal and traditional belief, messiah is implied.) *Prophet,* as opposed to *messiah,* is the clear choice of words. As for prophets, Israel had its share long before the first century. Moses may have been referring to any one of them, or perhaps all of them, knowing that other great individuals would surely follow in his footsteps. In regard to Israel's greatness, it had already become a powerful nation a thousand years earlier, although later it was eclipsed by the Babylonians, then the Persians, the Greeks, and ultimately the Romans. Without any specific references to Jesus in the Hebrew scriptures, it is only through individual interpretation that one can say Moses was writing about a coming messianic mission attributable to Jesus. If this interpretation is not the case, what then was Jesus referring to by saying that Moses wrote about him?

Esoterically, Jesus was referring (in scripture) to the concept of what it means to be "Man," as Moses expressed it in the story of the Garden of Eden and the laws that were put forth in the rest of his works. Only in this way does Jesus's identification with Moses have meaning.

Through the stories and the laws, Moses was explaining and instructing people how they should act in unity with God. It was his hope that the people would put God's natural, altruistic laws "in their minds and write it on their hearts."[2] Since such a condition is not easily achieved (it is a personal and inner quest), creating a physical system of laws would serve as a guide to living, with the hope that it would lead to enlightenment. Jesus himself suggests that this is the case by claiming that he did not "come to abolish the Law or the Prophets . . . but to fulfill them."[3]

At its heart, the Law is a moral code focused on the relation of man to the divine. In instituting a theocracy, Israel required that everyone follow the Law or face a penalty for noncompliance. Jesus was teaching that, through the understanding of his message, men and women would naturally follow the Law because it would become their nature, a nature that already exists, although dormant in every person.

Because of its esoteric nature, this message is difficult to grasp; it is something the individual must do alone: "Seek and you will find. Knock and the door will be opened." These words evoke an introspection of the human soul and a search for one's true identity, a search for God. Yet the people believed that by following Moses's laws and ordinances in an exoteric (external) manner, they would be righteous. They missed the point entirely, and did not receive Moses's doctrine in its true sense and meaning, just as Jesus claims: "You diligently study the Scriptures because you think that by them you possess eternal life."

As an Egyptian prince, Moses introduced the concept of the Man Cosmos into the traditional story of Hebrew origins, applying divine meaning to an otherwise ordinary legend. Jesus, in his wisdom and brilliance, attempted to reestablish the original intent of Moses's writings—abstract (spiritual) origins as the vital factor. Moses wrote about the eternal, abstract (spiritual) nature of man. He conveyed this message not just in Genesis, but also in Exodus, another esoteric and spiritual story.

Earlier writings, such as George Robert Stowe Mead's (1863–1933) *Fragments of a Faith Forgotten,* support this idea. Mead elucidates a scriptural exegesis ascribed by Hippolytus (presbyter of the Church of Rome at the beginning of the third century) to the Peratae, an unknown Gnostic school. Concerning Exodus, Mead writes:

> Thus then they explained the Exodus-myth. Egypt is the body; all those who identify themselves with the body are the ignorant, the Egyptians. To "come forth" out of Egypt is to leave the body; and to pass through the Red Sea is to cross over the ocean of generation, the animal and sensual nature, which is hidden in the blood. Yet even then they are not safe; crossing the Red Sea they enter the Desert, the intermediate state of the doubting lower mind. There they are attacked by the "gods of destruction," which Moses called the "serpents of the desert," and which plague those who seek to escape from the "gods of generation." To them Moses, the teacher, shows the true serpent crucified on the cross of matter, and by its means they escape from the Desert and enter the promised land, the realm of the spiritual mind, where is the Heavenly Jordan, the World-Soul. When the Waters of Jordan flow downwards, then is the generation of men; but when they flow upwards, then is the creation of Gods.[4]

Unmistakably, there were members of the early church that viewed the Exodus story as a parable in the Christian tradition. Such an interpretation supports Jesus's statement that Moses wrote about him. The message is esoteric, in that the elements of the story refer not to actual events but to a personal quest for each individual. Moses, in his disillusionment with Egyptian theology—possibly a result of Akhenaten's failed attempt to establish a new monotheism in Amarna—identified Egypt with the mental state of physical existence, the passions and perspectives of the body. The Exodus, through water and into the desert, represents the break from mental consciousness and a search for the Self, of which Jesus reminds us that the kingdom of God is "within" and not a physical kingdom to be sought in the immediate world.

This interpretation of the Exodus story also supports the minimalist view that some, if not all, biblical stories represent historicized fiction created for social, political, and spiritual purposes, esoteric truths that cannot be easily explained through rationality. Whatever the case may be, there are more important questions relating to the nature of biblical prose and its ambiguity.

Creating a philosophy/religion, as the canonical authors did, serves no purpose except to create social discord and disagreement, as history attests. Why would the so-called truth be hidden in a parable? If it were the intention of the Christ, his disciples, and the authors of the Christian scriptures to provide mankind with the truth and a way to relate to God, why did they not put it into simple terms? Why would a teacher hinder his audience by speaking in parables? Was there a strange need to create a mystery? Or did the prophets and inspired teachers not know how to express themselves?

This difficulty in Christ's message rests not in Jesus's ability to speak, but in the nature of the message itself, which is really the philosophy of living nature—a nature that is inherently subjective. The essence of the message can be understood only through esoterism and the symbolic, the vital part of the human mind that relates to the world of the abstract. Plain language simply does not prompt the curiosity of the listener, which is why, in literature, the parable can be so effective. The parable serves as a way of evoking esoteric knowledge through the symbolic, and is an apt way to convey an abstract truth as it relates to the immediate world. Thus, divinity is revealed only to those who reflect on the nature of their identity, which in modern terms is often called "self-realization," or, in the original words of the authors of the Christian scriptures, to be "born again."

Christ's message seems to convey an understanding beyond the state of mental consciousness (ego), which is governed by time, into an area of existence where time does not exist (eternity) and where lies the true nature of mankind. In effect, the man Jesus becomes the Son of God by achieving a level of perception that resonates with the absolute state of

Being. In the words of the ancients, this is "becoming," or attaining a unity with the One (or the absolute).

Externally (exoterically), Jesus was a man in all respects. Esoterically, however, he really was the Son of God and had every right to claim as much, in the sense that his innate consciousness blossomed into the dominant state of his being, thereby infusing his mind with a deeper level of existence and understanding. A majority of his parables reflect this type of thinking. From the fifteenth chapter of Luke, this principle of self-realization through introspection is well illustrated in the parable of the prodigal son.

The Prodigal Son: A Parable of Becoming

There was once a man who had two sons. The younger son demanded his inheritance so he could leave for a distant land. Once there, however, he squandered all that he had. As a consequence, he suffered greatly. After coming to his senses, he decided to return to his father, ask forgiveness, and work as a hired hand.

While the retuning son was still a long way off, his father saw him and was filled with compassion. So the father ran out to meet his son, threw his arms around him, and kissed him. The son said, "Father, I have sinned against heaven and against you. I am no longer worthy to be called your son."

But the father said to his servants, "Quick! Bring the best robe and put it on him. Put a ring on his finger and sandals on his feet. Bring the fattened calf and kill it. Let's have a feast and celebrate. For this son of mine was dead and is alive again; he was lost and is found." So they celebrated.

The older son was infuriated. "Look! All these years I've been slaving for you and never disobeyed your orders. Yet you never gave me even a young goat so I could celebrate with my friends. But when this son of yours who has squandered your property with prostitutes comes home, you kill the fattened calf for him!"

The father pleaded with him, "You are always with me, and everything I have is yours. But we have to celebrate and be glad, because this brother of yours was dead and is alive again; he was lost and is found." Now, his father loves him more than before.

This story explains a process by which Man, perfect in His abstract nature, becomes flesh, which is imperfect or "evil," then returns to conscious perfection. After the son returns home, his state is as it was before, yet slightly different. He has gained an added dimension of understanding. His mental consciousness (ego), the desires that led to the ill-fated journey, has been subjugated to his now realized innate consciousness (the Self). When the son leaves home, he identifies himself with the physical world, representing the descent from abstract to concrete. After seeing that bodily identification results in death and that his true nature lies in the abstract, he returns to his father, which symbolizes not only a return, but also a heightened level of perception. He becomes enlightened to his true nature and the true nature of all men.

Within the Christian tradition, we would say that he received God's gift of eternal life through Christ. The interpretation above, however, is the same as this one, but stated in more accurate terms, Christ being the abstract and archetypal man.

Here, "the kingdom of God is within you" is your innate consciousness (the Self). It awakens and blossoms and, in doing so, prevails over your mental consciousness (the ego). It replaces the identification with physical form with identification with the abstract (spirit). Consequently, the observer, who is the perceiver beyond form (who you really are), becomes stronger and your mental formations, as a reaction to the environment, become weaker, evoking a new perspective on all things. In its absolute meaning, this emergence of the Self is a death and a resurrection—death not to the physical world and its cultures but to your attachment to it; the resurrection is not of the body, but of the innate quality (the Self) in which all mankind participates. This concept of the abstract Self—the observer who is the receiver of all perception and the maker of all decisions—as the true nature of Man resonates throughout the Gospels in a unique way. According to the Gospel authors, more than any other term or phrase, Jesus referred to himself as "the son of man."

The Son of Man

The heart of Christian doctrine is that Jesus was exclusively God incarnate, the Son of God. Ironically, with only a few exceptions from the Gospel of John, Jesus never referred to himself as such, but instead chose "son of man." In the Gospel canon, he did so thirty times in Matthew, fourteen in Mark, twenty-six in Luke, and eleven in John.

Theologians have various theories about why Jesus referred to himself as the son of man. Some interpret his use of this phrase as an attempt to conceal his true identity as the Son of God, suggesting that saying "son of man" was a clever way of saying "Son of God." Others believe it was meaningless, and is nothing more than a reference to one's self, like using the English word *I*. They believe that Jesus, who spoke Aramaic, would never have designated himself as the "son of man" in any messianic or mystical sense, since the Aramaic term never implied that meaning.

The phrase *son of man* can have different meanings in various contexts, such as "the son of man is lord even of the Sabbath day," from Matthew 12:8, which simply conveys that a person has mastery over the Sabbath. In other passages, it is used in the sense of "that person," or "myself," which was common during Talmudic times. So when Jesus says, "The son of Man hath not where to lay his head" (Matthew 8:20), he means "I have nowhere to lay my head." When he speaks of his future suffering and betrayal (Matthew 17:22), "son of man" has nothing to do with a messianic title. But this conclusion ignores the fact that no one but Jesus made this phrase the exclusive manner of self-designation.

The author of the Book of Ezekiel used "son of man" ninety-three times, often with heavenly overtones. For him, it was how God addressed the prophet. It this sense, "son of man" conveys the idea that a chasm stands between God and the prophet, but it also implies that Ezekiel was considered to be the ideal man. Some scholars think this view is unwarranted, and that the Hebrew term *ben adam* is merely a cumbersome but solemn and formal substitute for the personal pronoun, possibly because of Assyrian and Babylonian usage. The author of Daniel, however, clearly uses the term in a metaphysical way:

> I saw in the night visions, and, behold, one like the Son of man came with the clouds of heaven, and came to the Ancient of Days, and they brought him near before him.[5]

In this passage from Daniel, "Son of man" *(bar enash)* has a peculiar use. Daniel envisions "one like the son of man coming with the clouds of heaven and appearing before the Ancient of Days" to receive "dominion, and glory, and a kingdom, that all people, nations, and languages, should serve him: his dominion is an everlasting dominion, which shall not pass away, and his kingdom that which shall not be destroyed."[6]

Bible commentators seem to agree that Israel is meant where the word *kingdom* is used, but differ in their understanding of "Son of man," which some think depicts a personification of the people, or a concrete personality representing Israel, such as the messiah or the guardian archangel Michael. Most scholars believe the messiah is being alluded to.

Given the context of the scripture, there can be little doubt that "son of man" refers to a metaphysical being. "Son of" is used not only to suggest a relationship, but also to bestow a special designation on the individual being described. According to archaeologist and art historian Ernst Emil Herzfeld (1879–1948), the use of the phrase *son of* in the ancient world carried a distinction other than family reference. Specifically, in Old Babylon, it meant heir or successor to royalty.[7]

Despite all this debate, in the Christian scriptures the argument that by "son of man" Jesus is referring to himself simply as a man becomes indefensible in Matthew's account of the trial before the Sanhedrin:

> Then the high priest stood up and said to Jesus, "Are you not going to answer? What is this testimony that these men are bringing against you?" But Jesus remained silent.
>
> The high priest said to him, "I charge you under oath by the living God: Tell us if you are the Christ, the Son of God."
>
> "Yes, it is as you say," Jesus replied. "But I say to all of you: In the future you will see the Son of Man sitting at the right hand of the Mighty One and coming on the clouds of heaven."
>
> Then the high priest tore his clothes and said, "He has spoken blasphemy! Why do we need any more witnesses? Look, now you have heard the blasphemy. What do you think?"
>
> "He is worthy of death," they answered.[8]

Clearly, Jesus is not hiding from the high priest that he is the Son of God, while equating the high priest's words with his own, "son of man." Except for a single use in Acts and another in Hebrews, and two instances in Revelation, however, "son of man" is never again used by the Greek writers of the Christian scriptures. One possible reason for its lack of use is the Greek language. According to the Catholic Encyclopedia, in the Greek version of the Hebrew scriptures (the Septuagint), *son of man* is always translated without the article *ho huios tou* and is simply *anthropou* (man, human being), where it is used as a poetical synonym for man, or for the ideal man. For example, in Numbers 23:19, we find that "God is not as a man, that he should lie nor as a son of man, that he should be changed."

In the Christian scriptures, the phrase *son of man* appears with an article, rendering it as *ho huios tou anthropou,* "the son of the man." Yet the consensus of Greek scholars is that the correct translation, as opposed to a literal translation, is "the son of man" and not "the son of the man."

Years later, possibly to clarify and avoid confusion concerning who Jesus was, in their own writings the founders of the early Christian church chose an alternative, and they chose "Christ." (He was also called Jesus, Christ, the Messiah, the Savior, the son of Man, and the Son of God.) This would dispel any ambiguity about whether Jesus was a man or a deity. Yet they chose to be true to his words in the Gospels themselves, depicting him as continually using the unique expression "the son of man" as his only form of self-reference. After his death, why did his disciples, people he had lived with for three years, almost completely ignore the expression? It is a mystery, but we are not without clues.

Jesus, the Archetypal Man

Observing that various cultures across the world held similar, deep-seated ideas, the Swiss psychologist Carl G. Jung (1875–1961) believed that all mankind shared a single collective consciousness, a set of ideas and concepts that lie deep within each individual's mind. According to Jung, all people share these basic, unlearned concepts, which can be spontaneously expressed. He called them "archetypes," and he saw these innate concepts as the unconscious images of human instinct. Over time, accumulated ancestral experiences composed these archetypes in the collective subconscious of mankind. As a result, individuals inherit the archetypal propensities of the collective human being.

Archetypal is a reference to a fundamental, structural thought or image that pervades people's mental activity. It is usually subconscious and always transpersonal. According to theoretical physicist Fred Alan Wolf, there seem to be universal archetypes in the myths and images of all cultures; he suggests they are formed during deep sleep. Also, according to Wolf, it appears that when we dream, we give structure to the world around us; our consciousness provides form to our surroundings. Interestingly enough, from the time the fetal brain starts to develop in the womb, the unborn human being spends close to eighteen hours a day in dream sleep.

Jungian psychologist Elizabeth Boyden Howes believes that "the son of the man" was an archetypal image Jesus used in describing his spirituality. According to Howes, this pattern of thought functions to transform the Self, rather than simply being symbolic of the Self.[9] In this way, Jesus viewed himself as "the son of man" relating to the archetype of the Self. In essence, he was using "the son of man" to describe the primary image dominating his life. Since this imagery operated through him, others could also find this essence of Self. It is why he always insisted "The kingdom of God is within you," since its essence *is* the Self. What this process describes, in our modern-day language, is the existential Self at work in a physical life, which is perceived by some as God manifest in humanity.[10]

For others, particularly traditionalists, this type of psychological analysis may seem impractical, or perhaps an inappropriate way of viewing the passion of the Christ. Yet the story of the paralytic in the Gospel of Matthew provides supporting evidence. It is a unique insight into the inherent, fundamental function of "the son of man":

> Some men brought to [Jesus] a paralytic, lying on a mat. When Jesus saw their faith, he said to the paralytic, "Take heart, son; your sins are forgiven."
>
> At this, some of the teachers of the law said to themselves, "This fellow is blaspheming!"
>
> Knowing their thoughts, Jesus said, "Why do you entertain evil thoughts in your hearts? Which is easier: to say, 'Your sins are forgiven,' or to say, 'Get up and walk'? But so that you may know that the Son of Man has authority on Earth to forgive sins. . . ." Then he said to the paralytic, "Get up, take your mat and go home." And the man got up and went home. When the crowd saw this, they were filled with awe; and they praised God, who had given such authority to men.[11]

In this passage, Jesus states, "[T]he Son of the Man has authority on Earth to forgive sins." Whereby Matthew then reports, "[T]hey [the crowd] praised God, who had given such authority to men." It is an astonishing statement, since one would expect the author of Matthew to use a title such as Christ, or at least his name, instead of "men."

Interestingly, the claim for authority was referring not only to Jesus, but also to his disciples. So "the son of man," or, more appropriately, the essence of this Being, was not restricted to Jesus but included his disciples, and, taking the concept a little further, anyone else engaged in the process of becoming whole.

The idea of the individual partaking in the existential human being is seen in the paralytic himself. Crippled, unable to walk, the paralytic man engages in an action, considered first in his mind. This action was an unconscious decision, a reaction, based not on a cerebral process but rather on an instinctive, intuitive, or spiritual quality he expressed while identifying and sharing in the same identity as Jesus. The power that was in Jesus was the same power that was in him. This principle of the Self, as the source of transcendent power, is set forth repeatedly: in Matthew 9:29, 15:28, and 21:21; Mark 2:5, 5:34, and 10:52; and Luke 5:20, 8:48, 17:19, and 18:48.

The Self is not confined to any particular culture or civilization; it is a natural part of all human beings. Even the Roman centurion at Capernaum, in Luke 7:1–10, who likely had little or no knowledge of the Hebrew scriptures nor any concept of their ideas, believed in the same way, instinctively emanating the same faith that others had. It was an event that inspired Jesus to say, "I tell you, I have not found such great faith even in Israel."[12] He explains this principle of Self as the source of power, in no uncertain terms, to the ruling religious elite of his day. God exists potentially within each and every person:

> Once, having been asked by the Pharisees when the kingdom of God would come, Jesus replied, "With your careful observation, nor will people say, 'Here it is,' or 'There it is,' because the kingdom of God is within you."[13]

Further support of this shared or collective consciousness is also evident in circumstances where Jesus could perform no miracles. Those who knew him since childhood, who had lived with him in his hometown of Nazareth, knew him as the son of Mary and Joseph, a carpenter and nothing more. They could not identify with the archetypal Self of Jesus that was recognized by others.[14] In both Mark's and Matthew's accounts, Jesus could do no miracles because of a lack of faith in the people themselves. The logic here is, if he were in fact God incarnate, performing miracles would be possible anywhere, regardless of anyone's thoughts.

The Gospel according to John, a more philosophical account than that of Matthew, Mark, or Luke, carries with it a weight of deliberation with insight, although it is at times obtuse. With phrases like "On that day you will realize that I am in my Father, and you are in me, and I am in you,"[15] it is clear that the nature of existence, of "being," is the subject, but it is confusingly expressed, since it is impossible for anyone to exist in anyone else. In fact, John's Gospel's esoteric approach nearly disqualified the text from inclusion in the canon, but it was saved by phrases such as, "I am the way and the truth and the life. No one comes to the Father except through me."[16] With no answer key provided, readers are left to their own perceptions and methods, or to those of a teacher, in interpreting such statements. There are clues, though, peppered throughout John's account of Christ's message, leading to an underlying theme: An entity exists to which all human beings are connected, and it endows them all with the potential to be "sons of God."

> But as many as received him, to them gave he power to become the sons of God, even to them that believe on his name: Which were born, not of blood, nor of the will of the flesh, nor of the will of man, but of God.[17]

What John alludes to in these verses is easily overlooked. The name of Jesus, and thereby the person of Jesus, was never intended to be the focal point of his ministry, teachings, and the ensuing Christian religion—which is why he chose to refer to himself as the archetypal son of man. John also tells of a private conversation between Jesus and his disciples, where this archetypal "Man" is clearly explained: "I am in my Father, and you are in me, and I am in you."[18] It is, perhaps, the most powerful and telling text in the Christian scriptures, as well as the most enigmatic. At a glance, it defies common sense. With the understanding that Jesus was portraying himself as the archetypal man, however, the power to become "sons of God" is in the process of identification with Jesus—how and why those that believed on his name (believed without actually seeing him—just hearing about him) could become one.

What the author of John's Gospel is referring to is easily understood as the intangible part of all of us that is the essence of life. Jesus's description of the spiritual nature of mankind is phrased in such a way ("I am in my Father, and you are in me, and I am in you") as to indicate that Jesus is introducing the concept of a collective consciousness. "[Y]ou are in me, and I am in you" can mean nothing other than that all people who identify with Christ share in a single consciousness, the consciousness of mankind, which is not bound to a physical existence. It is an apt description for the title "son of man."

The perplexing question is, how does identifying with Jesus—connecting to the collective consciousness of Man—give power to men to become sons of God? The answer lies in John's introduction:

> In the beginning was the Word, and the Word was with God, and the Word was God. He was with God in the beginning.
>
> Through him all things were made; without him nothing was made that has been made. In him was life, and that life was the light of men. The light shines in the darkness, but the darkness has not understood it [or the darkness has not overcome].[19]

In this passage, the subject—the Word—is traditionally understood symbolically as Jesus, or the unmanifest Spirit that was later incarnated as the man Jesus. It is a poetic, often quoted scripture, and it brings with it an air of mystery, setting the tone for the chapters to follow. Yet it is an important point, in that the author of John's Gospel helps clarify the concept of Logos as it relates to the early Christian point of view. The Logos is Jesus, but it is presumptuous to assume that John was referring to the physical man himself. Within the philosophical context of the Logos, whose tradition reaches back thousands of years to ancient Egypt and the Shabaka stone, it is clear that the author of John's Gospel is attaching Christ's message to a much older tradition, a tradition with its roots in Egypt's ancient mystery school.

Logos and the Tradition of the Mystery School

Rudolf Steiner, in *Christianity as Mystical Fact* (1902), takes Philo's Platonic view that the "Son of God" is best described as wisdom born of man. This wisdom, or world-intelligence, lives in the human soul and contains the intelligence that exists in the world. For Steiner, the Logos that John writes about also appears in the book in which Philo writes, ". . . has been inscribed and engraved the formation of the world."[20] Logos also appears as the Son of God following "the ways of his Father . . . looking to the archetypal patterns which that Father supplied."[21] Steiner, who believed Philo spoke of the Logos as being Christ, writes:

> For since God is the first and sole King of the universe, the road leading to Him, being a king's road, is rightly called royal. This road you must take to be philosophy . . . the philosophy which the ancient circle of ascetics pursued in hard-fought contest, eschewing the soft enchantments of pleasure, engaged with a fine severity in the study of what is good and fair. This royal road then, which we have just said to be true and genuine philosophy, is called in the Law, the utterance and word of God.[22]

For Steiner, the Logos referred to an intermediary between God and the universe and the manifestation of the divine principle in the world. It is both divine reason and reason distributed in the world, including the mind.

It is with this intellectualism—the development of Greek thought and reason over the centuries—that John endows his introduction to his Gospel. The Word (the Logos) has nothing to do with a birth of flesh, but rather to that of the observer, the archetypal man, into the realm of three dimensions, and its rebirth through mankind as a realization and awareness of its own nature.

John's introduction to the Gospel is a statement of becoming and of identification with what Steiner called the "Unspeakable Unity." The essence of Jesus—the Logos—is portrayed as a thing that is identical in respect to itself, and that returns toward itself. It is the definition of self-awareness with the understanding that one nature has to specify itself in another nature—that is, as the Hermetic says, "as above, so below." The essence is simultaneously Origin, Word, and With. This means the same thing as the opening to the Gospel of John: the essence of "Man" has always existed, and by the Word being spoken the universe was created. Herein is the heart of what was to become, three hundred years later at the Council of Nicaea, Christianity's doctrine of the Trinity: unity revealed, brought about through the intervening agency of a collective consciousness.

John also declares that an identity exists between the inborn, collective consciousness that defines Man and biological life in the broadest sense of its definition: "Through him all things were made; without him nothing was made that has been made. In him was life, and that life was the light of men." In essence, John is describing the Man Cosmos. Indeed, John is saying that *man is the cosmos*. Through esoterism, the heart of the Christ's message is revealed, and no longer is the suspension of reality a requirement to believe.

The Esoteric Message of Christ

According to traditional Christian doctrine, a little over two thousand years ago, Christ was born of a virgin, and at the age of thirty he began his ministry. After three years of teaching, he was convicted of blasphemy and crucified by the provincial Roman government. Three days later, he was miraculously reanimated and rose from the grave. All this was God's master plan for mankind's salvation. Since the Virgin Mary conceived him, he was born free from the bonds of original sin. As a result, he never sinned, and his trial and execution therefore served as atonement for all sin: past, present, and future. Anyone who believed this received eternal life. In the tradition of the Hebrew ritual sacrifice, he was the ultimate sacrificial lamb. Yet he was more than a man; exclusively, he was God incarnate.

Viewed exoterically, this message of Christ falls into the paradox of natural law suspended. In effect, it is superstition. Conception without sexual intercourse was impossible in Jesus's time, as was the reanimation of life once deceased. But this popular interpretation of Christ's message has been transmitted through the centuries to modern times, immersed in language translation difficulties and misunderstandings.

Viewed esoterically, each aspect of the Christian tradition—the virgin birth, salvation, resurrection, and Christ's return—relates to a viable and sophisticated understanding of life's mystery as expressed symbolically. It is the same message that has been shrouded in myth from time immemorial, a secret wisdom. This secret wisdom of all ages is not so much a secret as it is masked by the dispensation of dogma, but through a scientific understanding, it ultimately leads to the sacred.

Salvation

Given all the crime and atrocities that occur every minute around the world, there can be no doubt that evil exists. Its origin, however, lies not in a sovereign being who wages war against all that is good. Evil lies in the mental consciousness of human beings, where ego is alienated from the vital understanding of life's source, where knowledge becomes know-how and the human mind, with its material desires, becomes the imitator of nature's creative work. Evil is selfish; desires run amok and are the opposite of salvation.

In the 1973 film *The Exorcist,* the human-possessing demon appears as a relevant force in the war between good and evil. It is also a relevant fear factor within the psyche of mainstream America. It is unlikely there will ever be a scarier movie. The secret to its blockbuster success, besides the award-winning special effects and cinematography, is that it touched a deep nerve in our Christian tradition. During the fifteen hundred years of the church, demons were thought to be nonhuman spiritual beings that existed outside the realm of the human mind and had the ability to possess the human body. In more fundamentalist circles, this is still believed to be true. Yet where these demons came from, and how they operate, is (at best) vague, relying on flawed interpretations of the Hebrew scriptures.

The general belief is that spiritual beings existed prior to the creation of mankind. Known as angels, they were God's first creation. There was a war in heaven, and a third of these angelic beings were cast out of heaven for their rebellion. This, as the story goes, was the creation of evil. These angels-gone-bad were imprisoned on Earth and allowed to torment mankind. Loyal angels, on the other hand, who were still devoted to God, remained as God's messengers and servants.

Do angels and demons really exist? Some would say that they do. If such a concept were to be based on historical texts, however, there would be little evidence to support such an idea. A thorough reading of historical documents, such as the Greek philosopher Iamblichus's *Theurgia or The Egyptian Mysteries,* written in the second century, dispels such notions as a mere misunderstanding of philosophical concepts.

In describing the ancient Egyptian *theurgia,* which means "the science or art of divine works," Iamblichus clearly explains the role of different natural principles, the Egyptian neters, translated as "gods" in Greek. As for the demon (spelled *daemon*), it was and is a way to speak of physical principles that particularly relate to the vital functions of the body, man as well as animal:

> For the personal daemon does not "preside over specific regions in us," but simply over all at once. He pervades every principle about us, in the same manner as it was assigned from all the orders [of intelligence] in the universe. For it also seems proper to thee to remark as follows:

> "That there are daemons placed over specific departments of the body, one over health, one over the figure, and another over the bodily habits, forming a bond of union among them, and that one is placed as superior over all of them in common." This very thing thou shouldst consider as proof that the authority over everything in us is vested in one daemon alone. Accordingly it is not right to define "one daemon as guardian of the body, another of the soul, and another of the mind." For if the living person is one individual and the daemon manifold that is placed over him, the notion is absurd. Certainly the ruling powers everywhere are single rather than those that are ruled. But it is still more absurd if the many daemons ruling over special departments are not akin, but are to be classified apart from one another.
>
> Thou also declarest that there are contradictory characters among them, saying "some daemons are good and others bad." Evil daemons have no allotment whatever as guardians, and they are never classified in opposition to the good, like one party against another, as though having equal importance.[23]

The idea of a demon was a way to describe the characteristics of a person's body: its health, its habits, and even the figure or shape of the body itself. They could be either good or bad. Iamblichus writes, however, that these characteristics are never classified in opposition to the good.

Iamblichus continues explaining how demons impart divine principles:

> This daemon, therefore, is present as exemplar before the souls descend into the realm of generated existence. As soon as the soul chooses him for leader the daemon immediately comes into charge of the completing of its vital endowments, and when it descends into the body, unites it with the body, and becomes the guardian of its common living principle. He likewise himself directs the private life of the soul, and whatever the conclusions we may arrive at by inference and reasoning, he himself imparts to us the principles. We think and do just such things as he brings to us by way of thought. He guides human beings thus continually till through the sacred theurgic discipline we shall obtain a god to be guardian and leader of the soul. For then he gives place to the superior, or delivers over the superintendence, or becomes subject, as a tributary, to him, or in some other way is servant to him as to an Overlord.[24]

In Iamblichus's retelling of Egyptian beliefs, the demons are nothing more than the vital endowments of the physical body; they are the "common living principle" of nature. Evil demons would, therefore, be bad things, such as illness and disease. But it is clear from the ancient texts that demons were not considered unique and conscious entities poised to invade the weak-minded person. In a sense, however, we really are born with evil tendencies, meaning that we must rely on our mental faculties to survive and are slaves to the ego and its desires. In that respect, mankind *is* in need of salvation. But how does salvation come about?

The answer lies not in the suspension of reality in the tradition of the exoteric Christ, but in the understanding of mankind's nature and the symbolic essence at the foundation of our being. In realizing that the person is an observer (the Self), and that consciousness is the biological vehicle for perception, a person becomes conscious of his or her own abstract existence or Being. The person's sense of identity moves away from the ego (mental consciousness) and toward the essence of the observer, the Self, which is best described by what could be called innate consciousness. As a result, the person becomes conscious of his or her identity in the eternal, abstract world. Put another way, a person's unique consciousness becomes conscious of itself as a part of the collective consciousness that is Man. This is salvation, that the individual becomes a "son of man." It is the ability to look past material, worldly desires and to view reality as a product of the abstract (spiritual), where every one of us is connected. The individual's perspective changes, and the feeling that accompanies this change is that he or she has been "born again."

In creating the physical universe, the absolute quality that is the source (God) manifested itself as quantifiable objects. So every atom in the cosmos is a discrete quantity of God's quality. Add to that the principle of growth, and consciousness begins to develop, taking on increasingly complex forms through the stored memory of DNA. The goal of this process is to develop into a form where an awareness of identity—a relationship of God with Himself through the experience of the material world—can be achieved. Every person first identifies with the processes of the brain, which is mental consciousness that knows itself only as form. At this level of awareness, there is a need to identify with external things, such as physical appearance, possessions, work or career, social status, knowledge, education, relationships, personal and family history, and belief systems, as well as political, nationalistic, religious, and other group identifications. These are things people strive for. Since the mental consciousness is aware that it will one day die, however, every person also lives in constant fear of the form's destruction. This psychological projection of the self as the ego, with death as the ultimate fear, is the seat of dysfunction and emotional pain—the regret for yesterday and the worry of tomorrow.

Although curiosity, as a quest for truth, is sometimes a motive, suffering is typically how a person ceases to identify with form (the body) and awakens to innate consciousness. One's identity (innate consciousness) is manifest as flesh at conception, but innate consciousness is lost because of the complex biological form's growth requirements. Years of experience in dealing with the immediate environment are required for the human brain to fill up with information, a necessary process for the mental consciousness to become successful. Through this process, what was once purely innate consciousness becomes subjugated as a result of the quest by the mental consciousness for the form's survival. Once you realize that your innate consciousness is the true Self, and that the mental consciousness is a projection of the mind's function, you have embraced the dominion of the observer and have entered eternal life.

Salvation is to know that you are part of the abstract One, the life force from which everything receives its being. With that recognition, you are timeless and inseparable from the source of existence. Embracing this brings a feeling of wholeness, fulfillment, and peace. It is also a state of freedom: it is liberation from fear, suffering, wanting, needing, grasping, clinging, compulsive thinking, and negativity, and, most important, liberation from the compelling psychological needs of past and future.

Jesus referred to this state of being as "the abundant life." Its essence is an enduring presence, not a passing emotional experience. It is to know God, but its most succinct label is love—what the Greeks referred to as *agape:* a spiritual, selfless type of love and a model for humanity. It is what Jesus refers to when he says you should love (agape) your enemies and pray for those who persecute you (Matthew 5:44).

Resurrection

The message of Christ is invariably linked with resurrection. Traditionally, this is interpreted to mean that, at some future point in time, all people who have ever lived will be resurrected to either eternal life or everlasting damnation. But Jesus lived and taught in the social and religious context of ancient Judaea, where life was governed by the Sanhedrin, which was made up of two parties—the Pharisees and the Sadducees. One distinction between the two parties is that the Pharisees believed in the resurrection whereas the Sadducees did not. This difference of opinion between Pharisees and Sadducees helps illuminate how the Gospel authors portrayed the concept of resurrection.

One day a Sadducee posed a hypothetical question to Jesus about marriage and the resurrection. A man died, so his wife married his brother, as was the custom. The Sadducee asked whose wife she would be in the resurrection. A strange question to ask for someone who did not believe in the concept; nonetheless, he asked it, and Jesus replied, but not in an expected way. The answer was so profound that no one dared ask any more questions:[25]

> The people of this age marry and are given in marriage. But those who are considered worthy of taking part in that age and in the resurrection from the dead will neither marry nor be given in marriage, and they can no longer die; for they are like the angels. They are God's children, since they are children of the resurrection. But in the account of the bush, even Moses showed that the dead rise, for he calls the Lord "the God of Abraham, and the God of Isaac, and the God of Jacob." He is not the God of the dead, but of the living, for to him all are alive.[26]

Although it appears as a single answer, Jesus is really addressing two separate issues—the question first about marriage and second about the meaning of resurrection. In his last statement he switches the perspective to what he wants to convey: There is no past and no future; to God, all are alive. From a physical point of view, this statement is false, so what might be the point? With the understanding that the immediate world is a quantification of abstract quality, the true nature of man is seen to be timeless. So Jesus is claiming that the defining quality of all human beings (at all times) continues to exist in the abstract. To understand what this means, we have to return to the concept of the Self.

The body is a physical, biological form, but when it is reduced to its most fundamental substance, it disappears into a world of mystery. The human body, as are all other animal and plant forms, is a single entity composed of biological systems: respiratory, digestive, neural, and so forth. These systems are composed of cells, which are composed of molecules. Molecules are composed of two or more atoms of a given element, such as water—two atoms of hydrogen bonded with an atom of oxygen (H_2O). If we further reduce the human body to the quantum level, we are left with the "magic" of the proton/electron attraction within the atom. At the subatomic level, solid matter (including your physical body) is mostly empty space because of the vast distances between the subatomic particles compared to their size. A further reduction of these particles results in the concept of energy, which is generally defined as movement. The nature of the substance that moves remains speculative. A logical conclusion, however, is that this movement is the source of everything physical in the universe, and its connection to the Self, the observer, is where the collection of all sensory (and nonsensory) stimuli becomes perception. Movement is the essential and vital life force that determines who and what Man is, resulting in the manifestation of biological consciousness. By definition, the source of this movement—the Self—is infinite and eternal.

Although we speak of two different worlds, the concrete and the abstract or spiritual, to describe and explain things, there is only a single reality, a reality that is abstract and permanent. If so, then why bother with a physical body, or a physical universe, for that matter? It is the only way for the Self to experience. Through the fragmentation of the Self into individual units (bodies), a construct is formed where interaction occurs. In essence, the physical body and the cosmos itself are symbolic of an absolute quality where the concept of body already exists. In this way, the physical body is symbolic of the true body, the abstract (spiritual) body.

It is in this relationship (between the abstract body and the physical body) that the transformation—*resurrection*—of Self takes place. When you realize that the essence of life is abstract, you identify your physical body with your abstract body. At the moment of realization, you step into eternity. Through identification with the mental processes of your brain, your abstract body is, in a sense, dead. You attach your identity to the immediate, material world. In recognizing and embracing your true nature, the abstract body is resurrected. This is what it means to become one with God. In essence, this awareness of the abstract body is innate consciousness remembering its origin and returning to its source. In contrast, as long as you identify with the physical, you have an externally derived sense of self.

The abstract body—the observer that you are—is formless, limitless, and timeless. It is the reason why twenty years may pass and you still feel as though the events of those days happened just yesterday. From the perspective of the abstract body, they did. When the body dies, identification with form ceases. But since to the abstract body identification with form is just an illusion, death is nothing more than the end of that illusion.

Christ's Second Coming

Another significant part of Christ's message is his return, or "second coming." With numerous theories as to how and when this will happen, it has been a matter of contention within the Christian community for almost two thousand years. It began when Jesus, referring to the temple, stated, "Not one stone here will be left on another; every one will be thrown down."[27] His disciples asked when would this happen, and, believing that the two were linked, when the end of the age would come. The destruction of the temple, which occurred in 70 CE at the hands of the Romans, passed without Christ's return or the end of the age.[28] Ever since, Christians have believed they are living in the "end times."

During modern times, the second coming has been a sensational topic steeped in the hype of literal biblical interpretation, particularly of the Book of Revelation, which is highly esoteric and unfathomably symbolic. At some future, unknown date, all Christians will be whisked off Earth—an event known as the Rapture—allowing the Antichrist to form a one-world government. Those people who remain on Earth, as the theory goes, will have a final opportunity to gain salvation through Christ, all the while suffering persecution from a religion-hating, Gestapo-style regime. The scenario has been the stuff of popular movies and books, such as Tim Lahaye and Jerry Jenkins's *Left Behind* series. Although fiction, the series' themes are based on what self-styled end-time enthusiasts call the Great Tribulation.

One of the more ridiculous and outrageous variations of all end-time theories is that the office of the pope is considered by some fundamentalists to be the Antichrist that will eventually unite the world under one theocratic government. Few (if any) scholarly theologians believe such nonsense. Nonetheless, Christ did mention his return. Just as other aspects of the Gospel message have been turned inside out, so has the concept of his return.

Again, beliefs force the suspension of natural, absolute laws. Once a person is deceased, it is impossible to return, let alone miraculously pull all believers off the face of Earth. What was written concerning Christ (after the fact) must have been influenced by personal perceptions as well as doctrinal beliefs. This confusion concerning the suspension of physical laws purportedly occasioned by Christ's return, as well as all the other misunderstandings about his message, is resolved once time is put in its proper place. Christ was never concerned about the past or the future. He realized that there is no past or future.

His teaching was always esoteric, which is why he relied on parables to convey truths. For example, in teaching in the temple and arguing what it means to be a descendant of Abraham, he told the Pharisees, "I tell you the truth, before Abraham was born, I am."[29] To his listeners, he was claiming his existence was equivalent to that of God. But this is a very superficial, exoteric point of view, and was not his intent. He did not mean that he already existed before Abraham was born, which would have meant he was still within the three dimensions of the physical universe bound by time and form. He meant that his true nature (his realized divine presence), an abstract quality without quantity (spiritual), exists in a nondimensional state outside the boundaries of time and space. For this true nature, there is no concept of time, regardless of who you are. The key to understanding what Jesus meant is that time is a concept that exists only for the mental necessities of form. There is no such thing as time in the abstract world. Only the present moment exists. It is always now.

With this truth in mind, Christ's "second coming" is no longer some magical, sensational, impossible event that is yet to come; it becomes instead the spiritual truth he had always tried to convey. It is a transformation of human consciousness from the mental to the innate, a shift from the perspective of time to the eternal present, from thinking to becoming the observer.

The mistake that most people make is to personify Christ. Christ is the God-essence in you, the abstract Self. It is the indwelling divinity, regardless of whether you are conscious of it. The awakening, realizing that you already exist in the eternal, is the return to unity, or "second coming."

The Good News

The Gospel (or good news) message of Christ, as it has been described here, is simple and natural. No single person is God, for God's nature is by definition unknowable. Yet all of us collectively, and everything that exists physical, *are God.* This is the mystery that has fueled religion and philosophy since the dawn of civilization. The universe exists solely for us to perceive, and persists only through our perception. The physical is nothing more than the manifestation of the abstract, the discrete fragmentation of pure quality into a vast quantity. Time does not exist except as a man-made concept for record keeping. At every moment, creation continues. Eternity is not something in the future. It is now, here, where life occurs. Upon realizing this, the individual identifies with innate consciousness, which quickly blossoms as the seat of all understanding. Fear, anguish, bad memories of the past, and worries about the future lose their grip. Only peace and happiness are left. Existence, then, is aligned with the abstract (spirit); and the physical form, the temporal, becomes only a provisional symbol of what is real.

If existence is determined through the material world, physical death is its finality. Those who choose to identify with the physical can never disconnect from the pain and suffering continually churned out by the processes of mental consciousness. This leads to a life driven by the ego and a constant struggle to feel superior at the expense of others. In effect, this is the definition of "sin" that everyone is born with. If this state is maintained until physical death, then true death occurs. The death of the body is true death for those who identify with the material world. For them, nothing exists in the abstract world.

So, in a sense, hell is real, as it exists in the minds of those who refuse to identify with the innate consciousness of the Self. They live in continual fear of what might happen and continual regret for what did happen, only to die and be no more. This condition may even continue after bodily death, leaving the person who identifies with mental consciousness trapped as a specter in the material world, continually reliving all the regrets and worries of life. Perhaps this accounts for the occasionally alleged hauntings where tragedy and crime once occurred.

When pain and suffering become unbearable, salvation is provided by one's own inner connection to God, the innate consciousness that is the kingdom of God. Its perception-altering influence is represented best by a single quote from a disciple's letter: *God is Love.*[30] It is what the Greeks call agape, a spiritual, egoless, selfless, charitable state and a model for humanity—divine harmony.

10

Secret Wisdom

Esoterism and Inspiration

> *[A] "twofold truth" [exists] that has filled the history of Christian religion throughout the later Middle Ages. There is the very disputable doctrine that "positive religion"—whatever form it may take—is an indispensable need for the mass of the people, while the man of science seeks the real truth [in] back of religion and seeks it only there. "Science is esoteric," so it is said, "it is only for the few."*
>
> WERNER HEISENBERG,
> 1932 NOBEL LAUREATE IN PHYSICS

Despite repression, and at times violent opposition, from the Roman Empire, during the first three centuries of the Common Era, Christian beliefs spread throughout the Mediterranean world. At the beginning of the fourth century, the fledgling Christian religion became the beneficiary of a new emperor, Flavius Valerius Constantinus, more commonly known as Constantine. On October 28, 312, before his victory at the Milvian Bridge, Constantine, in a mystical sense, envisioned the Christ. A year later, with the Edict of Milan, he granted Christians religious tolerance. By the end of the next year, Constantine ordered that Christians, as well as all others, have the right to freely exercise their religious beliefs. As a result, Christian leadership within society grew, and local bishops took a more aggressive approach in public affairs.

Constantine's son, Constantius, took to Christianity with a passion and opposed paganism with a vengeance. With a motto of "Let superstition cease; let the folly of sacrifices be abolished," Constantius closed all temples and ordered sacrifices forbidden under penalty of death.[1] With these new antipagan laws in place, Constantius set Europe's political and religious tone for the next one thousand years. With the collapse of the Western Roman Empire in the fifth century, the church inherited the stewardship of Western civilization, and a new era began.

The authority of the church provided vision for the Western world. Philosophic thought shifted toward the theological and the relationship between God and man. Any connection to God was viewed as a function of the human soul. The immediate world (nature) as the system, or part of the system, was no longer considered the primary reality. What took prominence was the interpretation of spiritual revelation based on the canon of scripture.

At times, the church was dogmatic and repressive, but in 1517, when Martin Luther nailed his Ninety-five Theses to the door of Castle Church, the Reformation began. New theological ideas became acceptable, and then a quest for knowledge, not bound by biblical constraints, inspired the more progressive thinkers of the day. At that time, ancient wisdom and philosophy were all but lost, except for classic Greek manuscripts. One text of particular interest is the *Corpus Hermeticum*—the Greek version of the ancient writings of the Egyptian Thoth. Forty-six years earlier, in 1471, the philosopher, philologist, and physician Marsilio Ficino had translated the Greek Hermetic texts into Latin. They became a source of fascination to Renaissance scholars, who generally believed that Hermes Trismegistus was an ancient sage, or possibly a semidivine being, who had originally composed the texts in Egyptian.[2]

With an alternative philosophical approach to the status quo, particularly pertaining to nature, these ancient Greek texts inspired some of the great figures who ushered in the Age of Enlightenment and helped give birth to the modern age of science. During the sixteenth century, physicist Johannes Kepler (1571–1630) took a renewed interest in ancient texts and attempted to reconcile historical non-Christian knowledge with biblical truth. His goal was to retain the legitimacy of ancient knowledge without abandoning his devotion to the church.

In an attempt to understand Pythagoras's Harmony of the Spheres, Kepler, who had already accepted Copernicus's heliocentric solar system, discovered the laws of planetary motion. For Kepler, the number of perfect polyhedra (solids bounded by polygons) being one less than the number of planets was no accident. In his quest to comprehend the solar system mathematically, he proposed that the distance between the six known planets could be understood in terms of the five Platonic solids. Each planet was associated with a sphere and was nested inside the next outermost sphere. So, according to Kepler, the five Platonic solids represented the five intervals between Mercury, Venus, Earth, Mars, Jupiter, and Saturn.

Although Kepler's attempt to place the planetary orbits within a set of polyhedrons never worked, through failure came the realization that the planetary orbits were elliptical, not circular. Nonetheless, Kepler was inspired by, and honored in his 1619 book *Harmonices mundi* (Harmony of the World), the scientific and philosophical knowledge of the ancient Egyptians:

> I am stealing the golden vessels of the Egyptians to build a tabernacle to my God from them, far far away from the boundaries of Egypt. If you forgive me, I shall rejoice; if you are enraged with me, I shall bear it. See, I cast the die, and I write the book. Whether it is to be read by the people of the present or of the future makes no difference: let it await its reader for a hundred years, if God himself has stood ready for six thousand years for one to study him.[3]

Although *Harmony of the World* is mystical in its approach by modern standards, in it Kepler derives his formula for the third law of planetary motion—the square of a planet's orbital periods is proportional to the cube of its mean distance to the sun.

Other great thinkers were also inspired by ancient knowledge. Throughout Europe, the Renaissance was synonymous with the revival of Platonism, the Kabbalah, Hermeticism, and Pythagorean concepts of philosophy and science. Scientists with Pythagorean interests included Nicolaus Copernicus, Galileo, and possibly Isaac Newton, as well as the polymath and rationalist Gottfried Leibniz.

One of the most revered of the early Renaissance figures was the artist and inventor Leonardo da Vinci (1452–1519). With the recent speculative notion that he created the Shroud of Turin (believed by some to be the actual burial cloth of Jesus) using early photographic techniques as a hoax to humiliate the church, his legacy of brilliance still tantalizes today's historical researchers.[4]

Leonardo's *Vitruvian Man*

Although Leondardo da Vinci painted a number of masterpieces, such as the *Mona Lisa* and the *Last Supper,* his journal sketch *Vitruvian Man* (also known as the Canon of Proportions or Proportions of Man) is one of the best-known drawings in the world, commonly used in numerous advertisements and logos, particularly in health care–related professions.

Vitruvius, a first-century Roman architect and author, believed that temples should be based on the proportions of the human body, since he believed the body was nature's model of perfection. Leonardo, who had read Vitruvius's *De Architectura,* blended art and science in sketching within the circle and the square the image of a man. The purpose of Vitruvius's work and Leonardo's inspiration was to create natural aesthetics by relating structure to nature. Although we will never know what Leonardo envisioned in his sketch, scholars suggest that he was

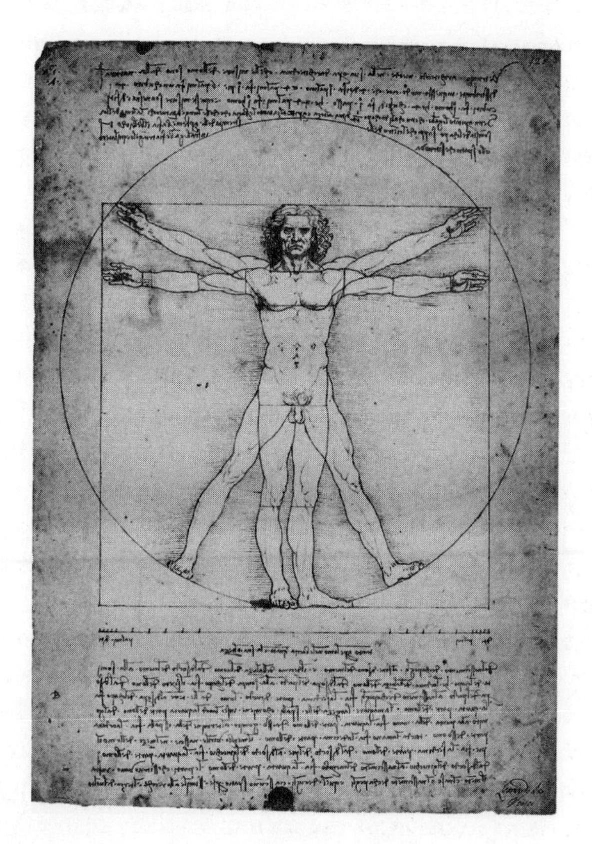

Fig. 10.1. Leonardo da Vinci's Vitruvian Man

depicting man as a microcosm, that man's body is an analogue of the functioning universe.

The human body placed into a circle and a square can be seen as a modern interpretation of Pythagorean philosophy, in which numbers symbolize an abstract quality as much as they represent quantity. The circle, with no beginning and no end, represents the divine abstract nature of Man, and the square, with its four sides, the concreteness of nature. The human body, symbolic of the circle and the square, is geometrically indicative of our dual nature.

Just as Pythagoras derived a relationship between music and number, in these sketches Leonardo related number to the human body. Proportion is a relationship between two ratios, and the golden ratio (1.618) is the key to natural proportion. Although the ratio varies slightly among individuals, the length of the hand is proportional to the forearm, as is the forearm to the combined length of the forearm and hand. So it is with the division of the body at the navel. The distance from the top of the head to the navel, in relation to the distance from the navel to the feet, is proportional to the distance from the navel to the feet to the entire height of the figure. In Leonardo's sketch, where the man's hands and feet meet the circle, a pattern is formed with four equally distant points emanating from a central (fifth) point, the navel—where the man's life began as a connection to his mother.

Just as Vitruvius used the natural form of man to design temples, Leonardo intentionally used the same natural ratios to proportion his figure. For Leonardo (the quintessential Renaissance man), this ratio was the key to not only artistic beauty but also the understanding that science itself was an esoteric adventure within the creative mind of man.

From Pythagoras's Square to Fermat's Cube

Even today, Pythagoras's mathematical legacy continues in the concept of geometric triplets. According to the Pythagorean theorem, the square of the hypotenuse of a triangle is equal to the sum of the squares of the other two sides, where the hypotenuse is the longest side opposite the right angle. Three, four, and five form a right-angled triangle, and Pythagoras's equation provides $x^2 + y^2 = z^2$. Pierre de Fermat, a seventeenth-century mathematician, wondered if there was a similar relationship in numbers for cubes, an extension of the idea of Pythagorean triples, where $x^3 + y^3 = z^3$.

At first the equation seemed to be unsolvable, but in the margin of his copy of Diophantus's *Arithmetica* (an ancient Greek text), Fermat scribbled the most famous note in the history of mathematics: "To resolve a cube into the sum of two cubes, a fourth power into two fourth powers, or any power higher than the second into two of the same kind, is impossible; have found a remarkable proof of this. The margin is too small to contain it." Unfortunately, the proof he allegedly wrote somewhere else was never found. Fermat's last theorem of Pythagorean triples remained unsolved.

But in 1986 at the University of California at Berkeley, Ken Ribet linked Fermat's last theorem to another unsolved problem that connects elliptic curves to modular forms, known as the Taniyama-Shimura conjecture. These modular forms are certain periodic holomorphic functions investigated in number theory. For British mathematician Andrew Wiles, who had been thinking of a solution to the problem of Pythagorean triples since his early teens, this meant that to prove Fermat's last theorem, all he had to do was to prove the Taniyama-Shimura conjecture.

On June 23, 1993, at a Cambridge mathematics conference, Wiles announced that he had solved the problem of Pythagorean triples in a 150-page-long proof. A *New York Times* article proclaimed, "At Last, Shout of Eureka! in Age-Old Math Mystery."[5] Pythagorean triples exist, and are "a triple of positive integers a, b, and c such that a right

Fig. 10.2. Computer visualization of Fermat's Last Theorem (courtesy of Stewart Dickson)

triangle exists with legs a, b and hypotenuse c. By the Pythagorean theorem, this is equivalent to finding positive integers a, b, and c satisfying $a^2 + b^2 = c^2$."[6]

Although modern mathematicians likely see in this proof nothing more than the brilliance and dedication of Andrew Wiles, it also validates the ancient marriage of science and esoterism. Philosophers such as Pythagoras believed that, in number, the divine order in nature was expressed. For Pythagoreans, an esoteric knowledge existed in the relationship between geometry and reality that made those who knew and understood it wise. More than numbers, this secret wisdom of how form arises from the abstract—from point to line to square and cube (now proved)—was the scientific and philosophical key to conceptualizing the mysteries of the universe.

Today's scientists are just as inspired as their Renaissance predecessors. Scientists' insight into the nature of things (their secret wisdom) is perhaps the true nature of man—the curiosity and the creativity. This observation leads to questions, which lead to speculation, that leads to experimentation and theory. In the end, such insight leads to a philosophical declaration about those grand assumptions that make science what it is and us who we are. After years of working with the scientific process and the quest to understand the fiber of reality, there is little wonder that Heisenberg wrote, "Science is esoteric."

Secret Wisdom

The importance of the Age of Enlightenment and its rediscovery of ancient knowledge was more about intuition and man's source of inspiration than it was about the ancient knowledge itself. For the Renaissance man, the study of ancient philosophical texts was a way to reconnect with the ancient tradition of the sciences and mankind's quest to understand the principles of nature. The secret wisdom of the Renaissance, as it was with the ancient traditions, was that *mysticism—esoterism—leads to science.*

Since the Enlightenment, the scientific method has become increasingly more influential in society, especially with the economic benefit of achieving higher degrees of industrial efficiency. Philosophically, along with its empiricism came a widening gap between the secular and the sacred. Mysticism, once at the core of scientific inspiration, lost its appeal in the formalized world of institutionalized disciplines. As a result, mysticism became synonymous with ignorance and superstition. For science, what could not be measured or observed was considered irrelevant at best. For religion, what could not be measured remained its trump card, and a matter of faith. Although it may appear that this set science and religion in opposition to each other, in fact both have seemingly carved out territories of usually peaceful coexistence where neither trespasses on the other. The dividing line for these two different approaches was (and is) *causality*—who or what is the cause of the universe. Reinforced by research into the structure and purpose of the brain and how evolution was responsible for its form, the dividing line grew into a chasm.

Mysticism and the Brain

As far as we know, the human brain, with its electrochemical network of neurons, is the universe's most complex structure. The most amazing aspect about the brain, and what we take for granted, is its innate learning process and a priori knowledge. This a priori knowledge makes it possible for us to learn complex, abstract ideas and skills from an early age.

Divided into two hemispheres, the cerebral cortex is the part of the brain that houses rational functions. Although both sides of the brain are involved in nearly every human activity, the left side of the brain appears to be the seat of language; it processes information in a logical, sequential order. The right side seems to be more visually based, processing information intuitively in a holistic manner. At one time, it was believed that the left hemisphere was the logical, verbal, and dominant half of the brain, while the right was the imaginative, emotional, and spatially aware side. But more recent studies suggest that this idea of brain functionality is oversimplified at best. No function is entirely isolated on a single side of the brain, and the characteristics commonly attributed to each side serve as a guide for ways of learning and understanding.

According to clinical neurologists Gereon Fink of the University of Düsseldorf in Germany and John Marshall from the Radcliffe Infirmary in Oxford, the evidence suggests that the left brain is organized on the basis of local bias or detail, while right-sided processing is more global. This is how the two sides of the brain complement and combine to form a full and true picture of the immediate world. The left hemisphere of the brain is the area where mental skills require us to act in a series of discrete steps, or fix on a particular fragment of what is perceived, such as stringing letters together to write a sentence. The right brain, in contrast, is utilized to form an expansive background picture whose panoramic understanding provides a general connectedness to the environment.

Despite the objectivity of the immediate physical world, the totality of brain functions composes a unique perspective that is inherently subjective for every individual. The beauty in this arrangement is that it creates an arena of abstract principles that serve as a knowledge base for conceptual analysis, which, in turn, serves as the foundation for the creative act in man. This process of how we create through thought, and then implement what we have created in the physical world as a construction or manufacturing project, is best described as esoteric. Up until now, how the esoteric or spiritual fits into the processes of brain activity has been a difficult phenomenon to grasp or study. Throughout history, cultures have developed religious beliefs based on the ideas and principles of the metaphysical to explain the natural processes of life, most of which they attribute to God. But how does one experience God?

Although this is a debatable and most assuredly complex issue, neuroscientist Rhawn Joseph theorizes that the capacity to experience God is primarily through the amygdala, a small, almond-shaped structure buried deep in the brain. Along with the hippocampus and hypothalamus, the amygdala makes up the limbic system (the first-formed and most primitive part of the brain). It is likely the source of emotions, sexual pleasure, and deeply felt memories.

According to Joseph, these tissues become highly activated when we dream, pray, or use hallucinogenic drugs. They enable us to experience a reality that normally exists in the mind's background, subordinate to the experience of the immediate world. Joseph, who earned his Ph.D. in neuropsychology, discovered through clinical and historical research that spiritual experiences are not based on superstition, but are instead real, biological functions that are part of our innate energy. Creatures that live exclusively in caves have not developed eyes, because, for them, light is not real. Likewise, humans would not have developed brain functions to process something that did not exist. We are hardwired for spiritual experiences because there is a real God to experience, however God is to be defined.

What can be said about the condition of the human experience and brain? As previously stated, there are two kinds of knowledge—what we know without learning and what we learn and develop over the course of our lives. Although some argue that a priori knowledge does not exist, the concepts of space and time are fundamental in our relationship to nature, but are not inherent in nature itself. It would not be possible to describe nature without these concepts. So, in essence, at the least these concepts are a priori. Space and time, as well as how the brain processes external stimuli, are the conditions for experience and perception, not the result. Learned knowledge comes to us through social means (primarily education) and is, for the most part, public or *exoteric.* Unlearned, a priori knowledge is *esoteric* or secret knowledge, but secret only in the sense that language is inadequate to describe its character, and that this esoterism is part of nature. This secret knowledge and, in particular, how we use the knowledge to be creative, either technically or artistically, can appropriately be understood as *secret wisdom.* For those who embrace this mystical yet objective reality, science ceases to be secular and becomes an aspect of the sacred. For Schwaller de Lubicz, this natural philosophy was the trademark of pharaonic Egypt and why he chose to call the Egyptians' blending of art, science, and philosophy "sacred science."

To explain how the human experience occurs and how the ancient Egyptians leveraged the understanding of such experience in the creation of their symbolism, Schwaller focuses on the symbol and the symbolic, and how it relates to a priori (or innate) knowledge.

The intricate processing of innate knowledge, more appropriately described by Schwaller as *innate consciousness,* operates differently from the processing of learned knowledge. Learned knowledge is a direct result of instruction and is based on the ego, which is part of *mental consciousness.* Mental consciousness is the part of brain processing that allows us to navigate the immediate world. It has colloquially been associated with the left brain. It is through the right hemisphere of the brain that the expansiveness of the panoramic is felt or perceived and the connection of the Self to the natural world occurs. It is within this connection of the Self to the natural world that innate consciousness exists and esoteric knowledge flows. In human experience, the functionality of innate consciousness is not external or literal; rather, it is intuitive, and its expression and comprehension are based on symbol and the symbolic.

Symbolism and the Nature of the Symbolic

Symbolism is a mode of expression and is distinguished from the symbolic, which can be thought of as a mental frame of reference or state of mind. In other words, symbolism is a technique used to express meaning through images, whereas the symbolic is a form of communication used to describe an object or phenomenon.

Symbolism and the symbolic can be auditory as well as visual. Any time an object has significance through memory (or imagination), a symbol is involved. The symbol immediately evokes its associated characteristics in the thought processes (or consciousness) of the observer. On the other hand, a symbol such as a trademark is meaningless unless its significance has already been explained to the observer. For example, disharmonic (eerie) music evokes the same irritating, chilling emotional response regardless of culture. In literature, the metaphor is a symbol in that it evokes meaning through analogy and allegory, and through the imagination, by replacing the given description with an alternative one. An example from literature would be *Animal Farm,* by George Orwell. Although about animals and a farm, the farm is an analogy for communist Russia. This is an example of exoteric symbolism, where the symbol and its evocation are objective. The symbolic occurs in the mind, whereas symbolism is an expression that has a physical existence.

Esoteric symbolism differs from exoteric symbolism in that it preexists in man's nature and is a part of innate consciousness. For example, without knowing anything about any particular grove of trees, a roaring sound or two glowing eyes peering out from within those trees elicits anxiety and fear. The sound of fingernails raking a chalkboard gives rise to goose bumps and a cringing expression. There are also the soothing effects of soft, melodious tones that lull us in the waking state, and the contagiousness of the friendly smile.

These types of reactions are the effects of esoteric symbolism. They are constant and real, and can be thought of as instinctive knowledge. They should not, however, be mistaken for the animalistic instinct for survival.

According to Schwaller, the esoteric symbol elicits an abstract response expressed physically, mentally, or emotionally. The effect of an inanimate symbolic object is the energetic reaction of the observer. For example, the atmosphere of a cathedral or church with its vaulted ceiling, columns, and stained-glass windows is related to the creative "magic" of spiritual action. When we enter such a structure, we have an emotional response of awe and reverence. Although it can be argued that such a reaction is a learned cultural response, we have a similar response to the majesty of such natural geologic features as Bryce Canyon and the Painted Desert.

The human response to esoteric symbolism can be understood as a relationship to the unknowable cause of existence, which is the essence of what we perceive as harmony, order, and beauty. Its effect is produced through unification between the setting (the nature of the cause) and the favorable circumstances of the moment perceived by the person. At that moment of unity, there is the basis for understanding causality and the original cause. The cause exists in everything and constitutes that which makes us alive—an inarguable point, since we are obviously here.

Any setting has attributes, or the perceived characteristics of a quantifiable group of objects or phenomena. Without exception, everything that is quantifiable also contains quality, an aspect of a priori knowledge that serves as the construct of experience.

To demonstrate how the experience of quantity and quality relate to cause, a brief digression into physics (covered in part 1) is in order. Through scientific analysis, we know for a fact that the origin of matter is energy. As long as life is considered to be solely biological and physical, we can dispute causality by viewing natural phenomena only as a quantitative sequence of events. Exoterically, through the mental consciousness, we see only polarized energy and nothing more. Movement of quality (the abstract, unknown substance) is energy, so it is also cause. Without cause, there can be no phenomenal effect. So the energy that makes up all phenomena must also be the cause. As a result, innate consciousness exists through the cause, and its materialization into the concrete produces the same cause in us.

It is possible that cause and effect are inseparable, meaning that there may not be a cause-and-effect relationship between whatever "caused" the cosmos and mankind to exist and our existence (the effect) for any experience or phenomenon. Perhaps things just *are,* and we are those things in a holistic way (cause as well as effect, spirit and physical). Nonetheless, we are accustomed to analyzing phenomena based on cause and effect. So with that understanding, it can be said that cause and effect are inseparable in time. Examples of this are rife in the natural world. No chemical reaction can liberate the elements during the moment of their transference from one combination to another. With the meeting of a sperm and ovum, for example, the crucial moment of conception is instantaneous. So is the germination of an acorn into an oak seedling. All natural phenomena appear to operate in this manner. Schwaller refers to this timelessness between cause and effect as the Law of Creation. Creation, therefore, is constant, and exists exclusively in the eternity of the present moment. Consequently, growth is our perception of creation when we experience one moment after the next.

Inorganic matter follows the same principle. Any unformed substance becomes form as a result of an activity balanced by an opposite activity. The effect of the activity is neutralization, which never occurs without the reaction of resistance. The effect of an activity is simply a chain of activities, where reaction becomes a new activity of the first effect as well as a new cause.

This chain of activities is perceived to be located in time and space, and from an exoteric point of view, that is obviously the case. Growth occurs over time, where time can be thought of as a quantity in duration through the union of parts (symbols of the universe), and that these parts (quantities) constitute a chain that has an apparent beginning and end. The result is the stability of phenomena, which is what science calls objectivity.

Eternity is the esoteric reality we are physically born into. Conceptualizing this is difficult because perception is a function of the past and the future, even though it always occurs in the present moment. As a result, through mental consciousness, we attribute an exoteric character to this esoteric reality. Remembrance and anticipation rule the mind.

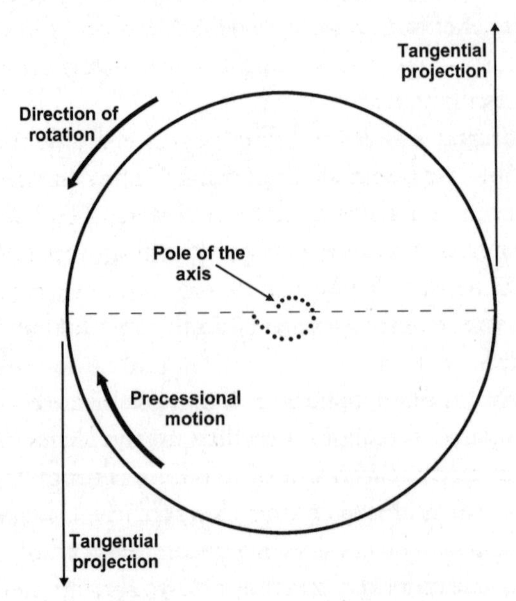

Fig. 10.3. Movement of a celestial body
(from R. A. Schwaller de Lubicz, Symbol and the Symbolic, *77)*

Even so, at the core of our nature, the symbolic serves as a transitory state unifying the world of the abstract and the concrete into a single reality.

In *Symbol and the Symbolic,* Schwaller illustrates this concept with the example of a rotating celestial body. A celestial body always rotates on its axis. The tangential projection of a diameter from the plane, at right angles to this axis, moves in opposite directions. At the body's center, directly on the axis, the tangential projection is canceled, resulting in no movement at the body's center, which is not possible. To compensate for this impossibility, the axial pole is displaced and moves in the direction opposite to the body's rotation.

In effect, this movement creates precessional motion (the gyroscopic motion of a spinning object, in which the axis of spin itself rotates around a central axis, like a spinning top). If the axial center were to stop while the body continued to rotate, the body would break up after reaching a certain threshold in velocity. This creates a mystery. The axis is not the abstraction of a concrete surface, nor is it possible to define it as cause or effect. The axis is an imaginary line. Nonetheless, it is a fact of the rotating body's mystical character.

The axis is not truly abstract, because it plays a large part in the mathematics and the behavior of the system (the rotating body). Nor is it truly concrete, because an axis is a line that is one-dimensional and cannot exist within a physical system. In effect, what Schwaller is saying is that *the axis is symbolic.*

Herein is an apparent contradiction between the symbolic and the historical. It exists, yet remains exoterically incomprehensible. If we view the symbolic as a transitory state, however, from the abstract to the concrete, unifying the discontinuity of appearance in the continuity of the present, then the symbolic no longer contradicts the historical fact. As a result, the historical fact appears as the exoteric character of an esoteric reality.[7]

In an exoteric frame of reference, the cause outside a system can never be demonstrated, which, in a duality, always leaves one pole as a matter of faith. In an esoteric frame of reference through the symbolic, its divided parts are unified, ending the problem of causality. The cause, which is apparently outside the system exoterically, is actually within it forever, united in the present. The deduction and conclusion is that creation is continuous, and a big bang cosmogony is no longer necessary.

The Exoteric and the Esoteric

As has been discussed, in human experience there are two ways to view the world: the immediate world of the exoteric is derived via sensation, and the abstract world of the esoteric is understood through the symbolic. The exoteric point of view is objective and creates our sense of detachment from everything else in the world. This externalized point of view creates the mystery of causality. The esoteric point of view is subjective and timeless, avoiding the problem of causality. Both perspectives exist simultaneously to create what we call *perception*.

Esoterically, the three dimensions of height, width, and length do not exist. Yet from within this abstract, nondimensional arena that is the seat of observation, spatial vision stems from the center and extends in all directions at once. We call this volume space, and within it quality (the abstract or spiritual) naturally forms itself into matter. Scientifically, form (the atom) is self-configured energy. This spatial becoming can be viewed as esoteric and a function of innate consciousness, or, to put it another way, as our consciousness of the continuous. Once spatial perception exists, the concept of time becomes our consciousness of the discontinuous, or our mental consciousness of the exoteric. The abstraction of substance (what constitutes electromagnetic waves in quantum physics) produces matter through action, and through the principle of bonding, the abstract becomes form. Scientifically, this becoming of form is through three principles: repulsion, attraction, and complementarity, which we find in the three charges of subatomic particles: negative (electron), positive (proton), and neutral (neutron).

Perhaps this is why the triplet of principles found in nature has historically been the basis for the simultaneous multiplicity and unity of deity. All three principles must exist to manifest the abstract as concrete, and to derive a quantity of objects where those objects can also be characterized as a quality. This process of quantifying quality is best described as symbolic. As such, all nature—the human body included—is symbolic of the qualities that make us who we really are. Sight, sound, odor, and taste, as well as touch, are all symbolic in the translation from the immediate world of biology to the abstract world of the observer, where mind and perception are the only reality. There is little wonder that symbol and the symbolic are an integral part of mankind's religious tradition. It is our nature.

Esoterism and Early Christian Symbolism

Everyone has seen the Pisces fish sign fastened to the trunk lid of a car. It is a display of Christian faith and allegiance by the owner—not a secret in today's society. Almost two thousand years ago, however, the fish symbol was a secret acknowledgment to others of one's faith, a way of safely communicating during a time of persecution.

The fish was not only one of the earliest symbols of Christianity, it was likely *the* first. The word for *fish* is an acrostic, in which certain letters, usually the first in each line, form a name, motto, or message when read in sequence. Each letter in the Greek word *icqus* (fish) represents another word; together, these words read "Jesus Christ God's Son Savior":

ΙΧΘΥΣ

This type of symbolism poses intriguing questions. Why did those first Christians specifically use "Jesus Christ God's Son Savior" and the fish when a number of other words, phrases, and symbols would have conveyed the same message? And how could it mask the mark of Christianity from the Romans? Surely those who zealously sought out Christians would eventually make the connection. The answers lie deep in a body of esoteric and scientific knowledge.

The ancient (Julian) calendar goes back to January 1, 4713 BCE, the day Julius Caesar decided was to be day 0 of year 0. This date was chosen because on that day important celestial bodies were converging on the number 14: Antares was 14 minutes from the horizon; the sun was at a celestial longitude of 14 degrees (13 degrees 43 minutes) into the constellation Aquarius; and the moon's velocity, which varies between 11 and 15 degrees per day, was 14 degrees (13 degrees 59 minutes) on that day.

Fourteen is the sacred number of Osiris, the Egyptian god whose body, according to myth, was hacked into fourteen pieces by his brother, Seth. Aldebaran, Antares, Saturn also played a crucial role in conjunction with the most sacred star in Egypt, Sirius, 19 minutes away and 19 degrees in its own sign. The sun, whose birth was being celebrated on that day, was in Saturn's sign of Capricorn and Aquarius. Julius Caesar chose January 1, 4713 BCE, as the start of his calendar to honor the birth of the sun at the Giza Plateau, and at the same time honor the Roman festival of Saturnalia.

The Egyptians associated the stars with the essence of divinity and marked the ages by the precession of the equinox. From 4000 to 2000 BCE, the bull (Taurus) was venerated; and from 2000 BCE onward, the symbol of worship became the ram (Aries). Around the time of Christ, the celestial transition from the Age of Aries to the Age of Pisces began.

There can be little doubt that the Christians' secret sign of the fish was born of this grand clock in the tradition of ancient science and wisdom. The Christian Savior was born on the eve of the age of Pisces (the fish) and sacrificed as a lamb (or ram—Aries). He was a fisher of men, and those he caught (actual fishermen), he turned into disciples and taught *them* to be fishers of men. Miraculously, on the Mount of Olives, Jesus fed a multitude by multiplying two fish (the zodiac sign for Pisces) and five loaves of bread. Even the story of Mary's virgin birth echoes Piscean polarity through its opposite sign of Virgo. As it turned out, the Christians' selection of the fish was an apt symbol to convey the idea of a new God who would reign throughout the age of Pisces.

Symbolic Fecundity of the Virgin Birth

Christ having been born of a virgin is one of the foundational concepts of traditional Christian doctrine. This concept, however, defies biological law. The esoterism of secret wisdom provides another interpretation.

In the manifestation of quality (abstract identity) as quantity (discrete identity as life), procreation is a result of creative unity acting upon its own state of disharmony. In other words, the One becoming Two created Man in the abstract. Two, the first number from the absolute, is female. Recall that in the story of the Garden of Eden, it was Eve who first ate from the Tree of Knowledge. In the symbolism of that story, it was the woman who first became a conscious being. So is it with the story of the virgin birth. Mary, the mother of Jesus, is only a virgin in the sense that Jesus represents the archetypal Man. As such, she was a symbolic virgin representing the cosmic concept of fecundity.

Explaining this concept is best done with the famous question of which came first, the chicken or the egg? The answer is the egg, for there is no first seed. The material of the egg has always existed as substance, but without form. There is no sexual intercourse in creation (sexual intercourse is procreation), so to represent the creation of Man symbolically, a cosmic virgin would convey that meaning.

The concept of the Divine being female has a long tradition.

Ancient as well as prehistoric cultures instinctively chose the female figure to represent and describe the formless and transcendental reality that gives birth to all life. They were worshipping the woman not as a goddess but rather as the life-giving principle of the creator symbolized in the role of the female. According to the renowned archaeologist Marija Gimbutas, prior to 5000 BCE, this was the prevalent worldview among all cultures. Only after the spread of a proto-Indo-European culture did cultures begin to project God as a male figure.

Egyptian Origins of Christian Symbolism

The literary metaphor (the esoterism) built into the story of Christ the Savior is becoming clear. Other symbolism of the Christian tradition also finds its heritage in an age when the science—the secret wisdom—of the Egyptians served as inspiration for the sacred. The symbolism of the cross and the nobility of the carpenter, as well as the true meaning of the nativity, relate to a hidden knowledge of the most ancient of traditions.

The Esoteric Meaning of the Cross

The earliest depictions of the Christ during the first century typically portray him as a shepherd carrying a lamb. The cross as a Christian symbol appeared several hundred years later, sometime after the beginning of the second century. One of the earliest-known references of the cross as a Christian symbol comes from the writing of the Montanist Tertullian (ca. 155–230). In *De Corona* (The Chaplet), at the end of chapter 3, Tertullian writes:

> At every forward step and movement, at every going in and out, when we put on our clothes and shoes, when we bathe, when we sit at table, when we light the lamps, on couch, on seat, in all the ordinary actions of daily life, we trace upon the forehead the sign.[8]

For the Christian, the original symbol of the cross was likely in the form of the Greek *tau,* the letter *T,* which later became associated with St. Philip, allegedly crucified on such a cross in Phrygia. The cross's first appearance in Christian art occurred in the fifth century on a Vatican sarcophagus as a Greek cross, with arms of equal length. What the cross represents in modern times, the crucifixion, did not appear until the seventh century.

Although the modern cross is depicted in the shape of a lowercase letter *t,* it is doubtful that Roman executioners used such a fixed, ready-made structure to perform their duties. Crosses used in this style of execution typically consisted of two pieces of wood: the upright, called the *stipes,* and a crosspiece, called the *patibulum,* which was carried to the site of execution by the condemned. Most likely, the crosspiece rested on the upright forming the shape of a capital *T.*[9]

According to biblical text, the execution of Jesus may have been atypical. The author of Acts refers to Jesus's execution as "hanging him on a tree" (5:30), as does Peter in his first letter: "He himself bore our sins in his body on the tree" (2:24). Being hanged on a tree meant that the individual was "under God's curse" (Deuteronomy 21:23). It was not a likely symbol for the burgeoning Christian Church. So, where did the cross come from?

The cross is actually a very ancient symbol with its roots in pharaonic Egypt. The ankh, the ancient Egyptian symbol for life, was likely adopted by the Christian Church, not because of its resemblance to the cross, but for its esoteric meaning of renewal and resurrection.

The Egyptian neter Neith presided over weaving (the manufacturing of fabric), symbolized by two crossed arrows. According to legend, she was endowed with the art of reasoning and the gift of discernment. Because of her, Thoth worked as counsel on behalf of the Great Ennead. Crossing, represented by weaving, symbolizes the idea that reasoning (science) stems from duplicity and from the comparison of any two notions—the hieroglyph for the concept of science, knowledge, or comprehension depicts a piece of woven material. For example, when the Egyptian medicine man says that he has "come forth from Sais with

Fig. 10.4. The Egyptian ankh

Neith," he is referring to a place where the practical aspect of medicine is taught.[10]

The cross was also symbolic of the pharaoh, who was always viewed as divine essence and the ferment (body or compound changing into another substance) of perfection. He was not a king in the traditional Western sense, but the culmination of Man. This process of fermentation is symbolized by the *heq,* a scepter in the form of a hook, similar to the shepherd's crook. This heq is used to gather what is represented by the *nekhakha,* the other scepter, in the form of a staff, where the three aspects of Being (existence) come in three waves. The neter Osiris also carried these two scepters in his aspect of destiny.

Crossed at the wrists, the scepter-bearing hands signify death; or, when opposed by two fists, they depict judgment. Double-crossing of the hands and the scepters always indicates resurrection for this life as well as a transcendent afterlife, which is the principle of renewal.

The principle of renewal or resurrection was also portrayed in the Osiris myth. He "is said to be born in his months as the moon . . . to appear at the monthly festival, to be pure for the festival of the New Moon. Barley is threshed for him, wheat [spelt] is reaped for him, and offerings of it are made at his monthly festivals and again at his semi-monthly festivals, as his father, Geb, commanded. He is told: 'Arise, O King, for thou hast not died.'"[11]

Osiris was also depicted being reborn as the germ, symbolized by the seed emerging from the base of his throne and then blooming into a lotus flower. The flower supports Horus's four sons, which are the four protectors of man's essential organic functions.[12]

The lotus, a prolific symbol in Egyptian art and architecture, represents the culmination of creation. The lotus is an aquatic plant similar to the water lily. It roots in the earth, but grows in and by means of water, with its round flat leaves lying on the water's surface, where they are nourished by air. The lotus blossoms by the sun's fire. Thus, the four elemental principles—earth, water, air, and energy (fire)—are all expressed in the lotus.[13] The lotus flower, in essence, represents finality, the end result of which is the divine nature.

Not only was the lotus representative of regeneration; it was also a symbol of the exalted. Egyptian carvings show that from the lotus arises a new leaf (air), a bud (coagulation), and new flowers.

According to Schwaller, the first constitutes a new Earth, a solar Earth, where the energy of the sun being transformed in a "living earth" is capable of exaltation. Philosophically, exaltation is in reference to the absolute. When the life force can no longer be contained in matter (bodily form), it returns to its spiritual source. This return to the spiritual source is the true meaning of exaltation.[14]

The Nobility of the Carpenter

Prior to his ministry, Jesus by tradition was a carpenter, and also the son of a carpenter. The Gospel of Matthew refers to him as the "carpenter's son" (13:55), and in Mark (6:3) he himself is referred to as "the carpenter." Although these two passages quite literally state that he was a carpenter, no one really knows for sure if, in fact, he was. It is *assumed* that he was. Given the theological importance of his ministry, this may not seem to be a terribly important point. There is another possible reason, however, why carpentry was supposed to have been his vocation.

During Egypt's Old Kingdom, Hesy Re, of the Third Dynasty (2600 BCE) was a practitioner of medicine and surgery. According to the New York Academy of Medicine's Smith Papyrus, he was chief of dentists and physicians of the early dynasties.*

The papyrus endows him with a great number of titles, two of which involve the term *carpenter:*

> The great medicine man (priest) hk of Mehyc the Ancient. Prophet of Min. Carpenter (of) royal Science, royal scribe. Grand Master of the Fly, Father of Min, Carpenter (of the) lioness. Great of the city of Pe, Chief of the guides (?). Great of the Ten of Upper-Egypt. Priest of Horus of Mesen of the city of Pe, Hesy.[15]

It is a curiosity that Hesy, the first of the ten most important personages with several priesthood titles, was also described as the "Carpenter of Royal Science" and "Carpenter of the Lioness." In a wooden panel (figure 10.5) depicting Hesy, an ax (the carpenter symbol) is inscribed beneath the term *royal science* (or *royal knowledge*) and above the term *royal scribe.*

According to Schwaller, "It is impossible to believe him a carpenter as

*The papyrus itself is 4.68 meters long with 21½ columns of cursive hieroglyphics actually containing the fragments of two books. One is anatomical, the Book on the Vessels of the Heart. The second is surgical, the descriptions of forty-eight cases, including injuries, wounds, fractures, dislocations, and tumors—the types of troubles that fall into the practice realm of a surgeon. These descriptions are arranged from the head downward, in order of severity within each group. The cases are given one of three labels, depending on the chance of successful treatment.

Fig. 10.5. Wood panel from the mastaba of Hesy, Sakkara, Third Dynasty (from R. A. Schwaller de Lubicz, Sacred Science, *plate 1)*

well, unless a particular meaning is attached to this title." Where Hesy is called "Grand Master of the Fly," "Father of Min," and "carpenter of the lioness," there is an unknown symbol comparable to the *nekhakha* scepter that Min carries in his raised hand. On the back of the lioness, the unknown symbol is found again.[16]

In the carving's fourth column of titles, Hesy is "Great in the City of Pe" and the double capital of the North (Dep and Pe). He is also the most eminent and the greatest of the "Ten of Upper Egypt," the first of the ten judges composing the high court that were sworn to the name of Maat (truth and justice). In the fifth column, he is called "priest of Horus of Mesen," a place where Horus wages his final battle with Seth.[17]

During the Middle Ages, a symbol similar to the carving's unknown symbol depicts the idea of sublimation (to refine or purify a substance) or high spiritual, moral, or intellectual worth. So this ancient Egyptian text may be understood as a sublimation of Min, depicted first by the fly, which represents the volatile (the bee is reserved for kings), then by the "father of Min" and the "carpenter of the lioness."[18]

Schwaller believes that the essence of all these gracious titles is reminiscent of the carpenter's designation, twice expressed by the symbol of the ax. As symbol, the esoteric carpenter is the one who crowns the house and executes the temple steps that reach into the heavens. When the ax (or any other tool) is used as a symbol, it likely refers to a principle or function as opposed to the worker who would use it. Such a noble man would never have been a common laborer or carpenter.

Also during the Middle Ages, carpenter guilds held precedence in processions with the raven as their standard, suggestive of the Hermetic tradition. According to legend, raven hatchlings are born white and not recognized by their parents until they turn black. In this raven, there is a conceptual relationship. The raven starts, as does the carpenter, by making a framework for the nest, and then finishes it with a covering.[19]

We can speculate that the carpenter of the Gospels refers to an esoteric tradition of nobility rather than to a common laborer, a literary symbolism evoking the idea that Christ creates the structure where all may find resurrection and renewal.

Secret Wisdom of the Nativity

The most compelling evidence for the tradition of secret wisdom can be found in a custom that has continued almost unchanged for over a thousand years, the Christmas nativity scene. Locked within its symbolism is the key to understanding the Gospel message, as well as the identity of the Christ himself.

The nativity scene is traditionally placed beneath the Christmas tree, where a stable houses the baby Jesus in a manger. At the infant's side are his parents, Mary and Joseph. The rest of the cast includes an ox, ram (or sheep), angels, shepherds, and what this enigmatic scene is so well known for, the mysterious magi from the East. It is possibly the most prevalent aspect of Christianity. Regardless of a particular brand of faith, it is commemorated by millions of people every Christmas. Yet few people realize its significance; it is cloaked in a way of thinking to which the modern mind is unaccustomed.

Although some may argue that the birth of Christ is not astrologically based, the principal feature of the story demands that it is. We know the herald of Christ's birth as the Star of Bethlehem, but it wasn't really a star, even though it appeared as one to the unaided eye. Until the seventeenth century, theologians believed it to be a miracle of God. On October 10, 1604, however, Johannes Kepler observed a new star in the sky and believed it was something unusual between Jupiter and Saturn, which themselves were only nine degrees apart. He believed this phenomenon was the same as what occurred just prior to the birth of Jesus. The next year, while writing *De Stella Nova in Pede Serpentarti* (About the New Star in the Serpent's Foot), Kepler came across the work of Polish astronomer Laurence Suslyga, who argued that Christ was born in 4 BCE. Ten years later, he published his conclusion that a triple conjunction of 7 BCE was followed by a massing, or grouping, of Mars, Jupiter, and Saturn in 6 BCE. Just as the conjunction and massing of 1603/4 had produced what looked like a new star, so the events of 7/6 BCE also produced a new star, the Star of Bethlehem.

Others believe it was a triple conjunction of Mars, Jupiter, and Saturn, but this event would not have been visible, because the planets were too close to the sun. Furthermore, it was not a conjunction; it was a massing. There was a conjunction of Jupiter and Saturn in 7 BCE, but the next year only a massing occurred that included Mars, Jupiter, and Saturn. A conjunction requires the planets to have precisely the same longitude. Another theory is that the Star of Bethlehem was a comet, or that it was the Shekinah Glory, the divine presence of God among humans.

There is no shortage of theories concerning the Star of Bethlehem. Yet it must be remembered that first and foremost, the nativity story conveys a message in which the importance of the Magi is paramount. It is also possible that the story has no basis in fact and was designed solely to disseminate a hidden meaning.

Nonetheless, these wise men from the East were holy men as well as scientists. In this event, they recognized a divine significance, and, since they were astrologers—astrology *was* astronomy in ancient times—it was, for them, scientific as well. Most likely they were from the region of Mesopotamia, which was at that time somewhat of a buffer zone between the Roman and Parthian empires. Why they came to Judaea has been a mystery now for over two thousand years.

In an attempt to understand these circumstances, Adrian Gilbert has put forth a cogent and compelling explanation of the nativity. In his book *Magi: The Quest for a Secret Tradition,* Gilbert proposes that the nativity is Jesus's horoscope, and that each element of the scene has a hidden meaning.[20]

For the ancient astrologers, July 29, 7 BCE, was a very special day. At that time each year, the sun would "rise" in the star of Regulus in the constellation of Leo. It was the King's Birthday position and in conjunction with the Little King or Lion Heart. At the same time, Sirius, which was the brightest star in the sky, would make its first appearance after a period of invisibility. In Egyptian mythology, Sirius's reappearance at dawn represented the goddess Isis emerging from confinement to give birth to her son, Horus, which was represented by the sun/Regulus conjunction. Osiris, Isis's husband, was represented by the constellation Orion. With these allegorical concepts in place, the nativity scene can be understood in its entirety.

The zodiac of the nighttime sky is represented by location, a stable where there is the ox and the sheep, representing the constellations Taurus and Aries. Bethlehem translated literally means "place of bread"—where the animals obtain their nourishment—and is located in the province of Judah, the Lion of the tribes of Israel. The baby Jesus in the manger represents the sun/Regulus conjunction in the constellation Leo (the Lion). Mary, the archetypal Isis, is represented by Sirius and gives birth to Jesus, who is the archetypal Horus. Her husband, Joseph, is, of course, Osiris, represented by the constellation Orion. The three shepherds represent the stars Procyon, Castor, and Pollux. Being in the hills, they are symbolic of a location north of the ecliptic. Since these stars rise before Sirius, like shepherds, they lead the way.

The most important elements, the three magi, refer to something completely different, and herein is the story's most hidden meaning. For the magi from the East, the Star of Bethlehem showed the way to the newborn king. This star was really the Jupiter/Saturn great conjunction that took place on and off for several months in the constellation Pisces. As previously discussed, even today, the sign of the fishes (Pisces) is symbolic of Christianity. The brightest objects in the sky, Saturn and Jupiter, would rise around 9:30 p.m. and remain visible until the brightness of the sun the next morning eliminated their glow.

Gilbert believes that the first two magi, Melchior and Caspar, with their corresponding gifts of gold and myrrh, stand for Jupiter and Saturn. Myrrh was used primarily for mummification, and Saturn is therefore represented by myrrh; Jupiter represents wealth, of which gold is an apt symbol. What about the third magus?

There was a third planet on that special day that was actually closest to the sun, Mercury. This was represented by the third wise man, and is symbolically the most telling. Balthazar represents Mercury, which means "the Lord's leader." As portrayed in traditional paintings, Balthazar is Jesus's closest attendant, kneeling at his feet. Aptly, his gift is frankincense, which is symbolic of magical priestly functions. Of course, the name Mercury is Roman. In Greek, it is Hermes, the Egyptian Thoth (or Djeheuty). Herein lies the secret. In the tradition of ancient pharaonic wisdom, what we have throughout this work referred to as secret wisdom, Jesus is initiated into the ancient tradition of the mystery school. Its great work being the *Corpus Hermeticum,* also known as the Writings of Thoth, it is the philosophy of living nature and the source of all major world religions, relating the principle of the Man Cosmos. Mythically, it is told through the life of Osiris, the archetypal Man who was murdered by his brother, Seth, and then resurrected into eternal life. What the authors of the gospels Matthew and Luke are saying is that Jesus—murdered by his religious brothers—also represents the archetypal Man in the tradition of Osiris.

In the concept of the archetypal Man's taking form and then being resurrected also lies the reasoning for Mary arriving in Bethlehem riding a donkey. Although this is not a well-known aspect of Egyptian mythology, Seth, in his donkey form, assists in the resurrection of the solar infant, as expressed at the Temple of Edfu. The art on its walls describes "transpiercing the donkey" on the Place-of-Transpiercing. Metaphorically, it is seen as an act of sacrifice that contributes to the maintenance of the cosmic order.[21]

As Jesus becomes an adult, higher gifts from the planets are realized; he meets up with John the Baptist and is baptized in the river Jordan. Further symbolism announces that he is the medium of the Christos, or, as John says, the Logos. Although Jesus's baptism is not as popular as it once was, the early church recognized its importance every January 6, when it celebrated the Feast of Baptism. Later, under the emperor Constantine, January 6 became Jesus's birthday. Then his birthday was moved to December 25 as a replacement for the Roman festival of Saturnalia. Yet the celebration of Christ's nativity, an indestructible icon of Western civilization, has continued unchanged for nearly two thousand years, hiding the literal key to the truth of the Gospels and the secret wisdom of pharaonic Egypt.

Given these historical and literary coincidences, the connection of the literature of the Christian scriptures to ancient Egypt is irresistible. There can be no logical conclusion except that the symbolism of the nativity was intentionally embedded in the story of Christ's birth, so that centuries or millennia later, those who had been initiated into the tradition of this secret wisdom would recognize its significance—a foolproof way to preserve the "Gospel truth," regardless of whatever dogma should arise.

It is possible that the leadership of the Western religious tradition has always known about this secret wisdom and embraced its values. The iconography and architecture of the Gothic cathedral tradition suggests that this may be the case. Adorned in a symbolic (and Hermetic) style and constructed in a very specific manner, cathedral architecture, and particularly that of the sanctuary, inspires a sense of awe and reverence in the psyche of the individual. Its ambience evokes a sense of mystery and a search within.

From one generation to the next and one culture to another—from the Egyptian to the Greek, Greek to Arab, and Arab to European—over a two-thousand-year period, the responsibility for the dissemination of secret wisdom rests primarily in a group of texts known as the *Corpus Hermeticum*. The content of these texts transcends cultural and religious customs; the provocative style of their discourse blurs the distinction between the rational and the irrational, as well as the scientific and the sacred.

11

Sacred Science

Ancient Wisdom of the Modern World

> *Knowledge left the temple first through the ideas of Greek visitors such as Thales, Pythagoras, and Plato; however modified by the sharpness of Greek concepts, it came to establish its presence at the roots of Western philosophy; a second path was through Egyptian Gnosticism and heretical Christianity; a third through the images of the Tarot whose cards, just as surely as the Emerald Tablet of Hermes Trismegistus, are said to portray the structure of the sacred science of the Egyptians.*
>
> ANDRE VANDENBROECK, *AL-KEMI*

During the Renaissance, scholars believed that the Hermetic texts were truly ancient in origin, dating to remote times of Egyptian civilization. Accordingly, these texts, known as the *Corpus Hermeticum,* were attributed to a man named Hermes Trismegistus, who is also the texts' principal character. During the first few centuries, Hermes was commonly known as "the Egyptian"' and was generally believed to be a real person, as well as the mentor of Moses. Although there is no evidence confirming that Hermes was, in fact, a living individual, the flavor of Hermetic literature lends credence to notion that he was.

In 1614, however, the French philologist Isaac Casaubon (1559–1614) proved that the *Corpus Hermeticum* was actually of a more recent origin. According to Casaubon, the texts themselves were from the second century CE and involved the works of a number of authors who lived in Alexandria. As literary works, they were of Greek origin with strong Egyptian influence, containing ideas common to early Christianity and other religious sects, such as the Gnostics. As such, Casaubon concluded that they were a compilation of first- and second-century religious ideas without any significant links to the remote past.

More recently it has been discovered that the use of the name Hermes Trismegistus predates the first century. During the second century BCE, a scandal broke out in Sakkara. Some men had been misusing the sacred ibis, the symbol of Thoth (Hermes), and may have been delivering empty jars that were supposed to have contained the mummified bird. The governing council of the ibis cult instituted a reform, which was recorded by the scribe Hor on June 1, 172 BCE. In Hor's notes there is a reference to "Thoth, the three times great."[1]

Nonetheless, despite Casaubon's dating of the texts, Hermeticism continued to influence and inspire the philosophical and scientific community of western Europe well into the eighteenth century. While it is clear that the Hermetic texts translated by scholars during the Renaissance were second-century documents, the popularity and growth of Egyptology during the nineteenth and twentieth centuries brought new evidence to bear on the history of Hermetic concepts. In 1881, at Sakkara, Egypt, the French Egyptologist Gaston Maspero (1846–1916) discovered that the subterranean chamber of the pyramid of Pepi I (the second ruler of the Sixth Dynasty) was engraved with hieroglyphics. Over the course of subsequent explorations, it was discovered that a total of five pyramids at Sakkara also contained inscriptions from the Fifth, Sixth, Seventh, and Eighth Dynasties of the Old Kingdom.[2]

In 1952, Dr. Samuel A. B. Mercer (1879–1969), professor emeritus of Semitic languages and Egyptology at the University of Toronto, published a complete English translation of these inscriptions as *The Pyramid Texts.* Although the Sakkara pyramids are generally dated to the late third millennium, between 2350 and 2175 BCE, Mercer acknowledged that the philosophical concepts recorded in the body of texts might have existed long before 3000 BCE. According to Mercer:

> Evidence of a date previous to about 3000 B.C. is seen in passages which reflect events and conditions previous to the union of the two Lands, for example, the hostility between North and South, before the time of the first king, Menes; in the mode of burying bodies of

> the dead in the sand; in the pre-civilized era reflected in the so-called Cannibal Hymn; and in the many references to the assembling of the bones of the deceased, passages which indicate a pre-mummification period.[3]

As for the collection's contents, Mercer categorized the inscribed "utterances" and "spells" (although the Egyptian phrase means "words to be spoken"[4]) into three topics: Solar Theology, the Religion and Myths of Osiris, and the Political Unification of Upper and Lower Egypt. He also concluded that the collection represented the following concepts:

- a funerary ritual of mortuary offerings, connected with the corporeal reconstitution and resurrection of the deceased king
- magical formulas to ward off harm and evil
- a ritual of worship
- religious hymns
- mythical formulas identifying the deceased king with certain deities
- prayers and petitions on behalf of the deceased king
- the greatness and power of the deceased king in heaven[5]

Although the Sakkara Pyramid texts do not appear to be a coherent self-contained set of beliefs concerning the king's journey into the beyond, among scholars there is general agreement that they are the earliest recorded religious beliefs anywhere in the world, and were likely representative of the secret knowledge that was later written on leather or papyrus scrolls.[6] During the Old Kingdom, this body of texts, composed by special priests called lectors, were kept in libraries and temples. Because of the extraordinary age of these books, however, none from the Old Kingdom has survived, and copies are scarce from later periods of Egyptian civilization.[7]

Ancient Egyptian religious customs no doubt changed over time, but the concepts in these texts were important enough to be incorporated into later writings, such as *The Coffin Texts* (ca. 2100 BCE) and *The Book of the Dead* (1240 BCE). What this implies is that the Egyptians' core beliefs were constant, even though their ritual and artistic expression changed with the progression of cultural trends.

Over the course of Egyptian civilization, various cults were dominant in certain areas at different times. Each cult focused on a specific neter, such as Re (the sun) at Heliopolis or Osiris (the resurrected king) in the delta. The traditions and rituals of these cults were generally secretive and reserved for the priests and nobles. So the knowledge and tradition of the temple remained within the confines of the temple. But after more than two thousand years of secrecy, with the influence of Greek intellectualism (particularly at Alexandria), a generation of priests and scholars were inspired to set forth on papyrus their philosophical and scientific views. Although it would be incorrect to say that the Ptolemaic period in Egyptian history created a religious doctrine in the *Corpus Hermeticum,* what the writers of the *Corpus Hermeticum* did do was preserve for antiquity the popular beliefs and philosophical mind-set of the people at that time. For the student of history, the interesting aspect of first-millennium Hermeticism is not so much when the Hermetic texts were written as how far back does the tradition they represent go. According to Hermetic researchers Timothy Freke and Peter Gandy, the hieroglyphics that make up the pyramid texts "contain doctrines that are identical to those expounded in the Hermetica."[8]

Hermetic Texts and the Early Christian Church

Beginning in the late first century, Hermetic texts began to circulate in the Roman world. By the second century, they were commonplace and generally accepted by the government and the people in general. With their philosophical and scientific base, they posed little threat to Pax Romana (the peace of Rome). Another movement, more imposing to traditional beliefs, was also under way, however, that needed grassroots support, a movement known simply as The Way (later to be known as Christianity). With their refusal to worship Caesar and accept Roman religious customs, followers of The Way posed a potential threat to the social stability of the Empire.

However unjustly, followers of The Way had become the target for persecution, first in Judaea by the Sanhedrin and then by the Roman government. After the weeklong fire in 64 CE that destroyed the capital city of Rome, the emperor Nero needed someone to blame. The strange new religious cult that grew out of Judaea provided a perfect scapegoat. Domitian, who reigned from 81 to 96 CE, followed suit. Although later emperors were less active in hunting down Christians, the secretive way of early Christianity was still deplorable to the Romans, and their laws reflected as much.

Despite persecution and anti-Christian laws, followers of Christ continued to gain ground as an up-and-coming religion. Although the efforts of early evangelists were effective to a degree in establishing new churches, another movement had already laid the intellectual and philosophical foundation for the Christic message, the Way of Hermes. Perhaps the secret of Paul's success was not so much in his ability to speak, but in Greek scholarship and the prolific works of the Alexandrian *Corpus Hermeticum*.

Although arising from historically different origins, the various and competing ideas of Christians and the philosophical concepts of Hermeticism were strikingly similar in the first century. For some leaders of the Christian Church, Hermes Trismegistus was perceived to be a prophet of Christ. This association of Christianity with Hermeticism continued as long as the church remained a movement without a formal doctrine.[9]

In *Adversus Valentinianos* and *De anima,* written in 206 and 207 CE, Tertullian of Carthage (ca. 155–230 CE) quotes Hermetic texts. He writes that the Egyptian Mercury declares that, upon death of the body, the soul keeps its unique identity and does not become a part of the "universal soul."[10] Didymus the Blind, a fourth-century Alexandrian Christian, quotes the *Corpus Hermeticum* word for word in his *De Trinitate,* interpreting the *Discourses of Asclepius* as a prophetic voice for the revelation of Christ. The word *Logos* occurs often in the Hermetica, and Didymus believed that it referred to the Christian Logos, the Son of God.[11] Cyril of Alexandria (? –444), who defended church doctrine against the Nestorian heresy, also frequently quoted the Hermetica.

According to research fellow Garth Fowden, of the National Hellenic Research Foundation in Athens, Cyril wrote vigorously, and had an impressive command of Hermetic material.[12] As did Didymus, Cyril also supported the Hermetic Logos as the Logos of Christ. Although the association of Hermes with Christ ended in the Western church at the end of the fifth century, even during post-Byzantine times (sixteenth through the eighteenth centuries), Hermes Trismegistus was still being painted on church walls in eastern Europe.[13]

Despite the favorable view of Hermeticism in the early church, a division grew in matters of doctrine that eventually consolidated power in the hands of those with a particular set of beliefs, what today is referred to as orthodoxy. The so-called Gnostics, who were, in particular, Gentiles who had converted, refused to believe that any man could be physically resurrected and that Jesus's teachings were of an esoteric nature based on knowledge. For them, Jesus did not have a mortal body, and the way to salvation was through the intellect. The primary Gnostic spiritual issue was not sin and separation from God; it was ignorance about the true nature of things—that it was impossible for the one true God to incarnate as an individual into the corrupt material world. The Gnostic concept of God was not anthropomorphic, as it is in Christian orthodoxy; rather, it was anthropocosmic in that God was the rudimentary essence of all existence. As a result, the Gnostic view of God could be known only through the abstract nature of man. For the Gnostic, it must have been a contradiction that God *was* the universe and at the same time a single individual, all other men and women excluded.

Hermeticism was different from Gnosticism, with the latter being more religiously oriented in its ideas. Yet because of their similarities, the demise of Gnosticism also meant the end of Hermeticism as a viable platform for Christian ideas. Nonetheless, the Christian message was so acceptable to Mediterranean cultures during the first few centuries primarily because of the Alexandrian intellectualism and the inspired writings from an amalgamation of Greek and Egyptian thought. During what is referred to as the Coptic Period, Egypt itself became a predominantly Christian country in the first three centuries of the Common Era.

The Egyptian Thoth as the Source of Hermetic Texts

Hellenism was politically successful in Egypt, but in cultural terms, Egypt, with its ancient traditions of more than two thousand years, remained Egyptian. Because the Egyptian culture was so progressive and cosmopolitan, the Greeks were eager to discover Egypt's ancient wisdom. As any foreign culture would, they translated, then interpreted those traditions and beliefs into their own way of thinking. From this blending of Egyptian and Greek scholarship arose the mythical figure Hermes Trismegistus.

The essence of this figure was a unification of the Greek Hermes and the Egyptian Thoth—Thoth being the Greek translation for the Egyptian god Djeheuty. The Greek Hermes, known for his wisdom, was the son of Zeus and the messenger of the gods. As the divine mediator, Thoth was also the messenger of the gods. As such, the Greeks identified Hermes with Thoth, and referred to the Egyptian city of Khmun, the center of the Thoth cult, as Hermopolis.

Thoth was known as the "heart of Ra," the "lord of divine words," and the "self-created, to whom none hath given birth, god one."[14] Usually depicted as a man with the head of an ibis or sometimes as a baboon, the Egyptian Djeheuty (Thoth) was a lunar deity who also represented wisdom and was believed to be the inventor of writing. In Egyptian art, he is often displayed holding a scepter and an ankh.[15]

Even during the earliest stages of Egyptian civilization, Thoth was venerated as the moon-god with a variety of functions linked to the principle of transformation, or death and rebirth. The moon was an ideal supranatural type for transformation, since it mysteriously appears from nowhere and is then transformed from a thin crescent into a full moon and back to crescent, only to once again disappear. The moon itself was not a god; rather, it represented the metamorphosis found in all nature—from tadpole to frog, caterpillar to butterfly—as well as the change in man from childhood to pubescence. In the pyramid texts, it is on the wings of Thoth that the king ascends to heaven.[16]

The principle of natural transformation was perceived as the essence of life and growth. Thoth became the counselor of Re (the sun), the true source of life and all other divine principles. During the night, Thoth was the ruler of the stars that marked the passage of time, and became "the lord and multiplier of Time, and the regulator of individual destinies."[17] His role was so great in Egyptian thought that he was deemed responsible for cosmic order, as well as society's institutions. As such, he was "the measurer" and the "measurer of time."[18]

Thoth presided over temple cults, the calendar, and legislation, as well as sacred rituals, the composition of texts, and the science and magic of the arts related to alchemy.[19] With these latter aspects, Thoth personified reason and logic and was considered the lord of knowledge, language, and science, the culmination of which was esoteric wisdom. Thoth was the creative power that creates with words, and "the Mysterious"—the essence of life that we still struggle to explain today.

But it was in his role in the afterlife as a guide to the departed and the judgment of the dead that Thoth was most famous. Through the Ptolemaic and Roman periods of Egyptian civilization, he remained popular. Consequently, newcomers to Egypt identified with Thoth through a native likeness from their home culture.[20] With the Greeks, it was Hermes who revealed the divine—the Logos.

The Hermes Trismegistus that was born of this Greek and Egyptian collaboration was distinctively human and hellenized. In the tradition of the Egyptians, the Greeks also identified their god Hermes with the invention of astronomy (astrology) and the sciences of numbers (mathematics and geometry), surveying, medicine, and botany, as well as the development of religion and government. In essence, Hermes was the author of all knowledge, human and divine.[21] Yet as great was his fame throughout the Roman world, he was still "the Egyptian," as Renaissance scholars correctly interpreted.[22]

The Egyptian perspective on Thoth was slightly different from what has been interpreted as the popular "classical studies" Greek version. Egyptians identified with Thoth in an esoteric way. He was an active and intrinsic part of nature that linked all humanity to everything else that exists. As such he was the "heart and tongue" of the gods,[23] the essence of mankind's unique role in nature. According to Geraldine Pinch, "Thoth was the god of wisdom and secret knowledge who invented writing and the different languages of humanity."[24]

Cyril of Alexandria writes that, for the Egyptians, Thoth represents the principles of being the "law-giver and an authority on astronomy, astrology, botany, mathematics, geometry, the arts and grammar."[25] For the individual, Thoth was the principal nature of attractiveness, nourishment, success, and a happy life. Implicit in the Egyptian understanding was the idea, according to Fowden, that "I know you, Hermes, and you know me. I am you and you are me."[26] In essence, Thoth was the archetypal Man in whom humanity and civilization was created, and not a living person.

Although the Greeks are credited with the popularization of Hermeticism, the source is unequivocally Egyptian and believed to be based on the *Book of Thoth.* According to Geraldine Pinch in *Egyptian Mythology: A Guide to the Gods, Goddesses, and Traditions of Ancient Egypt:*

> Most scholars now agree that the traditional wisdom of the Egyptian priests and their knowledge of Egyptian myth were among the elements that made up the Hermetica.[27]

The Hermetic texts follow the same literary pattern, as does the Book of Thoth, in which a dialogue between disciple and deity (or wise man) is the primary focus of the story. Pinch observes, "Several manuscripts ranging in date from the first century BCE to the second century CE preserve parts of an actual Book of Thoth."[28]

In one ancient Egyptian story, Setna (a prince), ignoring ghostly warnings, removes the magical Book of Thoth from its ancient tomb near Memphis. Consequently, Setna experiences terrible hallucinations until he returns the book to its rightful place. The punishment of Setna was a warning to the pursuers of divine wisdom: It should be used to gain personal enlightenment, as opposed to earthly power.

From the Old Kingdom through the New Kingdom, Thoth was an integral aspect of the Egyptian worldview. In the *Book of Two Ways,* during the Middle Kingdom, Thoth's mansion is a sanctuary for spirits using his magic to thwart the underworld's demons. In the New Kingdom's Underworld Books, Thoth directs the mystical union of Ra and Osiris. And in *The Book of the Dead,* he stands ready to record the judge's verdict. Egypt's tradition of Thoth grew into the belief that he had written forty-two books, which contained all knowledge mankind needed. This in turn grew into the literature we know today as the teachings of Hermes Trismegistus, Thoth the Thrice Great.[29]

Concepts of the Hermetic Texts

The *Corpus Hermeticum* is a collection of essays devoted to philosophically describing mankind, nature, the cosmos, and deity. Written in a dialogue style, the essays focus on the cosmogonic and cosmological and are distinctly a strong interpretation of the anthropic principle.

At the core of these texts is the role of man and man's relationship to nature and the cosmos. Mankind is viewed not as an infinitesimal and inconsequential part of the physical universe, but, the abstract essence of the cosmos itself. The immediate, observable world is explained as a manifestation of the abstract mind that can be comprehended only by understanding the nature of one's Self.

Creation of the Universe

In the first Hermetic essay, entitled "Poemandres, the Shepherd of Men," Hermes, the principal character throughout the essays, receives a vision explaining creation. In his vision, everything becomes light, and within the light darkness forms that is described as a coiling phenomenon, in "sinuous folds" like a snake. Representing matter (or earth), this "coiling" soon transforms into water. And then through Logos (Word), active light (fire) appears from within the water. Air also appears in the form of light and rises above the earth and water, which remain mingled together.

According to the text, the Light is the "Mind of God" and the Logos is the "Son of God." Although described as separate aspects of an abstract realm, these remain a single entity. Together they are described as "Life" and the source of the observable universe as we know it. They, mind and the Logos, are also described as energy in the form of light from which all matter arises and grows beyond all bounds.

The Concept of God

God is not so much a person in the Hermetic view as a conceptual understanding of what nature is. Only silence can express God's true name. According to the sixth Hermetic essay, God is self-cause and self-sufficiency—the meaning of the Greek word *good*—for all that is, and from God is available an infinite supply of "stable" energy. The essence of God is Mind, which can only be described as an unknown, nondimensional entity. From Mind comes forth Logos, representing the principles that constitute everything that exists. Also within Mind is the "Archetypal Form," which becomes the details of the physical reality, which exists prior to the creation of matter.

If such a thing as the essence of God exists, the Hermetic text proposes, then Mind is its essence. What that essence is, only Mind itself knows precisely. Therefore, Mind cannot be separated from God's essentiality.

In the Hermetic essay entitled "About the Common Mind," God is also defined as energy and power. Matter is separate from God so that it can be attributed to the quality of space. But, the Hermetic text asks, how else could mass be defined of whether it were not energized? Energies are parts of God, regardless of whether they are matter, body, or essence.

Materiality is matter's energy. Corporeality is the energy of all living bodies, plant or animal, which constitutes the energy of essence. This is God, *the All*. But it is because of reason (Logos) that there is adoration and worship. And there is only one way to worship God, which is not to be bad.

The Concept of Cosmos

Physical matter shares in the self-caused and self-sufficient. In this way, the cosmos is self-caused and self-sufficient. Logos, which comes forth from Mind, creates the principles of the cosmos. From Logos the abstract cosmos is born, and from the abstract cosmos, the physical cosmos is created, meaning that Mind (God) created, in God's image, another (second) Mind that is the abstract cosmos. This second Mind, also referred to as "Formative Mind," engenders itself into physical form and, as a result, matter exists and is set in motion.

Therefore, as a consequence of its immortal, abstract parent, the cosmos is a living immortal entity—a second God made in the image of God. Every thing that exists within the cosmos, man included, is immortal, since it is part of the cosmos.

Originally, matter was unordered. But through the principles of the abstract, the cosmos was formed into celestial bodies. In doing so, the physical universal, as a body, became immortal, existing in an ordered state.

In the second Hermetic essay, entitled "To Asclepius," the physical objects that make up the cosmos are defined as bodies that exist in the void of space. Since space is bodiless, the Hermetic author deduces it must be a godlike thing or God Himself. If space is God's creation, then it is substance. But if it is God Himself, it transcends substance. The author concludes, however, that space is "space," and God being "energy" contains all space. So the motion of the cosmos, and all material things—animals included—comes not from a source exterior to the cosmos, but from a source "interior to the exterior." This source is an intangible thing.

The cosmos is timeless and has always existed. Its existence is in its becoming, its evolution, and the genesis of quality and quantity, along with the totality of its purpose, is life. The universe is material, but since it is intelligible, it belongs to the realm of Mind.

The Concept of Life

To explain why conscious life exists, it is said that matter shares in Mind (abstract cosmos) as an equal to the Mind of God. In other words, God became form in the existence of the cosmos and became the archetypal Man. Thus, the anthropic principle is declared: Man's true nature is abstract, and the physical cosmos is a result of that nature.

As for biological life, after the creation of the cosmos, the planet

Earth became the focal point of the archetypal Man. For the (abstract) Man to become (concrete) form, Mind willed itself to live in Earth's waters to attain the image of Man's fairest form.

As a result, form became animated, even though initially devoid of reason. As Brian Copenhaver observes:

> [Earth] was the female. Water did the fertilizing. Fire was the maturing force. Nature took spirit from the ether and brought forth bodies in the shape of the man. From life and light the man became soul and mind; from life came soul, from light came mind, and all things in the cosmos of the senses remained thus until a cycle ended (and) kinds of things began to be.[30]

According to the Hermetic text, this is the mystery that is hidden: Man (abstract) embracing nature brings forth the wonder of life. And, through the harmony of opposites, Man became male and female, as are all animals, in order to multiply. Man, being aware of the body, is the third life made after the image of the cosmos:

> The cosmos was made by God and is in God but that mankind was made by the cosmos and in cosmos.[31]

Within Man there is also a mind, so he also has the ability to conceptualize the bodiless, self-existing Mind. Yet many men are unaware of this, because Darkness (unknowing) is the basis of material form that was fashioned through water.

Man as the Archetypal Soul

For mankind, God is intelligible, meaning humans can conceptualize and contemplate God's existence. Thought always occurs beneath the thinker's essence. So, being below God's essence, man can think of God. On the other hand, since there can be nothing above God, then God can never think of Himself. Therefore, God thinks of Himself as being nothing else but what He thinks.

In the Hermetic worldview, only Mind and Reason (Logos) are formless, self-embracing, and all containing. Mind, as a result of reason, is the principle of man. Reason lies in the soul, which lies in the spirit, which lies in the body.

Mind and Reason are also Man's Archetypal Soul, where God

is Cause and the creator of Light, Mind, and Spirit. Man's form, the material body, is created from water and earth as a result of this dispensation. The body, however, whose formation is a mystery, is temporal, since it is subject to change, just as the bodies within the cosmos are subject to change. Upon the dissolution of form, man must surrender life, leaving the body void of its energy. The body's senses return to their source and resurrect as nature's energies. Understanding this mystery in the common nature of man's "Being" is gnosis.

In the third Hermetic essay, "The Cup (or Monad)," Man is defined by the ability to reason, which is *of* Reason (Logos) and Mind. All men do reason. In general, men do not endeavor to understand the source of this ability to reason. Those men who do understand the nature of the source, Mind, do so by choice. Without Mind, men reason in a fashion similar to that of animals, through their feelings and impulses, and do not concern themselves with things worthy of contemplation.

Without Mind, men are concerned only with bodily pleasures, believing that this is the nature of existence. The seventh essay, "The Greatest Ill Among Men Is Ignorance of God," identifies the body without Mind as living death, where these bodily pleasures overwhelm man's consciousness, preventing him from understanding.

A person is not born with this notion of bodily attachment, but slowly grows into it as adulthood approaches. An infant's soul remains part of the cosmos soul, since it is not yet "fouled" by the pleasures and passions of the body. When the infant matures, its soul is drawn into the body and soon forgets its inheritance in the Self-existence of the cosmos. This forgetfulness soon becomes vice. The result is that man focuses on the trials of a material world and thinks to himself: I do not know what to do; ah, wretched me, I am devoured by all types of ills; poor me, I neither see nor hear! Such are the cries from the chastised soul who identifies with the ebb and flow of corporeal life. At worst, man is transformed into a vehicle of outrage, violence, and murder.

Gnosis and the All Soul

To have Mind, the body must be hated—not in a literal way, but there must be a recognition that the body (and material world) is not the essence or source of life but rather the expression of it. By not identifying with the material world, the nature of the Self becomes apparent. As the source of all things is a mystery, so is the Self. With the identification and recognition of the Self as an unknown abstract entity, the source (Mind) becomes the true understanding of reality. Achieving this understanding is gnosis. Not only does gnosis show piety to God, but it also defines Man as God. Identification with the abstract source of all that exists is called Oneness.

This Oneness is a mystery, since the source and root of all things is indefinable yet must exist for Man to exist. Nonetheless, the source and root of all things is understood by ordered nature, such as the sun, and the moon's course, or the stars, and is also understood in the miracle of conception. The Hermetic asks: Who fashions the man in the womb? And answers: Only Mind can contemplate these things.

> This god who is evident to the eyes may be seen in the mind. He is bodiless and many-bodied; or, rather, he is all-bodied.[32]

"On Thought and Sense," the ninth Hermetic text, argues that sensory perception and thought, although different, are really the same phenomenon. In animals, sensing is "at one" with nature, but in men sensing is "at one" with thought. Sensing requires thinking just as much as thinking requires sensing. This relationship is from the Mind that conceives all thought.

> Divinity comes to be by god's agency, understanding by the agency of the mind. Understanding is the sister of reasoned speech, or each is the other's instrument. There is no utterance of reasoned speech without understanding, nor is there evidence of understanding without reasoned speech.[33]

The cosmos is itself a divine creative power. For the single "sense and thought" of the cosmos is to create all things and then bring them back into itself. As such, the cosmos is the "Organ for the Will of God" and the cycle of dissolution and renewal.

In "The Key," essay ten, God's energy is described as God's will, and through the contemplation of self-causality and self-existing (Greek "good"), the goal of man is that his soul should become like God. In the sense that every soul comes from a single soul, the All Soul, man's soul is made like God. The essence of all men and women is contained in a single abstract entity that experiences through the fragmentation of its essence into discrete parts, what we refer to as individual persons.

The physical expression of this All Soul also includes the totality of nature and its continual building of more-complex forms. Some souls, which were creatures that moved close to the ground, change into creatures that live in the water. Souls of water things change to land dwellers, those that live on Earth change to things with wings, and souls that live in air change to men. The Hermeticist is addressing biological order and the evolution of life! As for mankind, the human soul is the first step toward deathlessness.

Science as a Sacred Endeavor

If a soul enters the body of a man and persists in its vice (its ignorance), being blinded by the body's passions, it does not share in self-causality and self-existence, and dies. As a result, it returns to the path that leads to creeping things. In contrast, the man who embraces virtue achieves gnosis, which is very different from sense. Sense is brought about by the physical cosmos and has mastery over man, while gnosis is the goal of science *(episteme),* and science is God's gift in Man.

> The virtue of the soul, by contrast, is knowledge; for one who knows is good and reverent and already divine . . . One who says little and hears little. He fights with shadows, my son, who wastes time on talking and listening to talk. One neither speaks nor hears of god the father and the good. This being so—that there are senses in all things that are because they cannot exist without them—yet knowledge differs greatly from sensation—for sensation comes when the object prevails, while knowledge is the goal of learning, and learning is a gift from god. For all learning is incorporeal, using as instrument the mind itself, as mind uses body. Both enter into body, then, the mental and the material. For everything must be the product of opposition and contrariety, and it cannot be otherwise.[34]

All knowledge (science) is incorporeal because the instrument it uses is the mind, just as the mind employs the body. For our perception, all things must exist out of thesis and antithesis; and this must be, or else things would not be recognizable by mind, and material things could not exist.

> Energies are like rays from god, natural forces like rays from the cosmos, arts and learning (science) like rays from mankind.[35]

Energy acts through the cosmos. And through nature, the rays (the elements) of the cosmos act upon man. In turn, Man acts through the sciences and arts.

The System of the Universe

The dispensations of all things are derived not by the means of two, the cosmos and Man, but through the One. God contains the cosmos and the cosmos contains man. So the cosmos is God's Son—man, as it were—the Cosmos's Child, the Hermetist believes. The dispensation of the universe is through the nature of the One who pervades all things through the Mind. Without Mind, the soul can neither speak nor act. Mind does not endure a sluggish soul, however, and leaves such a soul tied to the body. A soul such as this does not have Mind, and therefore should not be called a man. Man is a divine animal and is measured not with the rest of lives of things on Earth but with the lives above in heaven, which are called gods.

If we must boldly speak the truth, states Hermes, the true "man" is even higher than the gods.

> Therefore, we must dare to say that the human on earth is a mortal god but that god in heaven is an immortal human.[36]

In the eleventh essay, "Mind unto Hermes," from God there is existence of everything Aeon (eternity), which is the timeless and spaceless realm of ideal being. Aeon creates the cosmos. Cosmos creates time, and from time there is becoming, which means to become conscious in form. The energies of God are Mind and soul. The energies of Aeon are lastingness and deathlessness. The energies of the cosmos are restoration and obliteration. The energies of time increase and decrease, and are that of becoming. Becoming is in time, which is in the cosmos, which is in Aeon, which is in God. Therefore, the source of all is God. The essence is Aeon (eternal existence), which is in matter (the cosmos). As a result, the cosmos will never be destroyed, for its source is Aeon, meaning that it is eternal and indestructible.

> Eternity [Aeon], therefore, is an image of god; the cosmos is an image of eternity; and the sun is an image of the cosmos. The human is the image of the sun.[37]

Aeon preserves the cosmos (order) by necessity, or by foreknowledge, or by nature, or by whatever else a man supposes, says the Hermeticist. All is this: *God energizing.* Life is the union of Mind and soul. Accordingly, death is not the destruction of those that are at One, but instead the dissolving of their union.

Knowing God

Because it is impossible to conceptualize nothing, think of yourself as deathless and able to know all arts, all sciences, and all ways of life. Become higher than any height and lower than any depth, then collect into yourself the senses of all creatures and all the elements. This is knowing God:

> See what power you have, what quickness! If you can do these things, can god not do them? So you must think of god in this way, as having everything—the cosmos, himself, [the] universe—like thoughts within himself. Thus, unless you make yourself equal to god, you cannot understand god; like is understood by like. Make yourself grow to immeasurable immensity, out leap all body, outstrip all time, become eternity and you will understand god. Having conceived that nothing is impossible to you, consider yourself immortal and able to understand everything, all art, all learning, and the temper of every living thing. Go higher than every height and lower than every depth. Collect in yourself all the sensations of what has been made, of fire and water, dry and wet; be everywhere at once, on land, in the sea, in heaven; be not yet born, be in the womb, be young, old, dead, beyond death. And when you have understood all these at once—times, places, things, qualities, quantities—then you can understand god.[38]

The Hermeticist asks, Is God unseen? Who is more manifest than God? For this single reason, God made all things, so that through them, all men may see God. This is the good of God, God's virtue: that God may manifest through all that is. Nothing is unseen, really, even things without a body. Mind sees itself in thinking, God in making. So it is Mind, when it exists in man, that is the concept, definition, and essence of God. For this reason, some men are gods and their humanity is near to being divine. It can be said that gods are immortal men and that men are mortal gods.

Such a statement can be made because the soul is in the body and Mind is in the soul. Reason (Logos) lies in the Mind, which lies in God, who is the father of all of these. So reason is Mind's image, and Mind is God's image, while the body is the image of the form, and form is the image of the soul.

For each composed body, there is a certain number. Without number, structure cannot exist, cannot be composed or decomposed within the life of the cosmos. From Mind, the concept of the unit gives birth to number and increases it and, through decomposition, takes number back again into Mind. The cosmos is composed of matter, which is the image of the mightier One. It is the conserver of the will and order of the Father, and is filled with life. So the lives that exist within the cosmos never die but rather are dissolved as compound bodies. Dissolution of the body is not death; it is the disintegration of its constituent parts so that it may be renewed. The Hermetic asks: For what is the activity of life? Is it not motion? What in the cosmos is there that does not move? The answer is: There is nothing that does not move.

As a whole, the cosmos is not subject to change, although its constituent parts are. Nothing in the cosmos, however, is subject to corruption, or is ever destroyed. The perspective of men is what confuses the understanding of this principle. Genesis does not constitute life. Rather, it is the sensation of genesis that constitutes life. In the same way, change does not constitute death; rather, change constitutes forgetfulness. As a result, all that exists—all that lives—is immortal: matter, life, Spirit, Mind, and soul.

Furthermore, whatever lives is endowed with immortality through Mind. Most of all, as a recipient of God, and co-essential with God, Man is immortal. Life alone is God's partner. Whether through visions, signs of the day, or by all that exists in nature, through God the future is foretold: "Wherefore does man lay claim to know things past, things present and to come."

The Concept of Rebirth

Sensations of the body, as a result of the immediate observable world, create a perception of reality that clouds the essence of life and inflicts torments on the individual, of which, according to the "Secret Sermon on the Mountain," there are twelve. The first torment is fear of the unknown. The second is grief, and then intemperance, concupiscence (strong desire), unrighteousness, avarice, error, envy, guile, anger, rashness, and then malice. But under these are many more. Only by purging one's Self of these torments (things of matter) is divinity born from within, which leads to the awakening of knowledge of the Self and God.

The author of rebirth is the Son of God—the One Man—by God's will. The method of being reborn is not something that can be taught, nor is it something that can be seen. It occurs only in Mind, by seeing one's Self within Mind. When the Self is contemplated in the mysteries of the cosmos, the All is conceivable to man. This is the rebirth: that your point of view is no longer dictated by the body. Truth, then, takes on a new meaning. What is true is that which is never troubled, which cannot be defined, which has no color, nor any figure, which is not turned, which has no garment, which gives light; that which is comprehensible unto itself alone, which does not suffer change; that which no body can contain.

Within Mind, Man becomes a knower of the Self, as well as the commonality of Man's essence.

In Search of Pharaonic Egypt

After the rise of the Roman Empire (30 BCE), ancient pharaonic traditions and the use of hieroglyphics began to wane. By the end of the fourth century CE, these traditions had been lost in cultural and political upheaval. In 395 CE, the Roman Empire was divided into East and West. The Eastern empire (Byzantine) included Egypt. Consequently, nearly all pagan (non-Christian) temples in Egypt were closed. One exception was the Nubian-run Temple of Isis (the Osiris priesthood) at Philae, in the south of Egypt, which remained a viable cult center until the sixth century CE.[39] At this temple are inscribed the last-known Egyptian hieroglyphics, dating to 394 CE.

Although there is evidence to suggest that Arab scholars deciphered Egypt's hieroglyphics as early as the eighth century, in the West these sacred Egyptian writings remained a mystery until 1822. With the Napoleonic discovery of the Rosetta stone, a comparison of Greek text and Egyptian hieroglyphics became possible, and Jean-François Champollion was able to decipher the hieroglyphics. Egyptian hieroglyphics once again could be read, but their interpretation, particularly their theological elucidation, remained elusive. At face value, and within the Western tradition of thought, the deciphered hieroglyphics resulted in a confusing menagerie of gods and goddesses and an even more confusing theology of animal worship.

Since 1822, few scholars have endeavored to analyze and establish, in a coherent manner, the belief system that the ancient Egyptians espoused. Although one may argue that Egyptian beliefs were in fact primitive, the evidence for the wealth of their civilization spanning a three-thousand-year period is overwhelming. With many temples and palaces still existing today, not to mention the enigmatic structures of the Giza Plateau and all the intriguing artifacts housed in museums around the world, how can we accept ancient Egyptian civilization as a primitive society, technically or theologically? Clearly, the Egyptians' technical skills were sufficiently advanced to build the greatest legacy of all ancient cultures. Why would their philosophical or theological concepts exist in a primitive state? Any objective analysis would suggest that they didn't. Few scholars, however, would attempt any such analysis in the face of harsh criticism by their peers.

More than a hundred years ago, the British poet and Egyptologist Thomas Gerald Massey (1828–1907) took on such a quest in a series of books and lectures. For Massey, it was a huge mistake to assume that the "myth makers" of the ancient world fashioned the gods in the likeness of Man. The gods were created as a personification of natural forces and how those natural forces relate to the human experience—what Massey refers to as zootypes. For example, three of Egypt's earliest gods were Sut, Sebek, and Shu, represented in the likeness of the hippopotamus, the crocodile, and the lion, respectively. The lion is the representation of "force" exemplified by the wind, and how the wind creates swells in the ocean water:

> This power was divinised in Shu, the God of breathing Force, whose zootype is the Lion as a fitting figure of this panting Power of the Air. The element audible in the howling wind, but dimly apprehended otherwise, was given shape and substance as the roaring Lion in this substitution of similars. The Force of the element was equated by the power of the Animal; and no human thews and sinews could compare with those of the Lion as a figure of Force. Thus the Lion speaks for itself, in the language of Ideographic Signs. And in this way the Gods and Goddesses of ancient Egypt were at first portrayed as Superhuman Powers by means of living Superhuman types.[40]

With an outlook such as Massey's, a new meaning arises for the Great Sphinx of Giza that corresponds with the ancients' name of *shesep-ank,* meaning "living image."[41] The Sphinx was not a god to be worshipped, nor did it guard the deceased king. But with the head of the pharaoh and the body of a lion, it represented the divine purpose as Man himself, as the living image of God. There is no greater force in nature than man himself. The Giza Plateau was likely not originally a necropolis but rather a source of power for the ancient civilization that created it.

The hippopotamus—the cow of the waters—represented the Earth Mother and the bringing forth of life.[42] The crocodile was one of the earliest types representing the sun as the soul of life in the water, for the sun is the true source of energy for all biological life. Only in this ideographic way does the primordial and creative goddess Neith's suckling of young crocodiles make sense. She personified the wet nurse of all creation, whose child was the young sun god. In *The Book of the Dead,* Osiris in the netherworld exclaims, "I am the crocodile in the form of a man." In other words, Osiris is the soul of mankind, just as the crocodile symbolized the soul of the sun.[43]

The ancient Egyptian language of myth was based on the Egyptians' observations of nature, what we explain today through the various disciplines of science. What we perceive as myth was their language of science, and the reason why the Greeks, as well as Renaissance scholars, were so captivated with the Egyptian Mysteries.

The Egyptian Mysteries

The word *mystery* has been associated with Egypt for thousands of years. Although it is unknown exactly when the term became so closely connected with Egypt, the third-century Neoplatonic philosopher Iamblichus (ca. 250–325) is perhaps the reason. Iamblichus studied under Porphyry and wrote an influential treatise on Egyptian beliefs and traditions entitled *Theurgia or On the Mysteries of Egypt,* which was used and revered by Renaissance scholars such as Giordano Bruno (1548–1600) and Marsilio Ficino (1433–1499) and is still in print today. Iamblichus elucidates the metaphysics of nature in topics such as First Cause, Mind, Soul, Archons, and Demons, as well as Magic and the rituals associated with Egyptian temples. Despite Iamblichus's works, exactly what the Egyptian Mysteries were has been a point of conjecture and contention for modern historians and researchers.

Needlessly so, for the Egyptian Mysteries are the same mysteries that every one of us faces at one time or another: Who am I? Where did I come from? Why is nature the way it is? What will happen to me when I die?

Over the generations, there have been many different words attached to the mysteries. Most recently, those words are *soul* and *God,* as well as *consciousness.* For the ancient Greeks and Romans, the mysteries were *demon* and *genius.* For the ancient Egyptians, those words were *ka, ba,* and *ankh.* According to the Jungian psychologist James Hillman:

> These many words and names do not tell us *what* "it" is, but they do confirm *that* it is. They also point to its mysteriousness. We cannot know what exactly we are referring to because its nature remains shadowy, revealing itself mainly in hints, intuitions, whispers, and the sudden urges and oddities that disturb your life and what we call symptoms.[44]

What has been so baffling for scholars is what the ancient Egyptians were trying to express in their temples through their art and sacred writing (hieroglyphics), which were the essence of the mysteries. Although the deciphering of hieroglyphics by Champollion allowed a literal translation, taken literally the art and hieroglyphics were not easily understood by the modern mind. Without a proper interpretation, they were viewed as primitive.

After a decade of research in Luxor (ancient Thebes) at the temple of Amun-Mut-Khonsu, however, René Schwaller de Lubicz embarked on a scholastic venture to de-shroud the mysteries of the ancient Egyptians. In his book *The Temple of Man,* Schwaller provides an exhaustive study that clarifies in a scientific manner—through measures and proportions—the temple's axes and orientations, as well as associated medical and mathematical writings, which were the brilliance of ancient Egypt's philosophical and artistic traditions. Built into the architecture of the temple itself is the definition of Man and the story of the human experience, physically as well as spiritually. Physically, the temple describes the structure of man, from the importance of the femur in the creation of blood cells to the role of the pineal gland in the brain. The pineal gland was believed to be the "third eye" and man's link to the spiritual world. Even today, there is evidence that the pineal gland and its secretion of melatonin are linked to altered states of consciousness. Spiritually, the "temple of man" conveyed man's cosmic drama of, and its quest for, immortality.

We can conclude from Schwaller's analysis that the ancient Egyptian Mysteries are the same mysteries today's science is attempting to solve. For Schwaller, however, such mysteries lie beyond the realm of the physical and must be approached esoterically, in the tradition of ancient Egypt. Here's why:

When a fetus is conceived through the penetration of an ovum by a spermatozoon, matter develops. This is a type of becoming. For us, this is how the world appears, and we refer to this process as generation. This becoming is a quantifiable and observable fact, and it is the foundation of life. To explain why this is, we must ask where the energy for generation comes from in the first place. Scientifically, all that can be said is that there is a tendency for energy to be fixed into physical form, which requires an irrational source at the origin of all things.

This fixation of energy into form is a mystery. However, by embracing this mystery, esoterism exists and all of nature testifies symbolically to the source. With the mystery's embrace there is a return of the identity of the individual, the Self, toward the source, what Schwaller refers to as the resurrection of energy that has been fixed in bodily form.

Since physical form is corporeal, however, it is a "transitory and relative aspect of an activity and re-activity of two natures sundered and issued from a cosmic Oneness." For Schwaller, understanding this veracity leads to the conclusion that man is "final matter" and "the only true King,"[45] not in a sense of tyranny over other men, but as the culmination of nature and the end result of cosmic order. By embracing this reality, theology and science become a single expression of understanding, where the relationship of man and the universe is seen not as a materialistic struggle, but as a grand unified experience—the Anthropocosm. Such a worldview precipitates greatness for any civilization that lives by this dictum. According to Schwaller, such a civilization is found in pharaonic Egypt.[46]

Egypt's natural philosophy, disseminated from one generation to the next, became the source for the great religious tradition of Western civilization, an esoteric understanding later forced into the rigid confines of dogma (exoteric thought). More than a thousand years later, Egypt's natural philosophy helped inspire the age of science. As Martin Bernal has done recently with his three-volume work entitled *Black Athena,* and as Cheikh Anta Diop did in the 1950s with *The African Origin of Civilization,* Schwaller de Lubicz effectively demonstrated the advanced philosophical works of the ancient Egyptians even earlier with *The Egyptian*

Miracle, Symbol and the Symbolic, Esoterism and Symbol, The Temple in Man, The Temple of Man, and Sacred Science: ancient Egypt was the true source of Western science, civilization, and culture.

The Anthropocosm (Man Cosmos)

In the tradition of Gerald Massey, Schwaller quite convincingly suggests that the ancient Egyptians used zootypes to represent the principles of nature (Egyptian neters), and that these principles formed a coherent thesis explaining that all of nature is, itself, the essence of Man. As for the "royal" principle on which the ancient Egyptian civilization was founded, it has little to do with the Western feudalistic concept of the king and his kingdom. Although there certainly was a person who acted in a leadership capacity, the role of the pharaonic king was not based on a specific individual. Rather, the pharaoh was a symbol, a pretext for embodying the mythical and the mystical nature of Man—what the Greeks referred to as the Logos, and what Schwaller refers to as the "Horian Logos" (the Logos of Horus).[47]

For Schwaller, the Logos is the irrationality at the origin of things. Specifically, it resides in the language-oriented, creative capacity of man identifying with their source and the fragmentation of a homogeneous state of oneness into a physical state of individual uniqueness.[48] Logos refers to intellect and reason, but in its traditional sense it is "weaving." Two threads by themselves amount to nothing, but woven together, they produce form. This weaving together to produce form is the heart of the metaphysical concept of divine creation.[49] (In modern science, energy, or "strings," according to string theory, is what's being woven.)

In myth, Seth imprisons the Horian Logos (creative word) into physical form (in Earth) through the fall of man. The active principle of the abstract becoming form is Ptah, and represents the "fire in Earth." So Horus animates the king. But Horus must be delivered from his bodily prison in the same way the mortal's soul must be saved, and in the end he becomes the divine and perfect being. Therefore, Horus represents all phases of creation, from the becoming to the resurrection and return. In this way, the universal Horus is the divine Logos in all that exists.[50]

According to Schwaller, temple inscriptions explain this as a ritual where the royal fulfillment of Horus goes through phases as Horus becomes a glorified body. In the end, he becomes the "King of divine origin, almighty in things of created Nature."[51]

This royal progression of Horus through temple ritual is reflected in the mystic (godlike) names of the pharaohs, which represent the growth of the incarnate spirit that the king symbolizes. The temple does not humanize this principle; rather, it anthropomorphizes it. In the "theogamy" (marriage of the gods) chamber of Luxor, the spiritual birth of Amon-Ra is shown as the royal infant being baptized and named though celestial forces. The terrestrial father, the god, assumes the form of Thothmes IV, and Queen Mut-m-uia (Mut in the barque), the spiritual mother, becomes pregnant by him. Then Amon announces that the future (child) king will be Amon-hotep heq-uas, the hotep of Amon, as leaven of the rising flux. For Schwaller, the announcement is "the symbol of a historic phase in the Pharaonic opus."[52] Khnemu, the divine potter from Elephantine, announces the child's conception and fashions his form to be more beautiful than that of all the neters. With the assistance of celestial principles, the child is brought into the world to be nourished with the milk of the "heavenly cow," from which all beings have life.

Although we may suspect that such imagery must be relegated to a primitive belief system in the manner of today's horoscope devotees, the symbolism of this celestial story points to a coherent and highly refined cosmology, where Man is not destined by the fate of the cosmos, which is a physical fate, to death. In the philosophy we are speaking of, the essence of the human being is eternal and it is the fate of the cosmos that is determined by the human being—that is, Man *is* the cosmos. The story of the pharaoh's birth in the temple of Amun-Mut-Khonsu is symbolic and esoteric.

The story of the "becoming of the king" is that of the philosopher-king that Plato suggested in *The Republic,* a noble means of governmental leadership. The pharaoh was not a dictator or a tyrant. Nor was he a god; he was a representation of mankind's aim, the resurrected Osiris—the true king—who has returned to the source. According to Schwaller:

> The King, per ad, the great House (of the Neter), or Pharaoh, is the living image of incarnation and of return to the source of the divine Word; it is an ever-present fact which the King's persona makes tangible to the people. Such is the only true and great meaning of royalty as an idea.[53]

The Sed Festival and the Resurrection of Osiris

In one of the oldest festivals celebrated by the ancient Egyptians and continuing throughout Egypt's pharaonic period, the Sed festival is believed to have celebrated the king's ascension to the status of a god.[54] This festival, also referred to as a royal jubilee, is considered by Egyptologists to be the ritual reenactment of Menes' unification of Upper and Lower Egypt at the beginning of the third millennium BCE. In honor of the king, after he had reigned for thirty years he earned the right to celebrate this unique festival. For the remainder of his life, every three years the Sed festival would be celebrated.

Typically held after the annual inundation of the Nile, the Sed festival began with a sacrifice, and then the king was crowned with the white and red crowns of Upper and Lower Egypt. After running a ritual course four times (possibly alongside the Apis bull), the king was carried in procession to the chapels of Horus and Seth for his coronation.[55]

At Semneh and Abydos, Thothmes I recorded his Sed festivals. Except for a figure of the king, little remains at Abydos. At Semneh, however, Thothmes is depicted wearing the white crown and enthroned in a shrine before a number of standards—the foremost being Upuaut, Leader of the South and the Two Lands. In another scene, attended by the Anmutef priest, Thothmes also appears behind the shrine as Osiris, standing and wearing the red crown.[56]

The Sed festival of Amenhotep I was inscribed upon a slab in Karnak; it shows the enthroned king dressed as Osiris and bearing his emblems.[57] In the tomb of Kheru-ef at Assassif is a scene of Amenhotep during the morning of the festival. With his queen and daughters, he is setting up the *djed* pillar, a column with four disks in its middle. According to Wallis Budge, it represents the backbone or the body of Osiris, while the sacred cattle go around the walls four times.[58]

In the Sed festival of Osorkon III (883–855 BCE), as Osiris, Osorkon wears the white crown and stands with a stream of water pouring from his hands. Osiris the king personifies the Nile, symbolized by the crown of Upper Egypt whence the Nile flows—reminiscent of hymn lyrics from the time of Rameses IX: "The Nile comes forth from the sweat of thy hands."[59]

Another well-preserved depiction of Osorkon's Sed festival provides a detailed account of the ceremony. The king is dressed as Osiris, holding the crook and scourge (emblems of the god of the dead), and is carried in procession through the temple. Depending on the scene, Osorkon wears either the white crown or the red crown. In another scene his queen and princesses accompany him. As the center of attention, the king appears to be paid homage as a god. The figure of Upuaut of the South, carried by six priests, is immediately in front of Osorkon, the living Osiris.[60]

The oldest recorded Sed festival is from Pepi I, the second ruler of the Sixth Dynasty (2332–2283 BCE). Cut into the rocks at Hammamat, Pepi is shown in a double shrine. On one side, he is wearing the white crown, on the other side, the red crown. Below the scene is an inscription that states, "The first time of the Sed-festival."[61]

Exactly what the Sed festival signifies is debatable. According to Kent Weeks, professor of Egyptology at the American University in Cairo, "The Sed festival was a magical, religious, re-affirmation of the strength, and power, and prowess of the king."[62] What is certain is that the festival's title, "Sed," was in honor of the god (neter) Sed, who was represented by the form of the jackal, as well as Wepwawet (Upuaut), meaning Opener of the Ways.

Wepwawet was also Lord of the Sacred Land and one of seven gods who guarded the way to Osiris's throne.[63] Interestingly, Opener of the Ways is also believed to be the name of Thoth, who makes the joys of paradise available to those who obey the laws of Maat.[64] Wepwawet may have been the original (and archaic) title of the jackal god, who was later renamed Sed.

According to *The Pyramid Texts* and *The Coffin Texts,* by opening "a good path" through the hazardous territory of the afterlife, Wepwawet aided the spirit of the deceased to ascend into heaven. At Abydos, Wepwawet was identified with Osiris's son Horus, who aided the king in becoming the resurrected king of the underworld. As such, Wepwawet served as Osiris's spiritual guide.

The raising of the djed column in the Sed festival is one of the more common Egyptian symbols representing the principle of unchangeability; it was an integral part of the king's celebration. The djed column, in myth, represented the backbone of the murdered Osiris and, combined with the ankh, was also a feature of Ptah's (the patron god of Memphis) scepter.[65] The raising of the djed column symbolized (in ritual) the resurrection of Osiris. It was also an important feature in funerary rituals in general. Frescoes in temples and royal tombs often depict Horus raising the djed column as the act of raising Osiris from the dead. The djed column was often painted on the bottom of coffins, as well as placed on the mummy in the form of a golden amulet.[66]

Although it may seem that the ancient Egyptians purposely obfuscated their beliefs and rituals, such a notion is purely a modern point of view, a perspective far removed from dynastic Egypt. Their belief system was symbolic and complex, and its application involved a vast number of concepts and principles. Not knowing the precise meaning of those principles and how they were used in any given context is clearly a handicap in understanding the festival's meaning. Nonetheless, it is apparent that the Sed festival was a public rite of passage for the king, which has been traditionally interpreted as becoming a god. But what did becoming a god mean to the ancient Egyptians?

According to the Greek historian Herodotus (484–ca. 425 BCE), the ancient Egyptians had a supreme deity who manifested under various forms or names, but whose name was it unlawful to mention. Perhaps a better understanding of this unlawful name is that what the Egyptians were defining was unknowable and therefore could not have a name, a reality even today. God has no name except for that which man has applied. The ancient Egyptians appeared to realize this, for the pyramid texts state: "Unas says, 'O great god, whose name is unknown.'" Likewise, we find inscribed on the Re-ma stela: "His name is not known."[67]

One way that this unknown and unnamed god could have a name is in God's descriptive function as it relates to mankind. Such a condition is found in the mythology of Osiris. Murdered and dismembered by his brother, Seth, Osiris was resurrected into eternal life (symbolizing the spiritual manifesting as form) and became the lord of the underworld to judge the souls of the dead. In the Ennead, he was one of four principles brought into being by Earth and sky (Geb and Nut), and who became the husband of Isis. In essence, as the god of the sun, the god of the moon, and the god of vegetation,[68] and as lord of the underworld, Osiris represented life par excellence. As such, he symbolized the unknowable God who became the "living god" as a result of man's achievement, spiritually as well as physically. So anyone who had achieved a proper understanding of the "mysteries" (that existence is eternal and human being is the cosmos) in the Egyptian tradition was called Osiris.

This designation is also the epitome of the Sed festival, which Schwaller's refers to as a rite of "qualitative exaltation,"[69] in which "exaltation" refers to the characterization and quality of Man as the foremost principle of God manifest. As the leader of Egyptian civilization, with great achievements for the good of the people, the pharaoh personified and deserved such a designation. The king was the archetypal Osiris: not a god, but Man among all men.

Consciousness and Egyptian Philosophy

After an exhaustive, decade-long study of ancient Egyptian art and architecture at Luxor, Schwaller concluded that ancient Egyptians considered theogony, medicine, and astronomy as a single scientific discipline. Science was holistic and embodied the true creative power entrusted to humankind that resulted in the success and longevity of Egyptian civilization. Science was sacred; its aim was the perfection of existence.

Egyptian kings governed the land of the Nile in a pharaonic theocracy. But their form of government should not be confused with the less tolerant, tyrannical theocracies of later history. Through symbols and esoterism—the object of sacred science—a bond is maintained that ties the esoteric meaning of the myth (the basis of religion) to science in general. Philosophically, it also ties the king to the head of social organization. Benevolence and longevity were the results. For Schwaller, "This is the true meaning of a theocratic order, not to be confused with a royalty directed by a religious organism."[70]

Despite the recognition that ancient Egypt receives today for its monuments and temples, its brilliance in science and philosophy, as well as its understanding of the human experience, is often overlooked. In today's language of science, the subject of becoming, resurrection, and return is best described by the role of consciousness, its cause and effect. I see little difference except for the language used. Schwaller recognized the Egyptians' depth of understanding of the universe and was at pains to explain that they recognized and embraced the mysteries of life in a profound way.

Between the original cause and the effect, which in a sense is the final cause, is the immediate world we experience. The celestial bodies that populate the universe are the symbols of the evolution of consciousness and are "of thought" (in the sense of "I think, therefore I am" one can say the archetypal human thought and therefore the cosmos exists) for the archetypal Man through all experiences. The actualized knowledge of one's Self is still dormant in the accident of the "me." The life of the symbol (its esoterism) identifying with this life (its exoterism) defines reality. This is because the identification of the symbol with biologic life enables cosmic and historical events to persist in us from this moment forward. The life of the symbol is the experience of our consciousness, which is the consciousness of the individual (the human microcosm), summarized by the cosmic Man, of which we all are a part.

Potentially, or in fact, Schwaller writes, "[t]here is nothing in the world which is not in him [Man]."[71] Birth and death manifest the break in the eternal ring, which is the material form of man. The circle always exists, so only the natural aspect is affected by the break. This absolute circle is fixed and stable and is the essence of life that is the becoming and the return. The fragmentation of consciousness from the abstract into the concrete (birth), and then the return (death), gives us a lesser quantity of consciousness than the macrocosmic man. For pure quality, however, there is no size. In other words, size or quantity does not exist. An animal is not a microcosm; rather, it is an aspect of the microcosm and an intermediate state. Man's wholeness is a virtual wholeness; it becomes actual wholeness when consciousness is freed from mortal boundaries. This wholeness is consciousness in itself, with no further reflection, the circle without break: the entire cosmic phenomenon.[72]

As does all of nature, the human mind exists in a state of duality comprising innate and mental consciousness. That which we call the mind, mental consciousness, perceives only the immediate world, for that is its purpose. Innate consciousness, what we know as truth without instruction, is the fingerprint of absolute quality and our birthright as an abstract (spiritual) creation: how people perceive God as the essence of creation or the creator. Only through the resonance of innate consciousness with mental consciousness do we rise above the level of an animal. This rise is communicated from the innate to the mental through esoterism and the symbolic. Built into all nature is symbol, and that symbol evokes the innate consciousness in man, thus reminding him of his true nature. *Man is a totality,* and for that reason he himself is the cosmos. *Man is the cosmos.*[73]

Schwaller argues that philosophers can carry out systems of logic and analysis as far as they like, but to overcome the obstacle before them (cause), they will always be forced to resort to faith, or negation expressed through atheism, or rationally justified indifference. Any explanation of life can never be more than the circling of a central point that is logically undefinable. Such logic halts when reason acquires the certainty that this point exists.

In the Pythagorean science of numbers, the point is the One, representing the absolute—the concept of God—which is the mother/father of all that is. In truth, God is unknowable and undefined. It is only through the creation of Two that consciousness comes into being. Being made in God's image, Two, or Man, is that consciousness. But it is through Three, which is simultaneously created with Two, that we can relate to God. Through the creation of Two by One, the Self becomes the "I," the ego, which is the mathematical definition of the value One, as something in relation to itself. Next, with the original "I," the becoming of consciousness blends into the Self after all experience has been acquired. This is why nearly all concepts of God, such as the Christian Trinity, are described in the plural. Despite its irrationality (what does the Father, the Son, and the Holy Ghost really mean?), the Trinity is an apt way to express the creative principle of unity without explaining the complex relationships of number.

Sacred Science

In the tradition of the great physicists of the twentieth century, today's "new science" has embarked on a quest to redefine the human experience in ways reminiscent of ancient concepts: that there is no reality in the absence of observation. In the introductory episode of *Cosmos: A Personal Voyage* (1980), Carl Sagan asserts confidently, "[T]he Cosmos is all there is, all there ever was, and all there ever will be." I agree with that, but what is the universe, really?

Cosmologists often refer to an idea called the anthropic principle: To explain the existence of the universe, we must also explain our own existence. This line of reasoning falls into a circular motion of cause and effect and back to cause. Is the human experience the effect or the cause? Or is there really such a thing as cause and effect outside the concepts of science?

If the universe is all there is, what meaning would it have if it were not aware? Would the universe still exist if we were not present to perceive it? Questions such as these, spurred by some of the latest research into quantum physics, echo concepts from texts whose origins may be as old as mankind itself:

> The Mind, O Tat, is of the very Essence of God, if yet there be any Essence of God. . . . Wherefore we must be bold to say, that an Earthly Man is a Mortal God, and that the Heavenly God is an Immortal Man.
>
> HERMES TRISMEGISTUS

It is the role of science to provide explanations within the limitations of systems of measurement and instrumentation. Conversely, it is the role of sacred science to explain without limitation, and to embrace the intuitive function with which we humans are endowed. Sacred science is *not* humanism. Rather, it is the acceptance of cosmic mysticism—the fundamental principles of nature—which material science addresses only through assumption.

From a materialist perspective, science is no more than a cataloging of cause-and-effect relationships. But the true relationship between cause and effect (why we perceive, the greatest scientific assumption of all) remains a mystery. Perception, which is the effect, is abstract. Likewise, cause is also abstract. What lies in between, physical reality, serves as a means of relating cause to effect. Therefore, there is nothing in the universe that is not in man. At birth, man's material form manifests the break in the eternal ring of existence. This break is the essence of life. Quality's quantification into form, from the abstract to the concrete, produces man from what is eternal, the archetypal Man—the observer, the "Self" with no further reflection. The circle without break is the entire cosmic phenomenon. It is simultaneously cause and effect.

Evolution

The concept of evolution has been a source of confusion ever since Charles Darwin published *Origin of Species*. It need not be, for it does not address causality. The theory of evolution is based on mental consciousness through the specter of time, an exoteric point of view. To the Darwinian evolutionist, life is an accident created through the random cause of gene mutation, perhaps the deposition of bacteria from an errant comet. However important the ideas of Darwin, Lamarck, and Haeckel are, they are flawed, preventing the possibility of a correct understanding.[74]

A profound biological mystery remains unaccounted for in classical science. Over the course of a lifetime, every cell in the human body is renewed upward of ten thousand times. But despite the never-ending swapping out of physical parts, minuscule though they may be, our basic behaviors, memories, and sense of self maintain continuity.[75] There is a large missing piece in the theory of evolution. What holds the pattern of our identity?

In his 1933 book *Science and the Modern World,* British philosopher and mathematician Alfred Whitehead argues that the concept of evolution, which is founded on classical principles, is not consistent with materialistic assumptions:

> The aboriginal stuff, or material, from which a materialistic philosophy starts, is incapable of evolution. This material is in itself the ultimate substance. Evolution, on the materialistic theory, is reduced to the role of being another word for the description of the changes of the external relations between portions of matter. There is nothing to evolve, because one set of external relations is as good as any other set of external relations. There can be merely only change, purposeless and unprogressive. . . . The doctrine thus cries aloud for a conception of organism as fundamental to nature.[76]

The French scientist Lecomte du Noüy, director of biophysics at the Pasteur Institute during the 1930s, agrees:

> Evolution begins with amorphous living matter or beings such as coenocytes, still without cell structure, and ends in thinking man, endowed with a conscious [*sic*]. It is concerned only with the principle [*sic*] line thus defined. It represents only those living beings, which constitute this unique line zigzagging intelligently through the colossal number of living forms.
>
> Evolution, we repeat, is comprehensible only if we admit that it is dominated by a finality, a precise and distant goal. If we do not accept the reality of this orienting pole, not only are we forced to recognize that evolution is rigorously incompatible with our laws of matter, as we have demonstrated above, but—and this is serious—that the appearance of moral and spiritual ideas remains an absolute mystery.[77]

According to Noüy, what takes place to create the biological order of life and its hierarchical complexity is counter to the known laws of physics, particularly the second law of thermodynamics, which states that the disorder of a system (entropy) not at equilibrium will tend to increase over time. Energy spontaneously tends to flow from a state of concentration to a state of diffusion.

Noüy argues that there must be some unknown biological law to explain increasing biological order. Schwaller argues, as a growing number of scientists today also contend, that biological order and gene mutation are not random accidents but a result of consciousness, a fourth fundamental factor of reality, functioning within the environment. It adapts, grows, and brings about form from its surroundings. Genes are a blueprint and can be altered at will to meet the changing world around us. Mutations are therefore the result of an organism's perception, the rewriting of existing genes in an effort to overcome adverse conditions. The cause is consciousness, with the organism changing itself to accommodate a preexisting idea or belief.

Although such a concept is difficult to prove, since observation over long periods is not possible, other evidence suggests that this is the case. For example, when a human fetus reaches its third week of life, a heart begins to beat, *but no heart exists.* A single cell beats, and then recruits other cells around it to do the same. A few days later, blood of its own type begins to flow through the fetus's separate and closed circulatory system. Stem cells—cells that have not yet been designated as a certain type—have been an interest in the medical research community for some time. How these stem cells become a specific tissue or organ remains a medical mystery.

Exoterically, man is appointed once to live and then to die. But esoterically, as a conscious being, Man *is* and *will always be.* No one remembers his or her birth, nor is anyone likely to remember his or her death. Exoterically, the forward arrow of time ends for each man. Esoterically there is no arrow of time, for time is nothing more than a concept. For nature, the time is always now.

Now as Eternity

The classical and generally accepted concept of the universe's creation is that it began with the big bang. The difficulty with this model is that from a physicist's perspective, it is a disaster, because at the moment of the big bang, the laws of physics are irrelevant and inapplicable. More disturbing is that there is no "why" explanation for the big bang. Nor is there a what: What "banged"?

In his book *The End of Time,* physicist Julian Barbour marries the work of Einstein's general relativity and Schrödinger's wave mechanics into a unified view of the universe, what he refers to as the *quantization* of general relativity. According to Barbour, there is a strong argument that the universe is static and only the here and now exists. Nor is there a grand container encapsulating all matter. All that exists does so in an infinite nothingness.

According to Barbour's theory, the universe is the interaction of three wave functions. These three energies are the foundation of all that exists and even, he writes, "have a way of finding special structures: for example, they can create complex molecules like proteins and DNA."[78] Of course, wave functions do not physically exist; they are quantum principles that establish a ranking so that real things can exist according to our perception. In other words, the interaction of wave functions creates the universe, which contains particles, with and without mass, as well as a specific configuration of the universe at every moment, which we perceive as time.

Energy appears as mass when three or more waves intersect at a common point. When the wave amplitudes vary in an appropriate fashion and are in phase, a point emerges and a "localized blob" forms.

During the years 1925 and 1926, Erwin Schrödinger himself realized that such a wave event begins to look like a particle. This is important because it is what the physicist observes in nature: different waves of various wavelengths moving at different speeds.[79] In aggregate, they are our reality, and they combine to form everything physical. According to Barbour, the Wheeler-DeWitt equation, derived in 1967 by physicists John Wheeler and Bryce DeWitt, by its nature is likely to be the fundamental equation of the universe. It tells us "that the universe in its entirety is like some huge molecule in a stationary state, and that the different possible configurations of this 'monster molecule' are the instants of time."[80] As a result, quantum cosmology becomes a logical deduction of atomic structure theory, while also including the element of time. So time is nothing more than changes in the universe's configuration. Time does not exist except as a concept created by us to measure change. In classical physics, we could say that it is a complex of rules that govern the change.

Interestingly, this quantum point of view is in agreement with the concept of constant creation. Creation continuously occurs and did not occur as a result of the big bang. The time is always now.[81]

Barbour explains that each "now" has its own quantum intensity.

Using the analogy of a lottery, all the "nows" participate in this lottery and each "now" receives a number of tickets proportional to its intensity. Where there is an intensity of tickets, there is also a high probability for a "now" to be consciously experienced, with the following assumptions:

> All experience we have in some instant derives from the structure in one "now."

For "nows" capable of self-awareness (by containing brains and so on), the probability of being experienced is proportional to their mist (quantum wave) intensity.

> The "nows" at which the mist has a high intensity are time capsules (they will also possess other specific properties).[82]

As the "now" changes with respect to a direction, the appearance of dynamics arises from a static situation. The consequence is that two static wave patterns can, under appropriate conditions, be interpreted as time as we know it. As a result, the appearance of time and evolution arises from timelessness.[83] When we observe motion, the underlying reality is that our brains instantaneously process data corresponding to several different positions of the object that is perceived to be moving. In this way, the brain presents the observed data to consciousness and somehow plays them as a movie.[84] Despite this epiphany, Barbour contends that a more pertinent question remains: What is the role of conscious life, particularly man? Is a specially structured "now" capable of self-awareness?

Here, science must turn to the sacred for its explanation. Barbour, whom I view as a participant in the new science, a sacred science, believes that our conscious moments are embedded in the landscape of the abstract:

> Everything we experience is brought into existence by being what it is. Our very nature determines whether we shall or shall not be. . . . We are because of what we are. Our existence is determined by the way we relate to (or resonate with) every thing else that can be.[85]
>
> Things do not become, they are . . .
>
> I am nothing and yet everything . . . because there is no personal canvas on which I am painted. I am everything because I am the uni-

> verse seen from the point, unforeseeable because it is unique, that is me now.[86]

There is a coincidence in this quantum model of reality. Three numbers at each point of space describe reality's geometry. Again, we encounter the triplet (the trinity) as the fundamental concept of nature.

Ancient Wisdom and the Modern World

Today's leading edge of scientific knowledge, particularly quantum physics, rests at the brink of the perception mystery. From a physics point of view, the role of the observers that we are (as Barbour alludes to) appears as anthropocosmic as Schwaller's interpretation of pharaonic Egypt and as do the views of theoretical physicist Michio Kaku, cofounder of string field theory:

> If you had a super microscope you would probably see that an electron is not a dot but it is a vibrating rubber band, and so physics is nothing but the laws of harmony of these little vibrating rubber bands or strings. Chemistry is nothing but the melody you can play on these strings. The universe is a symphony of these vibrating strings, and the Mind of God that Einstein eloquently wrote about. The Mind of God, we think, is cosmic music resonating through eleven-dimensional hyperspace.[87]

Such a cosmogony is the deepest and purest meaning of the anthropic principle. Philosophically, it is clear that the new science embraces the same principles as ancient Egypt's natural philosophy from nearly five thousand years ago.

We arrived at this view of reality through scientific investigation, but how did the Egyptians arrive at their understanding of nature? Was it a good guess? Was it divine inspiration? Was it through scientific analysis? Or were they the inheritors of such a philosophy from a previous technology-based civilization? The latter two possibilities run counter to the grain of current thinking, yet they are the most logical candidates from among the four.

Although there is a tendency to regard religion and mysticism as the principal factors that shaped Egyptian civilization—the belief that the Great Pyramid was a "resurrection machine," for example—

such a notion is not compatible with any civilization, past or present. The principles of civilization thousands of years ago were the same as they are today. Each person specializes in his or her own trade, each company in its own industry. A workforce of thousands (or millions) creates the infrastructure for the good of all. Public projects propel civilization to new heights as the organized division of labor and specialization of function allow the best and brightest minds to engage in scientific research and development. Durable goods, monuments, and public buildings are the natural by-products of a technical society, as are philosophy and science.

12

Whispers of a Forgotten Technology

The Testimony of Atlantis

Could a pre-Pharaonic monument of great sophistication and a civilization whose antiquity makes our head spin possibly find room within current historical thinking?

R. A. Schwaller de Lubicz

We know very little about our own history. What we do know is that geneticists have been able to trace our modern human roots to a period between one and two hundred thousand years ago and to a place somewhere in the southern regions of Africa. We don't know how we became anatomically modern, or why civilization burst onto the world stage five thousand years ago. What were we doing for all that time before civilization?

Aside from hunting and fishing, the traditional answer is: relatively nothing, except for exploring and moving into all corners of the world. That in itself is a puzzle. How could primitive humans, as we are led to believe we were, manage to explore, navigate, and migrate into every corner of the world prior to the birth of civilization—was it aimless wandering?—and then build marvels of construction, such as Egypt's Great Sphinx and Great Pyramid, or the Baalbek Temple in Lebanon? Some things in history simply don't add up the way we want them to. They are anomalies, and we should attempt to explain them instead of explaining them away.

Although not as well known as the Giza pyramids or the Sphinx, the Baalbek Temple may be the greater mystery with its three oversized megalithic stones. The three stones, referred to as the trilithon, weigh somewhere between eight hundred and twelve hundred tons apiece and are placed together with precision. The largest block is sixty-five feet in length; fourteen feet, six inches, in height; and twelve feet thick. The two others are the same height and thickness but are slightly shorter. One block is sixty-four feet, ten inches in length, and the other sixty-three feet, two inches. There is a fourth quarried block, called Hajar-el-Hibla (stone of the pregnant woman), which was never moved from the quarry.

A wall characteristic of prehistoric cyclopean architecture joins the trilithon and forms the northern part of the terrace's wall. As Michel M. Alouf describes it in his *History of Baalbek,* "It is composed of nine stones, each measuring from 30 to 33 feet in length, 14 in height, and 10 in width. They form part of the same course and are on the same level as the six stones of the western side."[1]

Some scholars insist that the Romans were responsible for the massive trilithon and adjoining wall. Certainly, the Romans were expert builders and were responsible for the superstructure that makes up most of the visible remains at Baalbek. The size and weight of the three blocks that make up the trilithon, however, pose serious problems for a Roman origin. First, the Romans typically used smaller, more manageable blocks in their construction projects. In Roman construction, the use of such enormous stones is unprecedented. Second, moving the stones from the quarry to the site required an enormous amount of effort, even if the Roman winching system was used. Third, the trilithon and its joining wall are not really part of the Roman construction; they rest outside of that construction. So why move three stones that weigh in excess of eight hundred tons simply to create a facade? Fourth, the outer "podium" wall was left unfinished, which leads back to why the final twelve-hundred-ton block was left in the quarry. For whatever reason, the builders of the original cyclopean-style structure abandoned the project.

In addition to all this, according to Roman architectural standards during the first century BCE, the temple should have been placed at one end of a courtyard and been surrounded by it on all sides, which is not the case at Baalbek. The courtyard ends at the temple facade.

Taking the Ancient Egyptians at Their Word

Several years ago, Charlton Heston narrated a television documentary entitled *The Mystery of the Sphinx,* based on the research of John Anthony West and geologist Robert Schoch, on how the Sphinx was much older than mainstream estimates. For me, the compelling evidence was the undulating patterns of erosion where rainfall had melted away soft spots in the walls of the Sphinx's enclosure. According to the documentary, such features were not at all similar to patterns left by wind erosion. For such erosion to occur, the Sphinx must have been carved long before North Africa turned into a desert, when rainfall was still abundant. According to Schoch, the Sphinx was carved more than seven thousand years ago.

More than fifty years ago, René Schwaller de Lubicz noticed that the weathering of the Sphinx looked more like water erosion than wind erosion. "Except for the head, [the Sphinx] shows indisputable signs of aquatic erosion," he wrote.[2] As with all rivers, the formation of a delta occurs as soil is scoured from the land through the erosive forces of water and deposited at the river's mouth. According to geologists, the Nile emerged as a river system 130,000 years ago and deposited the first sediments of the delta 22,750 years ago; before this time, there was a maritime gulf where the delta now lies.[3] According to Schwaller, a great civilization must have existed prior to the delta's creation, since there is no other way to reconcile the Sphinx's existence and condition with geologic fact.[4]

The renowned twentieth-century psychic Edgar Cayce, during one of his psychic readings in 1933, claimed that somewhere beneath the Sphinx was a chamber that housed a hall of records from a prehistoric civilization commonly referred to as Atlantis. Although such readings (or channeling) are entirely subjective and are far removed from the works of science, in his 1991 seismic survey of the Sphinx's enclosure, Robert Schoch found such a chamber as Cayce described:

> I am not in the business to support Edgar Cayce or any other—I'll call them—psychic or prophet. I really knew nothing about him other than his name. We were able to model what was underneath the Great Sphinx. We found underneath, in front of the left paw of the Sphinx what I believe is a major chamber maybe up to twenty-five meters below the surface. Based on the regularity, it looks like it was human carved, and not only is it definitely there, but it seems to have

> something in it—the way it resonated, the way it rang—seems to indicate there's something in the chamber.[5]

We don't know when the Sphinx was built or by whom, despite the insistence of orthodox Egyptology that it was a Fourth Dynasty creation. Ancient texts concerning the Sphinx are rare. Only Pliny the Elder (23–79 CE) devotes a few lines to it after describing the pyramids:

> In front of them is the Sphinx, which deserves to be described even more, and yet the Egyptians have passed it over in silence. The inhabitants of the region regard it as a deity.[6]

Although Schoch's dating of the Sphinx remains controversial among professional academics, at least two other geologists who have also been to Giza agree with his conclusion, in part, that water was the source of erosion. Yet the conclusions of modern scholars who disagree with Schoch's analysis of when the Sphinx was carved remain hypothetical at best.

All the debate, interpretation, and various viewpoints of modern scholars seem to obfuscate the issue. Why not put a little confidence in what the ancient Egyptians themselves claim? The ancient Greek historian Herodotus did, and wrote that the Egyptians, "by their practice of keeping records of the past, have made themselves much the best historians of any nation of which I have had experience."[7] The ancient Egyptians were dedicated record keepers, and there is no mention of who built the Sphinx in their archives. What they do say about their history is that it reaches far back into remote times. According to the Papyrus of Turin, which is a complete list of kings up to the New Kingdom, before Menes (3000 BCE):

> *. . . venerables Shemsu-Hor, [reigned] 13,420 years*
> *Reigns up to Shemsu-Hor, 23,200 years*[8]

These last two lines in the king's list are explicit. So Egyptian history, from their perspective and according to their own documents, goes back 36,620 years.

The late-twentieth-century Egyptologist Walter Emery seems to have agreed in principle. He believed that ancient Egypt's written language was beyond the use of pictorial symbols, even during the earliest dynasties. According to his research, signs also were used to represent sounds only, along with a numerical system. At the same time that hieroglyphics had been stylized and used in architecture, a cursive script was already in common use. His conclusion:

> All this shows that the written language must have had a considerable period of development behind it, of which no trace has as yet been found in Egypt.[9]

Why should we not take the ancient Egyptians at their word? They are accused of being unable to describe history correctly, believing that gods, whose purpose was unknowable, controlled the existence of their civilization.[10] But with the understanding that the "gods" were a product of sophisticated and deliberate thought about how to describe the principles of nature, we have every reason to take them at their word. They themselves testify in their records to a historical context for an antediluvian Sphinx. But is there any evidence that this civilization held any type of advanced knowledge?

The astrophysicist Thomas Brophy thinks so. According to Brophy, megalithic arrangements at a place called Nabta Playa, in Egypt's southern desert, attest to sophisticated knowledge of astronomy that we, as a modern technological society, have only recently realized.

Discovered by archaeologists Fred Wendorf and Romuald Schild in 1973, the Nabta Playa megaliths are an arrangement of upright stones spanning twenty-five hundred meters. Like the spokes of a wheel, the stones radiate outward to the north and south from a bizarre sculpted rock vaguely resembling a cow. Six groups of stones, extending across the ancient basin, contain a total of twenty-four megaliths. The northern end of the arrangement of stones terminates at a small stone circle with two sets of stones in its center, three stones in each set. They represent the stars of Orion's belt and the stars of his head and shoulders. Wendorf and Schild, as well as Brophy, have identified this circle as a calendar.

According to Brophy's analysis, the calendar circle is a user-friendly star map of the constellation Orion, applicable between 6400 and 4900 BCE—not really all that extraordinary. But this is with reference to the northern arrangement of three stones within the calendar circle. The southern three stones within the calendar circle represent Orion's head and shoulders as they appeared on the meridian on a summer solstice sunset around 16,500 BCE. According to Brophy, the date of 16,500 BCE is symmetrically opposite the 5000 BCE depiction of the Orion's belt stars. In terms of the precession of the equinoxes, the dates are at the maximum and minimum tilt angle of the Orion constellation. For Brophy, this arrangement meant that

> the stone diagram illustrates the time, location, and tilting behavior of the constellation of Orion through the 25,900-year equinox precession cycle, and how to understand the pattern visually.[11]

That is extraordinary, but there's more. On further analysis, Brophy discovered that the northern group of megaliths represents the distance to Earth from Orion's belt stars, Ainilam, Ainitak, and Mintaka. (One meter at Nabta Playa is equivalent to 0.799 light-year.) Even more fantastic, the southern group of megaliths represents the stars' radial velocity, the speed at which they are moving away from Earth. How could this be, unless those who constructed the megalithic arrangement were thinking like astrophysicists?

It was initially believed that the center stone, which loosely resembles a cow (referred to as Complex Structure A), was a burial marker. The area beneath it, however, was excavated to no avail. The only artifact found was a strange maplike sculpture carved into the bedrock.

Brophy examined Marek Puszkarski's sketch of the sculpture, as well as Schild and Krolik's sketch. Over this sketch, he superimposed the location of the sun, and he found that the galactic center correlated to the direction of the spring equinox's heliacal rising of the galactic center in 17,700 BCE. Again, it was a match:

> Astonishing as it may be, the bedrock sculpture underneath "Complex Structure A" at Nabta Playa appears to be an accurate depiction of our Milky Way Galaxy, as it was oriented astronomically at a specific time: vernal equinox heliacal rising of the Galactic Center in 17,700 BC.[12]

Skeptical? Sure. But what are the chances of the megaliths being perfectly and randomly aligned in the way Brophy describes by a people without astronomical knowledge? According to Brophy, the chances of Orion's stars being aligned with the megaliths are less than two in a million, which is far beyond the requirement for accepting a scientific hypothesis as valid. Now, we have not only a historical context for a Paleolithic civilization but also *evidence* that the civilization developed advanced astronomical knowledge.

Without a historical context, the megaliths at Nabta Playa remain a curiosity, as do the megalithic stones that were quarried and placed as the foundation of the Baalbek Temple in modern-day Lebanon. The three stones that serve as the foundation each weigh an extraordinary sixteen hundred tons. As noted, a fourth stone of the same dimensions was left in its quarry, a few hundred yards away.

Sophisticated knowledge and methods of construction require the existence of a technically advanced society, which requires technical evidence that such a society existed. This is an important point, since skeptics require extraordinary evidence for extraordinary claims. It is my belief not only that the required extraordinary evidence exists but also that it is available for inspection by anyone who travels to Egypt and takes a tour inside the Great Pyramid of Khufu. The internal passageways and chambers of this pyramid are unlike any other interior design of any other structure ever erected.

Mystery of the Great Pyramid

Over the years, numerous books, ranging from the purely academic to the thoroughly mystical, have been written analyzing Egypt's pyramids, and in particular the Great Pyramid. All researchers agree that they "are astonishing in size and pleasing in their simple, perfectly harmonious form."[13] Zahi Hawass, secretary general of Egypt's Supreme Council of Antiquities, observes:

> Pyramids have magic and mystery. Their magic touches your heart, and when it touches your heart, you think about how they were built. Who built them?[14]

According to Miroslav Verner, director of the Egyptology Institute at Charles University in Prague, the pyramids "still challenge us to explain why and how they were built. And in many respects they remain a great secret of the past."[15] Even during ancient times, they were considered to be one of the Seven Wonders of the World. They have retained their reputation and are perhaps the most-visited structures in the world. So who built them, and why?

The standard theory, taught as a matter of fact and disseminated through educational institutions, is that Egypt's pyramids were tombs for the pharaohs. According to this theory, the earliest kings of Egypt were buried beneath a *mastaba,* a rectangular structure made of mud brick. The tomb consisted of a large open pit dug deep into the ground and partitioned into rooms, the center room being the burial chamber. Over the pit, a roof was built using timbers as the supporting structure. At the pit's edge, thick mud-brick retaining walls were built extending above the ground, creating a hollow space above the roof and below the top edge of the retaining wall, which was filled in with rubble, gravel, mud brick, or a combination thereof, creating a low, benchlike building.[16] In essence, the mastaba was a large, rectangular headstone.

During the late nineteenth century, the British archaeologist Sir William Flinders Petrie (1853–1942) excavated eleven royal tombs that comprised the tomb of the protodynastic king Narmer (ca. 3100 BCE), eight First Dynasty tombs, and two Second Dynasty tombs. During forty years at Giza, George Reisner (1867–1942) outlined the development of royal burials through the Fourth Dynasty. The Egyptian nobility and their families were systematically buried underneath mastabas around the pyramid.[17]

During the Third Dynasty (ca. 2800 BCE), however, Pharaoh Djoser decided that the standard mastaba wasn't enough. Thus, under the guidance of the architect Imhotep, Djoser built the first pyramid at Sakkara, known as the Step-Pyramid. It started out as a mastaba but through three phases of construction was transformed into a pyramid. According to the theory, in subsequent dynasties, each pharaoh had to outdo his predecessor, which resulted in more pyramids being built at Sakkara and the three massive ones on the Giza Plateau. As a result, Giza became a necropolis, and

> the site of the dead pharaoh's mystical transfiguration, re-birth, and ascent into heaven, as well as his residence in the beyond, from which he ruled over all the people of his time.[18]

Exactly how a pyramid worked as a vehicle for transfiguration, and how the king ruled his people in the afterlife, has never been fully explained.

Proof for the tomb theory is found in Nubia (modern-day Sudan), Egypt's southern neighbor. During the sixth and seventh centuries BCE,

Nubian kings were entombed in small, steep-sided pyramids. Painted on the walls of the burial chambers are Egyptian-style scenes of burial rites.

In conjunction with the tomb theory, there was a political reason for the pharaoh's tomb-building desires. The political ideology behind pyramid construction, according to the tomb theory, is that the people needed a project to bring them together as a nation. It is believed that every household throughout Egypt sent men and food, on a seasonal basis, in dedication to the king and his tomb. What better project than the largest tombstone known to mankind—mausoleum is more accurate—to rally the people and put a few dollars in their pockets? Everyone involved received compensation, Egyptologists now believe, which brings up an important point.

How much does a pyramid composed of 2.5 million blocks weighing between ten and fifty tons each cost? According to engineer Markus Schulte, of the global design and business-consulting firm Arup, the 5.9 million tons of limestone would cost $18 billion today. Add to that fifty thousand laborers working for ten years at a cost of $255 billion, plus 30 percent for general contracting costs, and the total bill for the pyramid would be $380 billion. Using modern techniques, the cost could be reduced to somewhere between $30 billion and $35 billion.[19] Such costs demand an answer to another question. Why would any society devote such a huge amount of resources to a mausoleum?

One possible answer is that the constituent Egyptian was naive and had little input into the management of public projects anyway. But without foremen and managers, no large project would ever be built. So the middle managers as well as upper management (those who oversaw the project's design and construction) were also naive and willing to follow the king's wishes. As a result, the Egyptians spent a vast amount of resources during the Old Kingdom's five-hundred-year existence, and in return received nothing of utility for their society—not a believable answer during any period of history.

Besides these rational problems with the tomb theory, there is another, more evidential problem. If the king was willing to spend his vast wealth on a tomb, why didn't he order a palace residence to be constructed during his lifetime? A royal residence for any of the pharaohs of the Old Kingdom has never been found. Furthermore, Egyptologists have not ascertained why one king built at Giza and another somewhere else, such as Sakkara or Dashur.[20] Consequently, the tomb theory does a good job of *explaining away* the Old Kingdom pyramids, but it fails to explain anything pertinent about those pyramids as they relate to ancient Egyptian society outside of Osiris resurrection mythology.

Would the ancient Egyptians really have spent nearly all their resources to construct a "resurrection machine"? Such an idea eludes the principles of civilization building, for it is unlikely that a technically sophisticated society would condescend itself as a result of farcical beliefs, and then make those beliefs the primary role of their civilization.

The Egyptians may not have delineated mathematical axioms as we are accustomed to doing today, but they were exceedingly proficient in their application of those mathematical principles. Both the Rhind and the Moscow papyri show that "they possessed sound practical knowledge and knew how to make the fullest use"[21] of those principles. Their repertoire of knowledge included a decimal system, the use of fractions, and the calculation of geometric areas for the rectangle and circle, as well as for the surface of a hemisphere; and volumes for pyramids, cylinders, and cones. Most important, they were aware of the relationship among the sides of a 3:4:5 (right) triangle—known to modern mathematics as the Pythagorean theorem.[22] Although there is no evidence to suggest that they theorized about pi, the evidence indicates that in practice, pi was used.[23] Such knowledge would be required for the undertaking of complex construction projects, but does not support the overall vanity of the civilization if, indeed, the pyramids were in fact tombs.

The most difficult of problems with Egypt's pyramids is an engineering riddle. How did the ancient builders create such a massive structure, in Robert Faulkner's words, "so finely dressed that it is barely possible to insert a playing card between adjacent stones"?[24] According to Ogden Goelet, Columbia University Research Scholar of Middle Eastern Studies:

> If we assume that Khufu reigned for fifty years and that his builders worked at a breakneck pace ten hours a day, one enormous block had to be added to the pyramid every four minutes or so—every day for fifty years, inexorably. Only the precision scheduling, rigorous planning, and careful organization of an efficient, honest, and clear-thinking bureaucracy could complete a project like this.[25]

The prevailing theory is that it was all done with a combination of ramps and lifting devices, in conjunction with a highly effective system of labor.[26] In recent years, there have been attempts to demonstrate that multiton stone blocks could have been moved with sleds under sheer muscle power. All these attempts have failed, which is an important piece of evidence in itself, and leads to the suspicion that the Egyptians of the Fourth Dynasty did not build the three Giza pyramids. (There is a fourth, smaller pyramid at Giza, but it is in a state of advanced deterioration.)

According to estimates, building a single large pyramid took twenty thousand men working in crews for a period of thirty years. Excavations by Zahi Hawass have discovered the remains of some of those who built the pyramids buried in a cemetery a half-mile from Giza.

If ramps and wooden lifting devices were used in construction, with a volume of 1,560,000 cubic meters to a height of 146.6 meters, the Great Pyramid's ramp itself is a modern marvel.[27] But there is no theory explaining how the ramp was removed once the pyramid was completed.

Besides the difficulty in determining how the pyramids of Giza were built, there is the mystery of the Great Pyramid's series of internal chambers, passageways, and shafts. There is little reason to suspect that so much work was devoted to the creation of three separate chambers, one of which was built from granite, simply for the purpose of entombing the king.

The Enigma: Interior Design of the Great Pyramid

According to the tomb theory, when the pyramid was being constructed, the king's burial chamber was originally the subterranean chamber that was cut into the bedrock below the pyramid. As construction progressed, however, the king or his architect decided that it should be moved to another location midway in the height of the structure. Later, it was moved again, to a third location above the second location, and that second location was then to become the queen's burial chamber. The logic in this construction theory fails to explain the large passageway, called the Grand Gallery, that joins the uppermost chamber to the middle chamber directly beneath it, which boasts a magnificent corbel vault ceiling built from enormous limestone blocks in seven layers.

On both sides of the Grand Gallery are low ramps that run the length of the passageway. Cut into these ramps are twenty-seven square openings, alternating from large to small, at regular intervals that correspond to right-angled niches in the gallery walls. Their function has been debated ever since their discovery. The prevailing theory is that a wooden structure was anchored in these openings (slots) to move construction materials or support blocks while the corbel ceiling was under construction. Admittedly, Egyptology recognizes that, so far, no theory accurately explains the curious slots built into the ramps of the Grand Gallery, which is but one curiosity among many of the Great Pyramid's interior design.

At the lower end of the Great Gallery, a narrow passageway leads to a corridor that descends into the bedrock underneath the pyramid. When it was discovered, this shaft was filled with rock and sand. Flinders Petrie believed it was an escape route for those who lowered the granite blocks into the ascending corridor after the king was entombed. But if that were true, the shaft could not have been filled from the top. Another theory suggests that it provided fresh air to the men who were excavating the underground chamber out of the bedrock, but that would mean that the underground chambers, as well as the shaft, were built after the Great Gallery.[28] Commonsense techniques of construction, ancient or modern, dictate that foundation and excavation work be performed first.

Near the entrance of the middle (queen's) chamber, the connecting passageway steps down and slopes, overall sixty centimeters, to meet the floor of the chamber. Why this is is unknown. According to some theorists, the original floor was granite and was removed by thieves or confiscated and used in the upper (king's) chamber.[29]

The queen's chamber itself is an enigma. Situated precisely on the pyramid's east–west axis, the room is built out of limestone blocks with a corbeled ceiling and sports a niche four and a half meters tall in its east wall. The purpose of this niche is unknown, but it is thought to be the spot where a statue of Khufu was placed. More bizarre are the narrow shafts built within the north and south walls of the chamber, which were originally tapered (and sealed) to a small hole as the shaft meets the chamber.[30] One idea explaining these shafts is that the queen's chamber was a backup burial chamber for the pharaoh in the event of the pharaoh's sudden death while the pyramid was still under construction. (The king's chamber also contains shafts.) After the king's chamber was completed, the shafts in the queen's chamber were sealed.

Finally, there is the king's chamber along with its antechamber—more baffling than the main chamber—constructed from red granite. The walls, floor, and ceiling are all carved from granite. Most interesting, there are a number of granite slabs built into the chamber's ceiling whose purpose is also unknown. In all, the ceiling of the king's chamber is composed of nine slabs of granite with a combined weight of four hundred tons.[31] It has been suggested that the extra granite was applied to support the weight of the pyramid. The queen's chamber, however, being lower in the body of the pyramid, supports more weight than the king's chamber, and it was not built with reinforcement slabs.

As does the queen's chamber, the king's chamber contains shafts in its north and south walls. Although the prevailing theory is that these shafts were constructed for the purpose of air circulation, the concept of a "resurrection machine" suggests another meaning. The soul of the king would ascend one of the shafts, believed to be astronomically aligned, on his way to becoming a star.

A Pragmatic Approach to the Great Pyramid

As a preliminary conclusion, to assume that the pyramids were tombs, simply because of the adjacent mastaba fields, is understandable. This is an idea we might come to at the beginning of an investigation, after a cursory survey of the area. Any in-depth study of the Great Pyramid would require much more, however. To estimate the project's complexity and the technology required to build the pyramids, there would need to be a feasibility study. There would also need to be mechanical (reverse engineering) analysis to estimate the pyramid's purpose, as well as a cost/benefit analysis projected for the society that performed the construction. Furthermore, historical context is required for the construction of the pyramids that relates to the historical progression of society, as we know it.

Feasibly, although modern technology is capable of building the Great Pyramid, the costs involved would be immense, like that of building a nuclear power plant, which takes somewhere between ten and fifteen years to complete. The estimate for building the Great Pyramid is somewhere between twenty and thirty years. The cost would be in excess of $30 billion. The average nuclear power plant costs between $3 billion and $5 billion.[32] Of course, we are using modern concepts of costs in dollars, and it can be argued that such an analysis does not apply to ancient Egypt. Those who commissioned the Great Pyramid project did not use slave labor, it is now believed, so the workers had to be paid in something. Today, money is used to acquire resources, which is more efficient than bartering. Nonetheless, in our modern system, resources are still being acquired and distributed just as they would have been in ancient times—money being the medium of transfer as opposed to actual goods. Accordingly, those who ordered the Great Pyramid had to pay for it. The fact that a pyramid costs a fortune (in resources), and that it required the cooperation of tens (or hundreds) of thousands of workers, demands that there be some type of civil utility provided in return.

The Egyptians built a vast number of tombs over their three-thousand-year history. Most of the royal tombs had been looted long before the excavations of the nineteenth and twentieth centuries. Nonetheless, there is a long history of what a royal Egyptian tomb looks like. The best source is the Valley of the Kings, as well as the aforementioned mastaba fields of Giza. What can be ascertained from the Valley of the Kings is that the tombs were highly decorated, houselike structures. They had hallways and chambers with level floors and steps for ease of use, similar to the way the earlier mastabas were sectioned off in house-like fashion. For King Tutankhamen's tomb, the only nonlooted royal burial so far discovered, the Egyptians filled the burial chamber with the deceased's possessions. The tomb was really designed to be the "house" of the deceased (the rationale for which we will encounter later on). Furthermore, the fact that real tombs were looted suggests that there was no homogeneous dedication to, or worship of, the king as a god.

As for the pharaoh being a god, such an interpretation of Egyptian culture is too simplistic and primitive, given the vast amount of knowledge and the complex way in which the Egyptians' beliefs were expressed and disseminated (a focal point in this work thus far). In fact, it is rather a demeaning view of the ancient Egyptians even to suggest that they believed that the pharaoh was, in fact, God. Nor does it match their social structure, technically or artistically. One of the more noble aspects of Egyptian society is that women had equal rights. They could own property or demand a divorce if they so chose. They could also rise to the position of pharaoh, which does not translate literally into English or any other language as "lord" or "king." The word *pharaoh* is derived from *per aa,* which means "great house."[33] Egypt was governed as a theocracy, but applying the Western concept of theocracy to the Egyptian version results in a confusing amalgamation of primitive mythology along with the evidence of the greatness Egyptian society achieved. Nor does it explain the ancient Egyptian understanding of *theurgia,* a way of understanding states of existence in order to purify and free the soul.

The reason, or reasons, why Egyptian civilization was so successful for so long also has to be explored. Nowhere in history is there a record of a tyrannical regime whose culture lasted very long. Look at all the empires that came and went during the course of pharaonic Egypt—Sumerian, Assyrian, Hittite, Babylonian, Greek, and Roman, just to name the more historically significant ones. Egypt's success lies in its technical prowess, as well as philosophical justice. On the same basis, a $35 billion mausoleum for a pharaoh makes no sense whatsoever.

With the understanding that the purpose of the Egyptian tomb throughout dynastic history was to be a "house" for the deceased, the Giza pyramids certainly do not qualify as tombs. Their passageways and chambers were not constructed for human use, nor to accommodate the burial rites. Nor were they haphazardly planned; they were designed and built to planned specifications to serve a specific purpose that had nothing to do with anything mystical. According to German engineer Rudolf Gantenbrink, who pioneered the exploration of the shafts in the queen's chamber:

> They did not embark on a reckless building spree but that the structure was already carefully planned before work commenced, with the consistent application of expertise that was still relatively simple for the period.[34]

For the African author and intellectual Cheikh Anta Diop, the Great Pyramid did not represent the "groping beginnings of Egyptian civilization and science, but rather the crowning of a culture that had attained its apogee and, before disappearing, probably wished to leave future generations a proud testimonial of its superiority."[35]

The Pinnacle of Ancient Technology

To determine objectively the Great Pyramid's purpose, any proposed theory has to describe every chamber and passageway in a coherent and holistic manner, just as its builders foresaw specific purposes for those chambers and passageways as they designed them. Under these requirements, most theories, such as the traditional tomb theory, quickly fail. There is a theory that explains every passageway and chamber, however, that was proposed in 1998 by engineering and manufacturing expert Christopher Dunn.

Dunn is no stranger to the Giza Plateau and the Great Pyramid.

Over the past twenty years, he has visited and revisited Egypt to examine and reexamine the Great Pyramid and its inner chambers to determine the meaning of its design and what it may have been used for. From a construction and manufacturing point of view, Dunn concludes:

> The Great Pyramid is the largest, most precisely built, and most accurately aligned building ever constructed in the world. To my mind it represents the "state of the art" of the civilization that built it.[36]

From the hieroglyphs to the granite boxes of the Serapeum (a large underground tunnel at Sakkara that holds twenty-one large granite boxes) to the intricate bowls and vases made in predynastic Egypt now housed in the Cairo Museum, Dunn has gone to great lengths to prove that the builders of the pyramids employed advanced machining techniques to achieve precision in the construction of their architectural and cultural treasures. As for the Great Pyramid, he has left no passageway or chamber unexamined in re-creating a model for ancient power. According to Dunn, the Great Pyramid was designed to transform vibrations, which are always present in the earth because of plate tectonics, and transform that vibrational energy into electrical power. Here's how the Giza power plant probably worked, according to Dunn's model.

The massive structure of the pyramid collected and funneled tectonic vibrations from the earth below, then up into the Grand Gallery, where resonators converted the vibrations into airborne sound. The sound traveled past an acoustic filter, which baffled all but a certain frequency, just before entering the king's chamber. In the king's chamber, the filtered sound vibrated the massive granite walls, ceiling, and granite stack above, converting mechanical energy into electrical energy. The king's chamber was filled with hydrogen gas produced from the queen's chamber, and the hydrogen absorbed the electricity, pumping its atoms into an excited state. Microwave signals were collected off the outer surface of the pyramid and directed into the northern shaft leading to the king's chamber. There, the granite box refracted electromagnetic radiation and, with oscillating crystals adding energy to the microwave beam, served to spread the signal inside the box as it passed through its first wall. Inside the granite box, the spreading beam would then interact and stimulate the emission of energy from the energized hydrogen atoms. Passing through the other side of the box, the microwave energy was then focused into an antenna device and exited the pyramid through the southern shaft, where it could be utilized.

What happened to the plant? According to Dunn, an earthquake triggered an explosion that rendered the plant inoperable, citing William Flinders Petrie's observation that the king's chamber had been subject to a powerful force that pushed the walls out over an inch. For Dunn, this was a prediction of his theory. Any hypothesis that correctly predicts an outcome, or a yet-to-be-observed phenomenon, is a powerful hypothesis indeed.

In 1999, Dunn returned to Egypt and found that the Great Pyramid had been thoroughly cleaned. It was a serendipitous trip for Dunn; the cleaning uncovered evidence that proved part of his theory correct. Dunn discovered that the Grand Gallery was made of smooth, highly polished granite, not limestone as was originally believed, and there were scorch marks on the walls of the Grand Gallery. There was also heavy heat damage underneath each of the corbeled layers, for a distance of about twelve inches. There had been an explosion within the confines of the king's chamber hot enough to permanently discolor granite.

M. E. Abdel-Salam, engineer and chairman of the board of the Egyptian Tunneling Society, agrees, in principle, with Dunn's analysis of ancient Egypt's engineering capabilities; that manpower and simple hand tools alone cannot account for the precision. According to Abdel-Salam, some technologies used by the ancient Egyptians "still astound modern artisans and engineers. . . . Even though the tools and machines have not survived the thousands of years since their use, we have to assume by objective analysis of the evidence, that something similar did exist."[37]

Although Egyptologists have generally ignored Dunn's power-plant proposal for the Great Pyramid, he has been invited to conferences as a guest speaker on numerous occasions, most recently with Paul Brenner of the Israeli Electric Company at the World Renewable Energy Congress at the University of Aberdeen in Scotland on May 22, 2005. Nonetheless, despite its apparent lack of academic support, no other theory explains all aspects of the Great Pyramid and ties them together in a meaningful way. The evidence of precision and sophisticated technology in the construction of the Great Pyramid, particularly its scope and design, leaves historical researchers with a nearly unsolvable mystery.

The Pyramid Tomb: More Guess Than Theory

According to the tomb theory, during the First Dynasty, pyramid construction began in Sakkara, ten miles south of Giza, where Djoser transformed his mastaba in phases into the Step Pyramid. At Sakkara, where there are several dilapidated pyramids, techniques of pyramid construction were developed. Years later, at Dashur, five miles south of Sakkara, pyramid development continued (with limited success) in the "Bent" Pyramid, and finally with complete success in the Red Pyramid. The Red Pyramid, two-thirds the size of the Great Pyramid, so named because of its reddish color, was (according to the tomb theory) the first true pyramid. With construction techniques now fully developed, a new site was chosen to build the greatest pyramids of all, at Giza.

There are a number of questions, problems, and inconsistencies with this scenario:

- If the Step Pyramid was the first pyramid ever built, where did Djoser get the concept of a pyramid?
- After the Step Pyramid, other pyramids built at Sakkara were adorned with hieroglyphic inscriptions, as previously discussed in the section on *The Pyramid Texts,* but no other later pyramids contain any inscriptions of any kind.
- Royal tombs throughout Egyptian history show a clear tradition of houselike interior design, such as mastabas during early dynasties and burials in the Valley of the Kings in later dynasties. The Giza pyramids were a temporary departure from this tradition. Why?
- The interior design of the Great Pyramid has no precedent or antecedent in Egyptian civilization, or in any other civilization.
- During the thousand years following the Fourth Dynasty, the Egyptians built nearly one hundred smaller pyramids. But they used a different method of construction, resulting in structures of substandard quality compared to the Giza pyramids. Today, these later pyramids are in a state of severe deterioration.
- It remains hypothetical that the ancient Egyptians had the technology required to build the three Giza pyramids, and, in particular, the interior chambers of the Great Pyramid. There is no evidence that they had that level of technology.

Given the difficulties listed above pertaining to the tomb theory, a more likely interpretation is that the three pyramids already existed on the Giza Plateau. When the early Egyptians came to occupy the plateau, they added to its existing structures. Not knowing what the three pyramids were, and recognizing that they were closed structures, they assumed they were tombs. As such, they turned the plateau into a cemetery, burying the dead underneath mastabas to the east and west of the northernmost pyramid.

This solves a number of inconsistencies. It provides a motive for Djoser to modify his mastaba into a pyramid, and for the early-dynasty kings to inscribe their religious convictions into the Sakkara pyramids. They were patterning their tombs after Giza, which they believed to be the tombs of their ancestors. It explains why pyramids built before and after were of a lesser quality. Old Kingdom pharaohs were attempting to copy the Giza pyramids but had no choice in techniques and had to use what was available to them. It also explains why there are two different styles of temple architecture on the Giza Plateau. The Sphinx Temple and the Valley Temple are built from large, rectangular blocks and are not inscribed with any hieroglyphs or other markings; they are referred to as cyclopean, which is what is found in Abydos at the temple known as the Oseirion.

In the same way the star-viewing diagram of the Nabta Playa megaliths suggests an advanced understanding of astronomy, the Great Pyramid at Giza provides the concrete example for the application of advanced knowledge of construction techniques. More important, if Dunn's theory is correct as to the purpose of the Great Pyramid—and there is no evidence thus far to suggest he is wrong in his analysis—an understanding of physics is also required on the part of its builders.

The Great Pyramid is the most distinctive structure ever built, an anachronism today as much as it was five thousand years ago. The mystery of its builders may remain a mystery forever. Yet we seek to fully understand its meaning as well as its purpose. If we ever do, maybe we will also understand the deeper aspects of what it means to be human and how our humanity relates to the divine, or perhaps how quantum physics may relate to both.

Understanding ancient mythology, whether it is Egyptian or that of any other culture or civilization, is an arduous task and open to personal interpretation, which is often dependent on the beliefs of the individual. When we are far removed from the period of myth creation and relatively naive to the keys with which the myths should be interpreted, it is reasonable to suspect that the myths are primitive rationalizations explaining why and how we are here. The theories of researchers that in myth there is scientific knowledge are often met with a roll of the eyes by naysayers and skeptics. There is just too much room, too much vagueness, in the storyline to support the notion that ancient traditions were based on technical and scientific principles. But as we have discovered in the brilliance of René Schwaller de Lubicz's analysis of Egyptian culture, interpreting ancient myth scientifically appears to be a valid approach for the researcher.

The mysterious and bizarre interior design of the Great Pyramid suggests that the structure was a device—according to Christopher Dunn, a device that produced power. The planned design and construction of such a device precludes the notion that those who built it knew nothing of chemistry and physics. Nonetheless, the principal difficulty with the theory of advanced prehistoric technology is that, just as there is today, there would have to be some form of written documentation somewhere outlining those technological principles, assuming, of course, that civilization has generally progressed in a linear fashion.

Quantum Physics and Dogon Mythology

If prehistoric civilizations had endured a catastrophe of global proportions, such as that which occurred at the end of the ice age, there would likely be no surviving documentation. Even so, those who survived the cataclysm would carry on the ideas of that civilization far into the future, and they would do so primarily through oral traditions. It might also be the case that their scientific language was far different from our own modern scientific terminology, which would impede an understanding of that documentation or oral tradition. Since the laws of physics do not change, however, those scientific principles would have to be the same regardless of the medium used for their dissemination. A case for advanced knowledge of physics is proposed by Laird Scranton in his research into the mythology of the Dogon and Egyptian culture as the founding symbols of civilization.

Linking the function represented by the symbol to the symbol's corresponding mythical storyline, and then comparing those symbols to the modern graphical rendition of the function they represent, shows a near-perfect match between ancient and modern concepts. Scranton performs such an analysis for a number of symbols, describing the structure of matter as well as reproduction and genetics. Although the skeptic may claim that these are mere coincidences, when do too many coincidences become a logical deduction? Laird Scranton appears to have far surpassed the coincidental. According to John Anthony West, Scranton "is showing in a way that is practically incontrovertible that the ancients had our science."[38]

Dogon Creation Mythology

Renowned for their mythology, mask dances, architecture, and wooden sculptures, the Dogon, who number approximately three hundred thousand, live in Mali's central plateau region of northeastern Africa. Dogon myth follows a similar pattern to that of the exoteric and esoteric aspects of Egyptian mythology. There is the exoteric, common knowledge with which the majority of the population is familiar, but there is also the esoteric, which leads to a deeper, more technical, meaning. For Scranton, Dogon mythology describes "the true underlying structure of matter, organizes it in the right sequence, diagrams it correctly, and assigns the correct attributes to each of its components." And "it does so within the explicit context of a discussion of the structure of matter."[39]

In the Dogon creation myth, before anything existed, a being named Amma, who is described as a spiraling motion, lived inside an (a cosmic) egg. At the center of this egg, the oval-shaped *po* seed—the smallest of things—was created. When the egg broke, Amma emerged in a whirlwind, and so did the world. According to Dogon tradition, the po, which housed Amma's creative will, emerged first "like a central air bubble."[40] Although inaudible and invisible, particles of matter were scattered in a deep, luminous motion. Thus, the opening of the egg created all the spiraling galaxies in the universe.[41]

The essence of the po is the image of the creator, representing the origin of matter. For some people, it was forbidden to speak of this, for "the beginning of things is Amma's greatest secret."[42] Although the po itself remained infinitely small, all things were created from it by a continuous addition of identical elements: "As Amma adds . . . the thing becomes large."[43]

Laird Scranton finds this creation story similar to how modern science describes the big bang. The "egg," before it bursts into life, is representative of a black hole, and the cooling process of the quark-gluon plasma that likely occurred just after the bang would have had a "central bubble" characteristic, just as the story claims. The po represents the hydrogen atom's emergence at the beginning of the universe. The Dogon also describe creation with an emphasis on "pellets of clay" and "spiraling coils," which Scranton interprets as an accurate description of energy particles and waves before they formed the primordial helium and hydrogen of the early universe.

Dogon myth also tells the tale of Earth's creation. Amma attempted to have intercourse with his wife, Earth, but to no avail. A termite hill rose up, barring passage and displaying its masculinity. So he cut down the hill and tried again, and from the union a single being was born, Thos aureus, the jackal.[44] Amma tried again to mate with Earth and this time too was successful, so the desired twins were born: "water, which is the divine seed, was thus able to enter the womb of Earth and the normal reproductive cycle resulted in the birth of twins."[45]

The twins were green, human from the waist up and serpent from the waist down, with red eyes and forked tongues. Their bodies were covered "with short green hairs, a presage of vegetation and germination." These two spirits, homogenous products of God, were called the Nummo, and, like God, were of a divine essence.

According to the Dogon, the Nummo represent water, and in their language, the words for *water* and *Nummo* are used interchangeably. The Nummo, or water as we know today, is the life force of Earth, which in Dogon mythology was molded from clay, which is also from water. In fact, for the Dogon, everything is from water: blood and stone, as well as copper and sunlight, referred to as Amma's excrement.[46] Speaking of the Nummo, "The Pair were born perfect and complete; they had eight members, and their number is eight, which is the symbol of speech."[47]

Although expressed in metaphor, the Dogon story of water as the essence of life is an apt scientific description for the basic principle of life on our planet. Why the Nummo pair had eight members is a technicality, not arbitrary. Scranton sees in the symbolism of the Nummo scientific significance in the numbers two (the pair) and eight, and for good reason. Water is a molecule composed of two hydrogen atoms and a single oxygen atom—H_2O. A hydrogen atom consists of a single electron held by its nucleus, whereas an oxygen atom is composed of eight electrons. So the Nummo, whose name is the Dogon word for water, are accurately described with *chemistry and physics* in mind!

The Dogon belief that copper and sunlight are also products of water (the Nummo) seems a bit of a stretch. But as Scranton notes,

"The rays of the sun could be reasonably seen as the excrement of the Nummo, since we know that solar energy is a by-product of the fusion of hydrogen atoms, which are in this sense of meaning the Nummo pair."[48] As for copper, which has twenty-nine electrons in its atomic structure, it is divided into four shells where its electrons are held in orbit. In its innermost ring there are two electrons, the second ring has eight, the third eighteen, and the fourth a single electron. The first two shells are a repetition of the numbers found in water (two and eight electrons), so one could say that copper contains water.[49]

Quantum String Theory and the Dogon

In Dogon tradition, there are 266 *sene* seeds (or signs) of Amma that combine to create the po. These seeds can be thought of as a prefiguring of categories for everything in the universe, inanimate objects as well as living beings.[50] According to Scranton, these signs are the conceptual foundation for the universe, based on the principle of vibration, and compare favorably to the definition of subatomic particle according to the string theory of physics.[51]

The Dogon honor these signs in an annual ritual before sowing seed into the fields. The head of the family clears an area in the "field of the ancestors" where the signs will be made. In the center of the field, he draws a circle using an upside-down basket as a guide, then creates a pile of stones. On the ground, he draws a small circle, roughly twelve

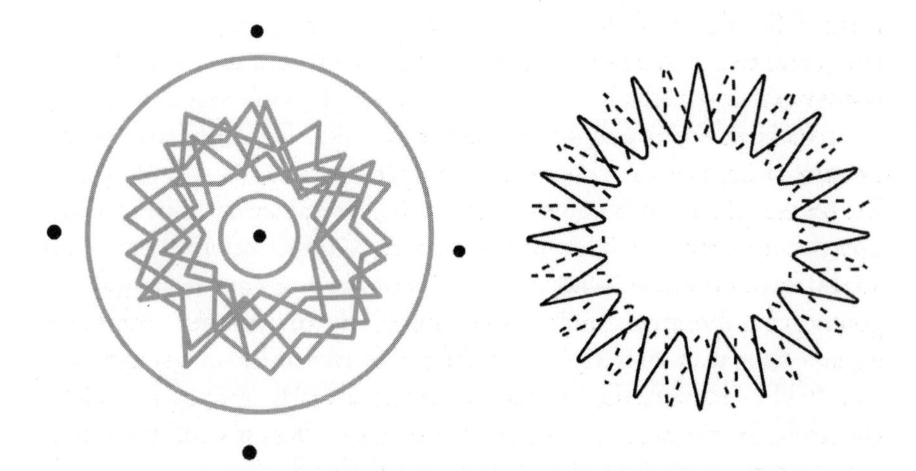

Fig. 12.1. Dogon field drawing (left) and quantum string vibrational pattern (from Laird Scranton, The Science of the Dogon, *75, 78)*

centimeters in diameter, inside the first circle, with a dot in the center. Then, throughout the day, he draws twenty-two zigzag lines around the inner circle to fill the outer circle's width with an intricate tangle of lines. This design that the family leader draws in the soil represents all possible signs.[52]

What is incredible about this drawing in the field is that French anthropologists Griaule and Dieterlen's diagram of it compares well with the theoretical vibrational pattern of an electron, according to quantum string theory. The Dogon refer to this representation in the field as a "thread" that has been made from the "spider of the *sene*."[53] Is this coincidence? Not for Scranton, for throughout the Dogon's history they have retained the image of what this "thread" should look like, which is a spiral or coil. Although this departs from what current string theory suggests a string should look like (a loop), the imagery is remarkably similar. The similarities between Dogon symbolism and modern physics continue.

From the work of the spider, the Dogon say that the four sene seeds are born. The first seed is the *mono,* meaning to bring together. The second is the *gommuzu,* meaning the bumpy. The third is the *benu,* meaning stocky. The fourth sene is the *urio,* meaning bow (as in bowing your head to honor your audience). On Earth, the four *yala* allocated to the *na* seed in Amma's womb pertain to the four elements: *na* (water), *gommuzu* (air), *benu* (fire), and *urio* (earth). Scranton suggests that these symbols refer to the four quantum forces: gravity, electromagnetism, the weak nuclear force, and the strong nuclear force. Further support for his thesis is that the behavior of a one-dimensional string, which stretches into a two-dimensional membrane through an increase in its string coupling constant, is uniquely similar to how the Dogon express spiraling components of matter.[54]

Even more support comes from the Dogon's imagery of life's development inside the seed. A series of seven symbols, referred to as the "drawings of the multiplication of the word of *po,*"[55] are said to be the successive appearances of seven vibrations. This corresponds with physics' M–theory, in which at every point in our four-dimensional spacetime there is a tiny theoretical space called Calabi-Yau space, where there are seven other dimensions. Although we cannot detect this Calabi-Yau space, this is where the strings exist and vibrate.[56]

In summary, the narrative of the Dogon creation myth agrees with modern scientific theory. It begins with Amma's egg (a black hole) and scientifically describes matter from the structure of the atom with its electrons, down to the massless state of strings. According to Scranton, "Dogon mythology seems to describe the true underlying structure of matter, organizes it in the right sequence, diagrams it correctly, and assigns the correct attributes to each of the components."[57]

Dogon Mythology as a Remnant of Ancient Egypt

"There are many persistent similarities between Dogon and Egyptian religious symbols and lifestyles,"[58] Scranton writes—similarities that, he suggests, are a result of modern Dogon culture being a remnant of ancient Egyptian culture. Although the ancient Egyptians often had alternative names for their gods—such as Amun the "hidden one," who was also called at various times Ammon and Amen, as well as Amen-Ra—Amun is phonetically very similar to Amma.[59] So is the Egyptian Nemu, the headsman of Osiris,[60] phonetically similar to Nummo, as well as the Sumerian deity Nammu, goddess of the fertile waters.

One of the earliest predynastic names for Amun was Menew, who was also called Min or Amsu. Again, Scranton notes that the name Menew is phonetically similar to Nummo, but more important are the hieroglyphs that represent Menew, the symbol for water, the sun, a clay pot, and a spiral. They are also key symbols in the Dogon creation story.[61]

Egyptian mythology also tells of a god named Khnemiu, who was four divine beings who wore red crowns and were to be found in the eleventh section of Dwat (underworld).[62] Another variant of the same name was Khnemu, the ram-headed god of creation and fertility. Khnemu, whose name literally means "molder," formed the great cosmic egg on a potter's wheel that contained the sun. Khnemu's hieroglyph was a simple clay pot, and his cult center was a town of the same name that the Greeks called Hermopolis. The ancient Egyptians called it the City of Eight and the "island of flame," which was the birthplace of the sun god. Just as the Dogon revere the number eight, and the twin or pair, so did the ancient Egyptians.

During Egypt's Middle Kingdom, the number eight was portrayed in the ogdoad, where eight entities composed a variation of Egyptian creation mythology. Although these beings were worshipped primarily in Heliopolis, aspects of the creation were combined with existing myths. As are the Dogon Nummo, each entity of the Ogdoad was a member of a masculine/feminine pair, and each pair represented an aspect of the primordial chaos from which the physical world was created.

Nun and Naunet represented the primordial waters; Kuk and Kauket, the infinite darkness; Hu and Hauhet, empty space; and Amun and Amaunet, the secret powers of creation. The gods were usually depicted as men with the heads of snakes (reverse of the Dogon), and the goddesses as women with the heads of frogs. They built an island in a vast emptiness where the cosmic egg was placed. From this egg came Atum, the sun god, who began the process of creating the world, just as the Dogon claim in their mythology.

Although over their history ancient Egyptian symbolism varied slightly in how creation was depicted, the principles of creation remained the same in the themes of earth, air, water, and fire. Similarly, the Dogon express these same concepts in their tradition of creation.

Perhaps the most fascinating comparison between Dogon and Egyptian symbols is the concept of a thread. In Egyptian hieroglyphs, the word for thread is *ntt–t,* which is also the word for weaving, and it was derived from the goddess Neith, who "was said to have woven the world on her loom."[63] Scranton finds that the string-intersection diagrams of theoretical physicists Paul Davies and John Gribben's *The Matter Myth* are contained in the hieroglyph *ntt.* And the last symbol in the hieroglyph for Neith matches the Davies/Gribben diagram *exactly* for the theoretical rendition of a more complex string interaction, a geometric form that results from a string/antistring pair eruption.

Although skeptics may claim that the Dogon's scientific knowledge comes from their twentieth-century contact with other cultures, why the Dogon would create such a myth in modern times is the real mystery. Furthermore, the Dogon mythology also asserts that the star Sirius is a binary star, calling Sirius B (the second star) Po Tolo, which means star-*tolo* and smallest-seed *po.* For the Dogon, this is their tradition and heritage.

In his book *The Science of the Dogon,* Laird Scranton cites example after example, many more than what has been covered here, that demonstrate a consistent relationship between Dogon symbolism and modern scientific concepts. He also explains the similarities between Dogon symbols and concepts and those of ancient Egypt in a truly groundbreaking way. In the Dogon creation myth Dada, the mother spider, creates the universe through weaving, just as the Egyptian Neith, the Mother Goddess, creates the universe through weaving. Given his thesis that a complex scientific understanding existed in ancient times, one is inclined to accept the notion that once upon a time there existed a civilization with one language and a common speech.

One Language and a Common Speech

A final piece of evidence that further corroborates the thesis of an advanced prehistoric civilization in the region of the Mediterranean comes from an ancient biblical text. The story of the Tower of Babel links the anomalous physical evidence—Nabta Playa, the Great Pyramid, and Baalbek, as well as the correlation between modern physics and the Dogon creation myths—to our historical traditions, oral and written. An unbiased look at the Tower of Babel story (Genesis 11:1–9) offers an interesting point of view.

On its surface, the Tower of Babel story is just another fable, in which God's will is forced on the world because God doesn't like what mankind has been up to. Whether the tower itself existed is not the point. The point is that the story was widespread during ancient times and important enough for Moses to include in the historical chapters of his texts, known as the Torah, or the Law.

In writing Genesis, it is my belief that Moses used the historical knowledge he learned from his royal upbringing in Pharaoh's house, as well as Sumerian and Akkadian traditions from his father-in-law, Jethro. Although Moses's actions were likely politically motivated in the interests of forging a common identity for his people, part of his goal was to establish a written history for the burgeoning nation-state of Israel. The other part was to create social laws. A common history is an important element for social cohesiveness, particularly if your ancestors are at history's center.

I believe it is doubtful he would have fabricated the Genesis stories. Most likely, he placed a Hebrew perspective on the history of the time that was, realistically, Egyptian history—although not exclusively about Egypt.

Moses may have been conveying the same story as Plato does in the dialogues of Timaeus and Critias—the story of a great civilization, destroyed by a cataclysm, told to his uncle Solon by the Egyptian priests at Sais. Moses was also Egyptian. Is this a coincidence? Probably not, Egypt really was the cornucopia of the ancient world.

In the Tower of Babel story, there are three important points:

1. Before the Babel catastrophe, "the whole world had one language and a common speech" (verse 1).
2. As a civilization, mankind was capable of planning and completing just about anything: "Nothing they plan to do will be impossible for them," God predicts (verse 6).
3. Those who survived the catastrophe were scattered throughout the world and were forced to speak different languages. The Lord "confused the language of the whole world . . . [and] scattered them over the face of the whole Earth" (verse 9).

According to the archaeological evidence as well as ancient texts, we know without a doubt that when civilization began (3000 BCE), there was no single language but rather a variety of languages and cultures. If so, what could the Tower of Babel story be referring to?

Although there is no mention of a catastrophe in Genesis, chapter eleven, it is implied. How else would a civilization with a common language become scattered over the face of Earth, other than to speak many languages?

In the insurance business, there is a clause that protects the insured from what are called "acts of God." These acts are, in modern terms, natural disasters. Five thousand years ago, people also attributed natural disasters to God. So what likely happened in the Tower of Babel story is that a natural disaster of immense proportions occurred. The civilization that existed was decimated, leaving only isolated pockets of survivors. Over many generations, those survivors struggled to rebuild civilization, but because of geographic isolation, unique languages developed in various regions.

The concept behind this story is entirely plausible. Today, astronomers talk about a Tower of Babel scenario, although they do not refer to it as such: a comet or asteroid impact, similar to the one that rendered the dinosaurs extinct. It has happened before, and astronomers assure us that it will certainly happen again; it's just a question of when. If it did, and 70 percent or more of the world's population died, what would happen to mankind? A likely result is that within four generations after the cataclysm, everything we know today—cars, phones, computers, and all the other niceties society has given us—would be a dim memory. By the tenth generation, who would care? Except for the tradition of telling the story of the civilization that once was, no one would care because that civilization would not be a part of anyone's memory.

The Kharsag Tablets

Did a catastrophe really occur that decimated the human population sometime in the remote past and inspire the Tower of Babel story? If so, did those who survived leave any trace that they had existed?

No one really knows for sure if such a calamity occurred, but a set of Sumerian cuneiform tablets discovered at the ancient city of Nippur by a group of American archaeologists in 1896 suggests there is more to the story than fable or folklore. The inscriptions on what are known as the Kharsag tablets tell the story of a group of people known as the Anannage who ventured into the mountainous region somewhere between the eastern shores of the Mediterranean Sea and the Zagros Mountains to establish an agricultural community.

When these tablets, now stored at the University of Philadelphia museum, were first translated, they were believed to be a Sumerian creation myth, and the personalities depicted in the epic stories were interpreted as gods. According to exploration geologist and historian Christian O'Brien (1915–2001), however, this religious interpretation was a product of preconceived notions; according to his analysis, it has had "disastrous results for the truth."[64] The nine epic stories on the clay tablets, he insists, which were written over a period covering the third millennium BCE, were clearly couched in secular terms and were important to the people who created them. "They give the impression," he says, "of being coveted library possessions that were copied in many places and over many centuries with sequential reprintings."[65]

According to the story, the area that the Anannage settled in was referred to as the Eden, or Edin, the Sumerian term for uncultivated plain, and the settlement that they built was named Kharsag, or alternatively Garsag. There were originally fifty Anannage who were democratically organized by a council of seven and led by Enlil (Father and the Lord of Cultivation); Ninlil (Enlil's wife-to-be and the Lady of Cultivation), also referred to as Ninkharsag; Enki (the Lord of the Earth or Land); and Utu, also called Ugmash (Sun Wisdom), whom O'Brien describes as an operations manager.[66] On occasion, the supreme leader, Anu, would join the council in an advisory capacity.

The Epics

In Kharsag Epic 1, the Anannage, those who are referred to as the Shining Ones by the native peoples because of their radiant appearance, settle in a fertile valley (the Edin) with a mountainous region.

The Anannage construct enclosed fields, plow the earth, and sow grain. They also plant three or four types of fruit trees, as well as other trees for shade. To ensure that the fields receive the proper amounts of water, they create a system of irrigation. Sheep and cattle are domesticated and housed in barns with running water.

As the settlement prospers with bountiful harvests, the local tribe's people—primitives who live in nearby caves—are invited in as helpers and partners. Later on, a reservoir is built higher up in the mountains to ensure a constant water supply. Houses are built from cedar, along with a community building, the House of Enlil.[67]

Epic 2 recounts the Anannage's decision to settle in the mountain valley. Ninkharsag addresses the council and argues for operating an experimental plantation for a short period. Although Anu disagrees with her, her reasons are convincing, so Anu agrees to this, to the cheers of the council members. According to O'Brien, "Ninkharsag's well-argued case was the turning point in the decision."[68]

The pilot plantation is so successful that the Anannage decide to make the settlement permanent, and they construct a maternity building for expectant mothers. Sickness spreads throughout the Anannage, however, from eating tainted meat. To remedy the situation, the Anannage decide to "burn the cedar-tree pests [of which] there are seven kinds," destroy "all the insects on the vines with a great light before sunrise," and cook food and clean more thoroughly.[69]

Such accounts, O'Brien argues, are more representative of our modern age than of our notions of prehistoric mankind. Knowledge of cooking and cleaning to ensure sterility suggests a level of scientific knowledge; it also indicates that the Anannage knew that a bacterium invisible to the unaided eye was the source of the epidemic.[70]

In Epic 3, the courtship of Enlil and Ninlil occurs before the construction of the settlement, just as the Anannage are settling in the valley. Originally, the text was translated as a description of the rape or seduction of Ninlil, but there is no suggestion that rape or seduction is actually the case. According to O'Brien, explicit lines saying the two kissed in darkness, then made love, demonstrate that it is a classic love story brought to its natural conclusion.[71] Then Ninlil requests that a "bright, high-sited house" be built, the Great House of Enlil, because she is pregnant. After its construction, Ninlil takes her place in Enlil's house:

Radiant, she stood by the Lord Enlil
Her heat rejoiced in him.
In the House of the Lord Enlil,
She stood proudly with him;
In the House of the Lord Enlil,
She proudly took his hand.[72]

In Epic 4, the Anannage council continue discussion as they plan the settlement. The harvests increase and construction continues. They build a wooden stable and a waterfall, and the lord Nannar plans to make a wooden sluice to water the land. Ninkharsag (Ninlil was her title after becoming Enlil's wife) plans for abundant planting and encloses the fertile land. Cedar trees are planted to create a windbreak and other trees are planted for shade and for fruit, although it is noted that some trees cannot be cultivated successfully in the orchards.[73]

A new character is introduced named Dungi. Although George A. Barton, the original translator of the tablets, refers to Dungi as an invented mythical king, according to O'Brien the name Dungi is more likely to refer to the teacher of canal construction, or a "reverence to the one who taught the art of digging the canals for the irrigation systems."[74]

In Kharsag Epic 5, the settlement is well under construction. The reservoir is completed and enclosed by the walls of a ravine. Houses are built from cedar, and work begins on the Great House of Enlil,[75] which is completed in Epic 6.

As with all other dwellings, the Great House of Enlil is built of cedar and stands in seclusion above the settlement on a rocky eminence. In the text, the Great House is called by a Sumerian term that translates as "sanctuary." Like the plantations, it is enclosed by a fence. As well as being the residence of the Enlil family, it features a nursery and a private boudoir for Ninlil. Those who live and work there enjoy the benefits of running water from the reservoir. As the leader's house, it also contains a council chamber as well as a banquet room, and is guarded.[76]

Now comfortable in their mountain homes, "the Principal Heroes planned justly; the great princes enjoyed discussing their strange mission."[77] According to O'Brien, the settlement in the mountain valley lasted for more than two thousand years but eventually succumbed to extreme weather that lasted for an unknown period.

In Epic 7, during the winter a great storm sweeps through the settlement. Many of the dwellings are buried under the snow, including the Great House of Enlil. The Anannage build stone fireplaces and seek comfort from the bitter cold in food and strong drink. According to O'Brien, exceptionally cold winters such as the story indicates, unknown in today's climate, were to be expected at moderate latitudes such as those of southern Lebanon. Possibly, it was "the last hiccough of the ice age."[78]

The climate grows worse, and in Epics 8 and 9, a great storm destroys the plantations as well as the settlement. According to the text, it appears that the Anannage have underestimated the power of nature and should have known better:

> *The fenced house was destroyed by burning . . .*
> *Disaster itself brought about by ignorance.*
> *Ignorance brought disaster.*[79]

The overflowing water from the thunderstorms destroys the dam. The Anannage hope in vain that the storm will subside and not wash away the crops:

> *The Settlement of Learning—the whole Settlement*
> *with its food-storage building, and its plantations,*
> *became marshland!*[80]

Ninlil "threw herself down in anger, screaming loudly."[81]

O'Brien believes that with the ruin of the settlement, the Anannage moved out of the highlands and onto the plain, which he describes as "a diffusion of Anannage influence from a nucleus at Kharsag into separate smaller kingdoms."[82]

The Creation of Mankind

In the traditional interpretation of Sumerian and Akkadian myth, the Anannage were the offspring and followers of Anu, the God of Heaven. Antu (meaning Earth) was a colorless female being who was the first to consort with Anu. Together they produced the Anannage.

Although the Anannage were considered to be gods (spiritual beings) by the Sumerians and Akkadians, there was a time when they lived in human form. According to the myth of Atra-hasis, the Anannage were the ruling class and forced the weaker Igigi to perform the works of irrigation and drainage necessary to maintain the fields. After thirty-six hundred years, the Igigi became tired of the work and confronted the Anannage, but agreed on a compromise to settle their dispute. From a mixture of clay and blood, Ea (Akkadian) and the Mother Goddess created the first humans to perform the work. Twelve hundred years later, mankind received forbidden wisdom and became too numerous and restless for the gods' comfort. The decision was therefore made to reduce mankind through pestilence and plagues. To hold the plagues in check, Ea advised mankind to withhold prayers and sacrifices from the other gods and to worship one god.

According to O'Brien, the proper translation for this Mesopotamian myth may reveal who these "gods" really were. The text in question is Akkadian, the story of Atra-hasis from Kharsag Tablet I, which was copied by the scribe Ku-aya during the reign of Ammi-saduqa around 1635 BCE from earlier material no longer in existence. (It was originally translated by Wilfred Lambert and Alan Millard.) It is O'Brien's opinion that the Akkadian scribe, in the middle of the second millennium BCE, fell into familiar traps that translators sometimes do when dealing with material of which they have little or no knowledge.

In the original translation of Ku-aya's story of Atra-hasis, the principal gods—Anu, Enlil, and Enki—were all that existed, but they decided to divide the territory of their domains into spheres of influence. Together with Ma-mi, Enki plotted mankind's creation from a mixture of clay and the flesh and the blood of a slain god.

In O'Brien's translation of this story, the Igigi were "lordlings," members of the Anannage whose job was to provide the work for the ruling council's plans. Since the work was heavy, it caused much stress "and the excess toil was killing them."[83] In a unanimous vote, the Igigi decided to confront Enlil about the situation. At this point there is a break in the tablet, but the story picks up again after the break. Enki has the floor and addresses his brother lords:

> While Belet-ili (Akkadian, Mistress of the Lords, Ninkharsag), the creator of life, is here, let her create offspring; and when they become men, let them bear the toil of the lordlings.[84]

Ma-mi (an alternative name for Belet-ili), the birth goddess, is sent for to create a *lullu* (Akkadian for man, referred to in the *Atra-hasis* text) so that the lullu can be assigned the tasks Enlil requires. O'Brien contends that the translation of "birth goddess" does not quite fit the context of the story, and that a more scientific term is needed. He believes that "biological expert" was meant.

The story continues with the Lady of Creation (Ma-mi) replying that it is not possible for her to do what is asked, and that she will need the help of Enki's skills because everything needs to be purified. He should prepare the material. Enki obliges, and orders that on three separate days, the Igigi will be dipped in a purifying bath for cleansing. After this, one of those dipped will have to be slaughtered so the Lady of Creation can mix the clay with the slaughtered one's flesh and blood.

This part of the story makes little sense. Why would they have to, or want to, kill one of their own? It also doesn't make sense that they would have to take baths. O'Brien claims that the translation of the words is the problem.[85]

The word for bath, *ri-im-ka,* is suspect, O'Brien claims. Its root is *rimku,* which does mean washing but does not necessarily refer to a bath. It can also mean pouring out, as in a draft. According to O'Brien, it is more likely that all these lords would have been giving blood (purifying drafts) on the first, seventh, and the fifteenth days of the month. Furthermore, it is not necessary to kill someone to obtain a sample of blood.

Another difficulty with the original translation is that a mixture of blood and flesh does not make clay. It could, however, make what we call today a "culture." As the story continues, this "clay" that the Lady of Creation mixes is placed in the wombs of "birth goddesses," which O'Brien translates as "foster-mothers," to produce life—in other words, babies.

So what was the clay mixed with?

According to the text, it was to be mixed with spittle: "The Igigi, the great gods, spat upon the clay."[86] The Akkadian term for spittle is *ru-tu* or *ru'u-tu;* but if this term was taken from the Sumerian, it could have meant a "conception-escape," which is a euphemism for the ejaculation of semen. The ejaculation of semen is actually "spittle" of a kind. O'Brien believes it is possible that the two terms are interchangeable.

So what was the clay?

The Akkadian word for clay was *tittu;* written *ti-it-tu,* which in the context of the story appears meaningless. If the word was copied from the original Sumerian without alteration, however, it could mean something else. Breaking the word into its syllables:

ti—to live, live, or living, or life
it—with
tu—bear, beget, or enter, or *tu(n)* portion or piece[87]

Consequently, it is reasonable to assume that the "clay," mixed with the Lord's sperm, was a "piece of life" or a "piece of begetting." It is more speculative to hold that this "piece of life" was an ovum. Nonetheless, such a conjecture is justified since, in the narrative, fourteen birth goddesses (O'Brien's "foster-mothers") are needed to carry out the project to its objective.[88]

According to the story, the events of this project occurred in the Bit Shimti, which Lambert and Millard translate as "House of Destiny." *Bit* was the Akkadian word for house, and since children, in a sense, are man's destiny, the word fits. But O'Brien claims the original Sumerian word may mean something else. In Sumerian, the syllables of Shimti mean:

si—see or look or be bright
igi—eye[89]

Overall, this means "a bright eye for seeing."

imi—clay
imi—embryo (from the Akkadian *ubarro,* which had that meaning; its archaic sign may have been a crude representation of a baby)
ti—observe or examine[90]

For O'Brien, the best combination of these closely compatible meanings is "bright eye for examining the life culture." The Bit Shimti may well have been the building that housed that essential piece of apparatus for completing the "creation of man"—an illuminated microscope.[91]

"Men of Renown and Heroes of Old"

The Shining Ones, the Anannage, settled in the region of southern Lebanon around 8750 BCE and lived in their mountain paradise for more than two thousand years. We know today of their doings because of their relationship with neighboring peoples, specifically the Sumerians, who may have received their knowledge and honored them and their mountain home as sacred.

The stories of the Anannage were also told in Akkadia, where the Anannage were known as the Elohim, recorded by the patriarch Enoch (in the Book of Enoch), and in the Hebrew scriptures, where they were referred to as "men of renown and heroes of old."[92] The strange introduction to the Noah flood story introduces them as the "Sons of God" who took the daughters of men as wives, but they are never mentioned again in the biblical canon. If they actually sired children, it precludes any notion that they were angels or other nonphysical beings.

In 1792 BCE, Hammurabi, the sixth ruler of Babilu (Babylon), created the first-known written code of law from Anu and Enlil's decrees by inscribing them in stone. Known as the Hammurabi Code, the inscriptions—composed of 282 articles—were erected on stelae in public places for all to see and championed the cause that "brought about the well-being of the oppressed."[93] The Assyrians apparently carried on this legacy, for one of the most magnificent treasures of their culture is the group of sculptures of the Elohim farming and caring for wildlife, on display in the British Museum.[94]

The ancient villages of Jericho, and particularly Çatal Hüyük, with its stylized, almost modern dwellings (see chapter 5) and its inhabitants' personal possessions, tell of a relatively sophisticated life that existed three thousand years before the official birth of civilization in 3000 BCE. And, in the tradition of the Kharsag agricultural settlement, the residents of Çatal Hüyük also revered the Mother Goddess, the Lady of Creation.

Like the settlement in the Kharsag Epic, Çatal Hüyük was inhabited for several thousand years and then was suddenly abandoned. It is convenient to dismiss the Anannage as an ancient story steeped in a primitive people's struggle to explain its own existence. But the anomalous town of Çatal Hüyük cannot be dismissed, nor can the mysterious thousand-ton blocks of Baalbek.

"Who knows to what degree of civilization and prosperity Baalbek attained in olden times?" asks Michel Alouf, Baalbek's curator from 1897 to 1934.[95] Baalbek, like Kharsag, was said to have been located in a mountain valley, in a fertile land overlooking a vast plain. With an ample supply of water, its fields produced the best grains and fruits. The mountains were covered with cypress, cedar, pine, oak, and juniper forests; flocks of goats and sheep grazed on the plain. These are sufficient reasons for Baalbek to have been one of the richest cities in the world, even in the most remote of times.

Although the ruins at Baalbek and the story of the Kharsag agricultural settlement remain a matter of interpretation, recent events in Iran appear to make at least some aspects of the tale plausible. According to archaeologists, the first domestication of sheep and goats occurred in the Zagros Mountains around 8000 BCE. Within five hundred years, those early herdsmen had established agricultural villages. By 4200 BCE, the prehistoric city of Susa was established on the Mesopotamian Plain to the northeast of the modern Iraqi city of Basra, near the foothills of the Zagros Mountains.[96]

In 2000, flooding of the Halil River uncovered numerous previously unknown tombs. During the last several years, a wealth of evidence has been discovered in the Halil River Valley that indicates a "large and vibrant bronze age society"[97] that existed east of the Zagros Mountains.

Hundreds of stone vessels, some intricately carved, have been found that are similar to vessels previously found throughout the region, from the Persian Gulf to central Asia, and as far north as Uzbekistan and Turkmenistan. One artifact in particular, a stamp-seal impression (made from a metal or clay stamp usually in wax), excavated by Yousef Madjidzadeh, was etched with an unknown style of writing.[98] Art historians who have had the opportunity to view some of the stone artifacts date their artistic style to the middle of the third millennium BCE,[99] with some artifacts said to be as old as seven thousand years.[100]

Near the town of Jiroft, there are also the remains of massive structures. One building, which has been partially exposed, measures one hundred by two hundred feet. According to the director of the Archaeological Research Center Masoud Azarnoush, there are at least three hundred mounds yet to be examined. Preliminary excavations hint that a city as large as four hundred hectares may exist beneath the soil. Azarnoush also believes that during the third millennium BCE, this unknown civilization was as important as the Nile and Indus River civilizations.[101] Madjidzadeh, who held the position of archaeology chair at Tehran University during the 1970s, declares that what has been found are the remains of "a distinct civilization" that is "characterized by unique monumental architecture and by its own form of writing."[102] According to the Iranian Cultural Heritage News Agency, the unknown civilization to which the artifacts and buried ruins belonged has been nicknamed "the lost paradise" by the experts.[103]

Before the Pharaohs

This emerging scenario also helps explain one of ancient history's most perplexing puzzles: the apparent sudden appearance of Cro-Magnon cultures in western Europe. This sudden appearance of the Cro-Magnon is established fact. In France and Spain, hundreds of caves display thousands of truly magnificent images.

Forty thousand years ago, anatomically modern humans arrived on the Iberian Peninsula and brought with them complex social structures, diverse symbolic behavior, and what they are most recognized for, art. The suddenness with which they appeared and the speed with which they replaced Neanderthal man are remarkable. From our perspective tens of thousands of years later, it is no wonder that contemporary authors write of a "creative explosion" or "human revolution" when referring to the western European phenomenon of the Cro-Magnon. But did such a "creative explosion" or "human revolution" really happen?

David Lewis-Williams, professor emeritus of cognitive archaeology at the University of Witwatersrand, in South Africa, thinks not. Although it is understandable that archaeologists and anthropologists describe the appearance of the Cro-Magnon in these terms, they should not ignore the evidence of cultural precursors from Africa and the Middle East.

According to Lewis-Williams, the overall picture looks much less explosive, and the seeds of the "creative explosion" are to be found in Africa and the Middle East. Africa contains the earliest evidence for the human revolution and is likely the source of modern man.[104] If so, the interesting question is why there are not more archaeological sites in North Africa and the Middle East dating to the ice age. If modern humans came from the southern regions of Africa and migrated to the north and west, shouldn't areas around the temperate climes near the Mediterranean Sea be hotbeds of Paleolithic artifacts? Common sense dictates that they should. Here's why.

Waves of migration always extend from an established base where the necessities of life are easily obtained. In other words, migration begins at home. For the prehistoric Cro-Magnon cultures, home was Africa. In and around the Mediterranean, particularly in northern Africa and the eastern areas that are now submerged by higher sea levels, a civilization would have been built as a consequence of northerly migration patterns. Prior to the sea level rising, the Mediterranean was likely a series of freshwater lakes similar to the Great Lakes of North America, sealed from the Atlantic by the Gibraltar dam. It would have been an ideal location for a civilization to flourish.

For tens of thousands of years, the people of this civilization would have thrived and developed unique technologies and engaged in trade, only to be decimated by some catastrophe around 9000 BCE. Although precisely what the catastrophe was has been debated for years, there is no question that a cataclysm occurred that resulted in the extinction of numerous species around the world. Its consequence was a new topography for the Mediterranean region, as well as the rest of the world. Man, of course, was among the species that suffered but managed to survive.

If the greater Mediterranean area was heavily populated from France to the Mesopotamian Valley to North Africa and the movement of this population was from Africa, then if the heart of the civilization was destroyed, outlying areas such as France would have survivors that to us seemed to appear from nowhere, according to the archaeological record. Furthermore, the remnants of this civilization would appear to us as anomalies, such as the Great Pyramid, the megaliths at Nabta Playa, and Baalbek in Lebanon, as well as the unknown culture responsible for the temples and other structures on Malta.

The survivors, those who lived in Egypt and North Africa, struggled to carry on, and with the Sahara Desert growing, they found themselves migrating to the Nile Valley, where water was continuously available. As the inheritors and keepers of ancient knowledge and wisdom, whom twentieth-century Egyptologists refer to as the "dynastic race," they rebuilt their civilization with the help of newcomers from the northeast who had also migrated into the Nile Valley, providing an increasing level of manpower. Dynastic Egypt was, in a sense, the rebirth of a prehistoric civilization that once was.

The Testimony of Atlantis

Although there is no evidence that a specific city named Atlantis ever existed, the historical and archaeological evidence, in my opinion, describes a civilization reflected in part by not only Plato's story of Atlantis, but the Tower of Babel story as well. Interestingly, ruins thought to be those of Atlantis have been discovered by various researchers throughout the Mediterranean and North Africa, including, but not limited to, Tartessos in Spain, the Atlas Mountains in Africa, the Lake of Tritons in Tunisia, Corsica, Sardinia, Thera, Sicily, Egypt, and Lebanon. In June 2004 Rainer Kühne, of the University of Wuppertal, in Germany, claims to have discovered the fabled city in a salt marsh off Spain's southern coast. Satellite images depicting ancient ruins appear to match Plato's description of rectangular structures surrounded by concentric circles.[105]

If there were only ruins, little more could be said about Atlantis except that it is speculation. But the engineering expertise required for the construction of Egypt's Great Pyramid, and the astronomical knowledge depicted in the arrangement of the Nabta Playa megaliths, as well as the mysterious Trilithon of Baalbek and the histories of the "first" civilizations, leads me to a conclusion. The idea that an ancient, sophisticated civilization existed is more than speculation. Based on the evidence, it is a valid theory, and there is more in the depths of our remote past yet to be discovered.

Professional skeptics are eager to dismiss Atlantis theories, sight unseen, as a continuing concoction in the tradition of Francis Bacon and Ignatius Donnelly. I would have to agree with them *if* the Giza pyramids did not exist, *if* the Great Pyramid's interior design was simple, *if* the megaliths at Nabta Playa didn't exist, *if* the Trilithon of Baalbek didn't exist, and *if* the histories, both oral and written, that were passed down from the Egyptians, Sumerians, and Akkadians that tell of a previous civilization didn't exist. If all these anomalies and histories about these anomalies didn't exist, I would agree with the skeptics.

Those who consider the profusion of Atlantis theories to be a syndrome or disease must wonder why Atlantis theories won't go away. The answer is simple: There's too much evidence.

13

The Path of Osiris

The Principles of Consciousness

> *To wish to find proofs of evolution by starting from organic evolution is to start from the wrong end, because only consciousness "evolves," or enlarges, and the physical—the corporeal—adapts itself to it. This is clearly to give to Consciousness the nature of Being, and to bodies and their characteristics the role of instruments at the disposition of this Being. This is the way to view Reality.*
>
> R. A. Schwaller de Lubicz

Typically viewed as superstition, the ancient Egyptian ritual of mummification and burial has long been a subject of fascination in the Western world. In 1922, with Howard Carter's discovery of King Tutankhamen's tomb in the Valley of the Kings, we learned that it was traditional for the Egyptians not only to spend a great deal of effort in making the deceased look lifelike, but also to entomb the mummified body with numerous personal possessions. For us, it is impractical to bury so much wealth with the deceased, as opposed to handing it down to an inheriting generation. The common interpretation of the ancient Egyptian burial practice was that the deceased needed all these things in the afterlife. Such a rationale is overly simplistic and fails to reach the heart of ancient Egyptian beliefs.

Ancient mortuary texts provide much of what we know of Egyptian beliefs, particularly the afterlife. It seems to have been their hope not only to transcend terrestrial life but also to achieve immortality as a heavenly star.

Egyptian Spirituality

For the ancient Egyptians, the two most important concepts concerning the human experience and the afterlife were the ka and the ba. Although not an exact analogy, the ka and the ba are what the Western religious tradition might refer to as spirit and soul. A third concept, the ankh, represented by the crested ibis, was immortality. Symbolized in Egyptian art by upstretched arms, the ka was believed to be the part of man's consciousness (the personality and inner qualities) that related to the immediate world. The ka is that part of us connected to the physical body: where the body lived, its possessions, as well as the people it associated with. The ka was the energy or spirit that emanated from the person. The ba, depicted as a winged human head or sometimes a human-faced bird, represented the part of consciousness that is immortal, the eternal force that causes all of nature to exist.

When someone passed away, it was the hope of the family that the person's ka and ba would unite. To accomplish this merging, the person's possessions were gathered together by the family and placed in the tomb along with the body. Food offerings to the ka of the deceased were interred with all the other goods. It was also believed that if the body were allowed to decay, the deceased's ka would be reabsorbed into the energy field that is the fabric of nature, making it impossible for an eternal union with the ba. To prevent this decay, the deceased were mummified.

When the ba and ka were joined in the afterlife, ankh, or immortality, would result as the fully resurrected and glorified essence of the deceased, reaching beyond the limits of an earthly realm and thereby achieving eternal life. For this reason, the tomb was referred to as the Per Ka (house of the ka) and the priests, who were in charge of it, the priests of the ka.[1] This is also why the temple was called Per Ba (house of the ba).

At death, the ka is separated from the body and naturally seeks a means to take form again. But if the person harbored regrets or violent desires, the ka would seek "any substance whatsoever, psychic in particular, borrowed from a living being, in order to return to a ghostlike shadow-existence."[2] Only if the ka were able to unite with its ba would the individual person have a continued existence through unity or "oneness" with its ba.

The ba, the animating spirit of nature manifested as physical form in Man, was the source of a person's ka, separate yet unified in a person's physical form. For this unity to continue into the next life, the ka would have to be transformed in the tradition of Osiris, whose essence was reincarnated as his son, Horus.[3]

Because of the modern popularized definition of reincarnation, in which a person's mind and memories are reborn in an entirely new body years, centuries, or millennia later, the Egyptian concept of reincarnation requires some explanation.

The Ancient Egyptian Concept of Reincarnation

The Egyptian concept of reincarnation was associated with the cyclical character of the natural world. Since mankind was a part of nature, mankind would also be subject to this cyclical character. The ancient Egyptians did not view the concept of reincarnation through a dogmatic system, but through an understanding of the human experience that was symbolically explained as the myth of Osiris. It is a story of natural principles, not people.

According to the myth, Osiris is murdered by his brother, Seth, dismembered, and his body parts scattered throughout the land. Realizing what has happened to her husband, Isis, who in most versions is aided by her sister, Nephthys, searches the land for Osiris's remains and buries each piece where it is found, except for his penis. According to one version of the story, fish have eaten it, so she replaces it with a wooden prosthesis. After all Osiris's body parts are found, Isis brings him back to life, but not in physical form, as might be expected. In the myth, both Isis and her sister are depicted as birds,[4] representing the volatile aspects of Man, the principle that is the source of life, the soul. So Osiris's resurrection is metaphysical, not physical.

Isis, destined to bear the child king, restores Osiris's libido as the "Hand Goddess aroused the penis of the creator to create the first life." She does so as a bird by using "her wings to fan the breath of life into Osiris,"[5] while Thoth and Anubis protect his corpse. Once resurrected, Osiris sires a son, but does so without a penis. After ten months, Isis gives birth to Horus, who was already king while still an egg.[6] Thus, the resurrected Osiris continues on, reincarnated in his son, Horus. Such a story is clearly metaphorical.

In less symbolic terms, Osiris, as a physical being, dies and his body returns to the soil. Yet the essence of his person (his ka) remains and is resurrected to seed life again through the ba, by way of nature's cyclical character. The role Isis depicts is the volatile or abstract (spiritual) essence of Man, which is the true power of life. Therefore, the new

Man, Horus, is the product of nature's mystical powers just as much as he is of procreation. Since the true source of Man's life is not really procreation, but the enigmatic Self that exists in a unified, nondimensional state, it is that source that incarnates time after time. *This is the original meaning of reincarnation: continual incarnation of a single entity.* This entity is the quality that is Man, the archetypal Man of Jungian psychology—the Son of Man—who is continually manifested as form through nature's process. The essence of the individual, the quality of his or her personality, returns to the prenatal state of unity and oneness to be born again. In the myth of Osiris, physical death holds no power over the resurrected Man, just as in Gnostic Christian belief.

The individual Egyptian could choose to believe or disbelieve in reincarnation. To believe simply means "Be good, be just, be charitable, and the sooner you will reap illumination," and not to believe means "Be evil, be cruel, be egotistical, whether you believe or not, you will pay."[7] The concept is inherently nonreligious and more a principle of life, even science, which is why the ancient Egyptian religion has continued to baffle those who approach it from an exoteric and dogmatic point of view.

The story of Osiris is about the transformation of one's thought (self-revelation) in the realization of nature's unity, told symbolically as the uniting of the earthly dismembered Osiris with his spiritual body; Osiris, who was the ruler of Earth (symbolizing the physical body), becomes united with Ra, the Lord of Heaven (symbolizing the abstract body). This explains the dual, but unified, nature of reality.

For the believer, life was a search for or path to enlightenment—the path of Osiris. This is the principle of Osiris: Resurrection is the awakening and acknowledgment of the Self that is at the heart of every person alive or who has ever lived. And it is the Self that reincarnates its character. The newest translation of the Papyrus of Ani (Book of the Dead) illustrates this principle of reincarnation:

> Yesterday, which is pregnant with the one who shall give birth to himself at another time, belongs to me. I am the one secret of the Ba-spirit who made the gods.[8]

But this phrase may be better illustrated in Budge's translation:

> I am Yesterday, Today, and Tomorrow; and I have the power to be born a second time. I am the hidden Soul who createth the gods.[9]

The individual is not that which reincarnates but rather the "one secret of the ba-spirit" or the "hidden soul." In modern terms, God is that which is being reincarnated.

Also from the Book of the Dead, it is clear that the principle of Osiris is, in fact, the basis for the ancient Egyptian theocratic style of government. In the final invocation of the papyrus, Ani (the scribe) tells the meaning behind the double crown of the pharaoh: "O white crown of the divine Form!"[10] he exclaims, meaning that the physical cosmos is the divine Man who incarnates as men. But the essence of men, Ani writes, is that they are eternal through the characteristics of feeling and perceiving, the principle of consciousness, which is the "lord of the red crown."[11] Horus, as Osiris, is "the Child" who has been reincarnated. With this understanding, the Sed festival is put in its proper context. The unification of Upper and Lower Egypt was not a military and political victory; it was a celebration of eternal life within the Man Cosmos.

The Egyptian Book of the Dead is also clear on this matter. Man is the sum total of all the neters. The deceased declares:

> *I am the holy knot within the tamarisk tree.*
> *I am Ra who establish those who praise him.*
> *My hair is the hair of Nu.*
> *My face is the face of Ra.*
> *Mine eyes are the eyes of Hathor.*
> *Mine ears are the ears of Ap-uat.*
> *My nose is the nose of Khent-sheps.*
> *My lips are the lips of Anpu.*
> *My teeth are the teeth of Khepera.*
> *My neck is the neck of Isis, the divine lady.*
> *My hands are the hands of Khnemu, the lord of Tattu.*
> *My forearms are the forearms of Neith, the lady of Saïs.*
> *My backbone is the backbone of Sut.*
> *My privy member is the privy member of Osiris.*
> *My reins are the reins of the lords of Kher-aba.*
> *My breast is the breast of the awful and terrible One.*
> *My belly and my backbone are the belly and backbone of Sekhet.*
> *My buttocks are the buttocks of the eye of Horus.*
> *My hips and thighs are the hips and thighs of Nut.*

My feet are the feet of Ptah.
My fingers and leg bones are the fingers and leg bones of the living uraei [serpent].[12]

Osiris, as the principle of man, is the Man Cosmos who changes form (evolves), but is "the only One born of an only One"[13] who moves about in his circular course. Every part of his body is "the member of some god,"[14] and Thoth, who represents the principle of knowledge, shields the body. The source that moves him forward is unknown, yet states, "I am yesterday, and my name is 'Seer of millions of years.' I travel along the path of Horus the judge. I am the lord of eternity; I feel and I have power to perceive."[15]

In this passage, the ancient Egyptian philosopher has equated the Self who feels and perceives with the unknowable All, who, in traditional terms, must be referred to as the creator God. Ani then describes those same characteristics as the essence of Man:

> I am he who is unknown, and the gods with rose-bright countenances are with me. I am the unveiled one . . . the only One, [son] of an only One. I am the plant which cometh forth from Nu, and my mother is Nut. Hail, O my Creator, I am he who hath no power to walk, the great knot within yesterday. My power is in my hand. . . . I am not known, [but] I am he who knoweth thee. I cannot be held with the hand, but I am he who can hold thee in his hand.[16]

During the Middle Kingdom, the themes of reincarnation and resurrection are also apparent in the Coffin Texts (ca. 2134–1782 BCE). In spell 330, the ancient Egyptian writes:

Whether I live or die I am Osiris,
I enter in and reappear through you,
I decay in you, I grow in you,
I fall down in you, I fall upon my side.
The gods are living in me for I live and grow in the corn that sustains the Honored Ones.*
I cover the Earth,
whether I live or die I am Barley.

---

*Neters, or principles of nature.

I am not destroyed.
*I have entered the Order,**
I rely upon the Order,
I become master of the Order,
I emerge in the Order,
I make my form distinct,
I am the Lord—of the Chennet (Granary of Memphis?)
I have entered into the Order,
I have reached its limits.[17]

And in spell 714:

I was (the spirit in?) the Primeval Waters,
he who had no companion when my name came into existence.
The most ancient form in which I came into existence was as a drowned one.
I was (also) he who came into existence as a circle, he who was the dweller in his egg.
I was the one who began (everything), the dweller in the Primeval Waters.
First Hahu (the wind that separated the waters) emerged from me
and then I began to move.
I created my limbs in my "glory"
I was the maker of myself, in that I formed myself according to my desire and in accord with my heart.[18]

These concepts of reincarnation and resurrection are deeply embedded in the questions concerning the nature of humanity and the nature of reality. In part 1 of this book, "The Mysteries of Modern Science" recalls that the essence of "man," from a physics point of view, is epitomized by the question of measurement, which, despite whatever tools or instruments are used, ultimately is the brain.

And there is sufficient evidence from physics experiments to establish that the world we observe is born out of our own observation. For this reason, consciousness and perception that are formed within the mind and brain are "the observer observing" (see page 62).

*The cosmos.

Consciousness is biological, but it is also cosmic in that the compounds, molecules, and other structures that make up our form have been traced to the creation of heavy elements from within the stellar cycle. Scientifically, a gap exists in our knowledge between the creation of the elements required for our form and the actual creation of that form. Effectively, there is only "creation" for the individual. We humans are our own creation. I am not talking about procreation but rather about the fact that we exist and to the best of the physicists' knowledge, philosophically speaking, we exist because of observation. If one chooses to believe that life randomly came into existence, then the universe is self-creating. Ironically, such a concept is the same as believing that an intelligence created the universe. If some "unknown desire to express" created the universe, from what could it create except itself? Again, there is self-creating taking place. The effect of this self-creation, conscious life, is perception. What enables us to perceive is the physical universe. So, between original cause and perception, both of which are abstractions, there exists the immediate, concrete world we experience. It is a giant step in thought, but in applying the anthropic principle to this sequence of cause and effect, the idea emerges that Man is not so much form as he is an abstraction, or that which perceives. Form is merely the means by which perception is attained and the act of self-creation is expressed. The faculties of memory required for self-awareness develop only after the second or third year of life; and after about the eighth year, awareness of thought and the question of "why am I me" emerges. These are surely the effects of biological growth. All the while, there is the observer observing, waiting for the proper development of the biological structures needed for experience and expression at its fullest. Only that which is a function of biological processes conforms to the concepts of age and time. The observers that we truly are know neither age nor time, and so there is a mirror to remind us of our place in the socially structured hierarchy of civilization.

It is within this aspect of cosmic consciousness that the observer exists, for it must be the generator of the system required for experience. Reality is energy, a wave function, as quantum physics tell us; but it is only the immense complexity of the brain transforming the energy wave into the observed particle that makes everything real. So, within Man, there is the eternal essence of life, which is the observer (the macrocosm), and the consciousness produced through biological form, which is individual perception, and all the qualities that make up personality (the microcosm). In all respects, the observer is a fundamental aspect of reality just as much as are the spatial three dimensions. Yet its true nature outside the concept of experience is unknown and can only be referred to as the source.

For the Egyptians, the observer—the source that is the cosmic energetic essence of life—was called the ba, and the individual's personality (energy as it is imparted to others) was called the ka. Resurrection was not a physical raising of the dead but instead a truth of nature itself when the individual embraced the scientific principles of reality, however esoteric, and the ka was identified with the ba. Reincarnation was not the guaranteed rebirth of an individual's specific personality as it relates to the physical world. Rather, reincarnation was the uniting of the ka (the energy of the specific person) with its ba, and the recycling of the ba (animating force) back into form. In other words, what reincarnates is what all people share by nature: the eternal, energetic source of life, the ba. Hence, the Egyptians prevented the destruction of form through mummification, thereby preventing destruction of the ka. True reincarnation, then, was achieved when the ka reunited with the ba after death, resulting in ankh, or eternal life. The general context in which reincarnation must be understood, however, is through an esoteric theory of evolution that pharaonic Egypt espoused.[19]

The Ancient Egyptian Concept of Evolution

Biological evolution and the enigmatic concept of speciation, in which one species genetically mutates into another, higher-ordered species, is purely an exoteric approach. The Darwinian theory of evolution is best described as a biological (mechanical) process for the building of form. Where Darwinism fails is in its exclusively exoteric point of view. Materialistic evolutionary theory does not address the esoteric, particularly creation, or what other process governs the sequence of development from microbe to man. It thus misses a vital aspect of evolution, particularly with the notion that an organism's interior drive creates an organ for a particular function, such as hearing, seeing, tasting, feeling, or smelling.

"Function" is perhaps the most abstract of all concepts, and, according to Schwaller, the least likely to have a direct influence on any physical entity. [20] Darwinism suggests that the relationship between the organism and its environment, and what we experience as an amalgamation of functions, can be reduced to a fact that function creates organ.

We have a tendency to view life as form (body), which creates an impasse in understanding our true nature. Any attempt to describe how the human body arose out of some primordial soup is an exercise in mental frustration. But approaching the subject from the perspective of function gives a deeper meaning to the principles of evolution that is *not at all* in conflict with the general idea of creation.

There is no question that our organs (eyes, ears, nose, and so forth) are tailored to carry out those functions, and that certain development of an organ does occur through its functional activity. What is not recognized in the creation of an organ is the effect of a living environment. Why would there be any drive for an eyeless organism toward the faculty of sight? For Schwaller, the answer lies in the vibrational construct in which we exist. (All things are reduced to atoms, which reduces to movement or vibration.) The energetic essence of this vibrational construct makes its imprint onto matter through the organism's reaction. Organic life is formed out of the symbol or image of an energetic impulse from the environment, which is not only a terrestrial environment, but a cosmic one as well. (The circumstances for life on Earth are dependent on its cosmic structure—the planet's distance from the sun, the placement of the solar system within the galaxy, and so on.) Whatever the function or principle we define as an experience, that function will possess an existing organ and make use of it at the level of consciousness presented by the organism.[21]

At one time, Earth was lifeless; but through some event or events, organic life became distinguished from the inorganic through functions such as respiration, digestion (assimilation), and reproduction. Such primitive functions are evident even in the simplest of single-celled creatures and persist throughout the animal kingdom, man included. Schwaller contends, "One can philosophically argue that everything that characterized man has been conceived since the first manifestation, and the sequence of evolution from the simplest forms is but an increasingly perfect organization of these characteristics."[22] The mystery is how this transformation from microbial form to human form was accomplished.

But if we recognize that man's natural state is abstract (spiritual), and that the sole purpose of the universe is a means of expression and experience, then we can also say that man is "prefigured" with the first manifestation of life. Identified with consciousness and from the point of view of cosmic consciousness (Egyptian ba), the manifestation of the human form and its individual energy (Egyptian ka) is little more than the imprisonment of cosmic consciousness in matter, resulting in that matter attaining consciousness seemingly all by itself. In recognition of this microcosm from macrocosm, it is the liberation of the ka (resurrection) and its identification with the ba (reincarnation) that is the grand quest.

Yet liberation (the motivating force that propels true speciation) is a cosmic function. The march of organic incarnation into consecutively more complex forms is what develops the "innate consciousness that each individual bears from birth."[23] During each successive stage of evolution, a new embodiment of a function dominates the organism, such as the eye with the bird or the sense of smell with the dog. With this progression of form and function in mind, human history becomes inseparable from the history of the animal kingdom, from the first single-celled creature to modern man. Although each step in the ladder of evolution represents the addition of an essential function, ultimately all functions are innate in man. For life on Earth, Man, of course, is the final function.[24]

According to Schwaller's pharaonic evolutionary theory, how a species evolves into another, more advanced form involves the destruction of that species and the decomposition of its form, where all that remains is mineral content of that form, what Schwaller hermetically refers to as "fixed salt." These fixed salts, which are never destroyed and are recycled through the food chain, provide the physical framework through the influence of cosmic energy (consciousness) for a new form.

From a purely competitive and adaptive point of view, it is noteworthy that catastrophism has recently been recognized as an important factor among the mainstream tenets of evolution. A case in point is the Yucatán asteroid impact sixty-five million years ago that ended the reign of the dinosaur and began the epoch of the mammal.

For Schwaller, simple chemical reactions are responsible for the transition from one form to another, more complex form. When a new chemical is introduced into a liquid chemical composition, which is receptive to combining with other bodies in a solution, a moment eventually arrives when one or another of these chemical bodies is molecularly liberated in transition between the first compound and the new one.[25] At this juncture, the evolutionary principle in nature can "consciously" produce mutations of genes and align the viable varieties of genes to form.

In the Hermetic tradition, it is believed that all organisms, plant and animal, are composed of a volatile part and a fixed part. Upon an organism's destruction—by fire, for example—the volatile is evaporated through chemical reaction, leaving a residue of ash. Within this ash is an alkaline salt, which, if sown into the earth, promotes the generation of form by its absorption into other life-forms. According to Schwaller, this is the true and esoteric meaning of the Egyptian legend of the phoenix, which rises from the ashes as an initiatory tradition "for revealing a mode of generation that the cellular seed [biological procreative] cannot realize."[26]

Salt and other minerals are elementary nutrient substrates necessary for life, and are important for an organism's structure and growth. The presence of salt in cells, which is necessary for cell metabolism, has been recognized by modern science for more than a hundred years. These cell salts, first identified by the German Wilhelm Heinrich Schuessler (1821–1898), vary in chemical composition and range from calcium fluoride to sodium sulfate. In all, Schuessler discovered twelve different tissue salts that were required for human health.[27] Thus, in essence, salt really is the inscription of life.

Every cell in an organism's body relies on energy from simple molecules, such as ATP and a related metabolite called ADP, in which salts are an important part of the processes of digestion and respiration for ordinary tissue. For example: The function of a brain cell (neuron) is to fire an action potential (AP) to another brain cell. (This is how the brain functions as the governor of autonomic processes—heartbeat, for instance—to physical movement of the body, as well as thought process.) Neurons act as information relay stations; the AP is an informational medium. When a neuron fires an action potential, it collects inducted ions from excitatory postsynaptic potentials (EPSP) and inhibitory postsynaptic potentials (IPSP). At this level, any change in the cell membrane potential is local. With all aggregate potential summed at the axon hillock, however, a cascade of sodium ion (Na+) influx perpetuates the AP down the axon to the terminal button.

Minerals such as salt also provide the strength that the bones need to support the weight of the organism's body. Between 60 and 70 percent of mature bone is mineral. The remaining 30 to 40 percent is an organic component composed primarily of collagen fibers and water. Within the structure of bone tissue, calcium salt and phosphate salt combine to form hydroxyapatite crystals, which give bone its incredible strength.

Schwaller believes that salt is the nucleus of inscription and the key to evolution. The inscription of information is from the consciousness that is Man into the physical world. The natural salt that is ingested becomes part of the body, and its use is altered, Schwaller believes. When the body dies and is buried after complete decomposition, only the salts remain. These salts then become part of the soil and flow through plant life back into animal life. Over a long period of time, these salts accumulate changes and serve as the basis for the evolution of form. Salt always survives, so it is the permanent part of Man's existence, and, in a sense, the ultimate physical presence. What is being inscribed is our existence, which is, fundamentally, knowledge and experience. DNA is a blueprint for biologically replicating what already exists, but it does not survive death of the individual. On the other hand, the component salts that constitute the body do, becoming a part of the soil after putrefaction and reentering the nutrient cycle, where it is ingested by plants, and the plants are then ingested by the animal. "The natural path of metempsychosis [reincarnation] is through the plant," Schwaller writes, "because during its growth, the plant is first to resorb [to absorb again] the fixed salt."[28] Salt, with its biological importance—in muscle movement, digestion, the regulation of osmotic cell pressure, and the firing of nerve cells—appears to be the secret interface for the biological manifestation of consciousness in physical form. From our brains to our bones, salt penetrates every aspect of our physical bodies, which in the end are nothing more than the instruments of consciousness. This concept of evolution, which Schwaller defines as evolving consciousness, is the essence of the anthropic principle and the sacred science of the anthropocosm:

> Biology shows us that all the vital functions that characterize man have been inscribed in living matter since the beginning, that all of nature is like a womb for the gestation of the human being. From the origin a perfect end is foreseen, just as the fruit itself is virtually contained in the seed. In other words, the final product is necessarily implied in the Cause. This doctrine, this understanding of evolution presents the role of consciousness as the determining factor for life. Consciousness alone evolves. It does so by enlarging the instrument for itself. It expands, enlarges itself passing from the physical to the vegetal to an emotional state, then to a mental, and, finally, by means of a mental consciousness of abstraction, to spiritual consciousness.[29]

Such a principle built into the order of nature also explains how individuals can transcend themselves even in their present state. The consciousness with which matter has been endowed is a sequence of the abstract (spiritual) identity becoming conscious of itself. The end result of this process is conscious immortality.[30]

Cosmic Evolution as the Science of Life

Scientists argue that, since we are made from stardust, we are literally children of the stars. Chemically, we are 65 percent oxygen, 18.5 percent carbon, 9.5 percent hydrogen, and 3.2 percent nitrogen, with less than 1 percent calcium, phosphorus, potassium, sulfur, sodium, chlorine, magnesium, iodine, and iron. We also contain trace elements of chromium, cobalt, copper, fluorine, manganese, molybdenum, selenium, tin, vanadium, and zinc. These elements are the results of a cosmic process.

Early on in the universe's life, the only matter that existed were hydrogen and helium. The heavy elements (such as iron, carbon, and calcium, for example) were created by exploding stars that later condensed into what came to be solar systems and planets, ours included. It is imperative to recognize that evolution begins not with the first life-form on Earth, but with the universal creation of matter at a time when physicists agree the laws of physics have no meaning. This is the significance of the cosmologist's anthropic principle, as well as the heart and soul of Schwaller's anthropocosm.

Michael Shermer, a defender of science and the editor of *Skeptic* magazine, agrees on at least this final point—that we are "children of the stars." In a recent issue of *Scientific American,* he defines spirituality, or the nature of the soul, as "the pattern of information of which we are made—our genes, proteins, memories and personalities,"[31] and concludes, "[s]pirituality is the quest to know the place of our essence within the deep time of evolution and the deep space of the cosmos."[32] For Shermer, science is spiritual, for it too seeks the answers about who we are and where we came from. In the article, Shermer quotes Carl Sagan at length:

> The cosmos is within us. We are made of star stuff. We are a way for the cosmos to know itself. . . . We've begun at last to wonder about our origins, star stuff contemplating the stars, organized collections of ten billion billion billion atoms contemplating the evolution of matter, tracing that long path by which it arrived at consciousness. . . . Our

> obligation to survive and flourish is owed not just to ourselves but also to that cosmos, ancient and vast, from which we spring.[33]

If we sprang from the cosmos, as Carl Sagan asserts and Michael Shermer reiterates, then to become a conscious function of the cosmos we would have to expect that speciation as a result of random genetic changes is not random at all but rather a self-referencing system where conscious organisms react to the environment to become self-conscious.

The neo-Darwinian theory of evolution has its difficulties. For example, natural selection—survival of the fittest—may just as well be a function of reproduction as it is of survival. Organisms with the ability to reproduce in greater numbers than their competitors and predators stand a greater chance of survival than organisms whose reproduction is poor, regardless of environmental survival capabilities. So, nature's selection process may be more sexual than survivalist.

There is also the problem of explaining the creation of animal species, such as the Cambrian explosion. Where once there were only bacteria, plankton, and algae, six hundred million years ago suddenly animal life appeared, which blossomed into even more animal life during the next seventy million years. An even bigger problem is how the first living organism emerged from inanimate matter.

It seems to me that the evolutionary approach to the emergence of biological life, according to science and apparent in the fossil record, is more of an assumption based on observation than it is a mechanism. This is not to say the model is wrong, but rather that it is incomplete.

Why should accidental genetic mutation result in more-complex (and healthy) organisms? Accidental genetic mutation is just as likely to be toward the less complex. Although it is certain that speciation has occurred, given the vast diversity of life today and the fossil record attesting to past ecosystems and organisms, the mechanism behind speciation remains elusive.

By approaching evolution through the principles of unity and consciousness, with consciousness as a fundamental characteristic of the universe, we find a viable mechanism for speciation. "Quantum" evolution, as proposed by molecular geneticist Johnjoe McFadden and physicist Amit Goswami, puts forth a theoretical framework for the mechanism of speciation that is in agreement in principle with Schwaller's concept of expanding consciousness.

To understand how this theory of quantum evolution works, we'll need to briefly revisit some important concepts of quantum physics. In chapter 2, we discovered that reality, at its most fundamental level, is a state of superposition of energy waves that we experience (a priori) as physical existence. How this perceived transformation of energy into particles occurs is through what physicists call state vector collapse, which is a characteristic function of the brain as a result of conscious observation. In state vector collapse, the superposition of waves becomes a single event in three dimensions. Yet the brain itself must be represented by a superposition of waves until the collapse occurs. This creates a paradox of circular logic. Without the awareness of the brain, there can be no collapse, and without collapse, there is no awareness for the brain. This back-and-forth logic continues for infinity.

To resolve the paradox, the two events must be understood as simultaneous and dependent. In other words, subjective experiences in the mind and the objective awareness of the brain are cocreators of the immediate world. But this is also a matter of perception based on our methods of questioning. Since we always see particles (objects) and never waves, it must be the case, as physicist John Von Neumann (1903–1957) proposed, that conscious observation terminates the infinite and circular reference between the brain and state vector collapse, and this is why consciousness cannot be an emergent phenomenon of the brain. If it were, the circular referencing would never terminate.

The direction further deduction must take is that consciousness is transcendent to three-dimensional space and nonlocal. As a fundamental characteristic of the universe, consciousness must create the objects we observe as well as the subjective perception we experience. For Goswami, the theater of consciousness serves as the container and director of Being:

> We have to see that consciousness is the ground of being; we have to see the whole reality as consciousness and matter as quantum possibility waves within it.[34]

Gene mutations are caused by the replacement of a single base nucleotide with another nucleotide (called point mutations), which is the result of either radiation or some other kind of chromosomal rearrangement of a quantum nature.[35] The motion of subatomic particles, electrons and protons, existing within the DNA double-helix structure causes mutations. For geneticists, this phenomenon of gene mutation is known as DNA base tautomerization, which is how chemists describe how a particle can be in two or more places at once. Protons that form the basis of DNA coding are not located in specific positions. Rather, they are spread out along the length of the double-helix structure. All the while, various positions for the coding protons correspond to different DNA codes. This means that DNA exists at the quantum level in a superposition of mutational states and as a molecule.[36] But how are these possibilities manifest as a change in form?

Only observation collapses the superposition of waves (selects a single choice from a set of possibilities) within the cell, which leads to the question, How and when does consciousness choose? The observable physical characteristics of the genetic change must first exist in this subquantum world of possibilities (what Heisenberg called *potentia*). Only after enough change occurs does physical choice arise from the possibilities, which involves a number of structures such as regulator genes, structural genes, RNA, and proteins.[37]

The possibilities of form (mutations) are inherited by subsequent generations, and as life continues, more mutations accumulate in the unexpressed world of possibilities. Finally, at some point, one or more new amino acids are created that change proteins. This leads to a new physical expression. The change in form still exists only as a possibility. However, when consciousness, in referring to its self (self-reference), selects from all possibilities, a new way of re-creating itself occurs. If the organism does not interbreed with the old, a new species emerges.[38]

In quantum evolution, the living cell is viewed as a conscious, self-referencing system, as biologist Bruce Lipton suggests in *The Biology of Belief*. Its choice of form, the basis for speciation, is derived from the propagation of uncollapsed superpositions of possible genetic states. This leads to the possibility that an archetypal form—a morphogenetic field (a field containing the information necessary to shape the exact form of an organism)—is necessary to contain the specific form that defines the species. When matching qualities of Consciousness and biological form between the archetypal form that exists as an aspect of consciousness and its manifestation takes place, then state vector collapse occurs.

In this quantum manner, evolution becomes Schwaller's creative principle and the defining characteristics of life—the creation of environment and the organisms as they relate and react to that environment. Only through conscious choice does evolution speciate, as well as maintain a direction for greater complexity and the order of biological life. At the same time, later (more-evolved) species, which have a greater gene pool to select from, have less need to adapt to the environment.[39]

We are made from stardust through the self-organization and self-reproduction of the cosmos. Yet self-organization and self-reproduction are natural extensions of self-reference and self-evolution.

The ancient Egyptians viewed the order of life in a similar manner, for, as Miroslav Verner puts it, the "noblest of men were believed to become stars and join the magnificence of creation; the life force for all that exists."[40] Behind ancient Egypt's symbolism of becoming a star is another way of expressing the natural principles of the cosmos we call evolution.

Principle of the Anthropocosm

In chapter 1, these questions were introduced as the questions everyone wants answers to: *Who are we? What are we? Where did we come from? Why are we here?*

There are many answers, more than six billion, in fact. Whether scientifically or religiously oriented, each person has his or her own answer, and regardless of the general commonalities, each answer is unique. Such is life, and as a result there are myriad theories and beliefs, from New Age crystals to orthodox religions, as well as unorthodox religions, materialistic philosophy, and Darwinian evolution, with various mixtures and matches in between. For those who take their beliefs too seriously, one person's religion becomes another person's heresy; the pontification of any one religion thus becomes a condemnation of anyone who happens to think or believe differently.

With a mystery at the heart of life, there can be no right or wrong theological or philosophical system. Moralities, altruism—the virtues of mankind are more a question of conscience and empathy than of belief. As is science, theology and philosophy are methods of explanation, which allows an initiatory experience to a deeper understanding of one's self and the cosmos. Ultimately, the answer to the questions—who am I, what am I, where did I come from, and why am I here—depends on the individual, and how he or she chooses to relate to the Self and to others, as well as to all of nature.

The sacred science of the ancients and its secret wisdom, which has been passed down through uncountable generations, and which the new science is discovering, brings everything together under the umbrella of a single, absolute truth: the self-aware conscious experience.

Our true identity is the Self, which is the observer within. And our true nature is abstract, a substance that can be referred to, through deduction, as nonpolarized energy. Without beginning or end, it is our role to experience and express and create.

All the various religious traditions throughout history, whether ancient or modern, approximate these truths. Although reality appears scientifically objective, it is so only because the scope of investigation is limited. What we are absolutely certain of and know as real—our conscious experience—is subjective, but is such that the stability and order of the universe are mistakenly interpreted as objective. The physical universe as we know it exists because we exist to perceive it. It is our perception of it that makes it real; otherwise, it is nothing more than a superpositioning of energy waves.

Characteristics of the Universe

The nature of the universe—physical laws, some of which remain a mystery—is limited by its own intrinsic characteristics. We can describe these characteristics, but we cannot alter them.

The universe is a vast expanse containing untold billions of galaxies. Our home, the Milky Way galaxy, is more than a hundred thousand light-years across, and we sit at the edge of one of its spiral arms. The light from distant galaxies that astronomers gaze at from their observatories left its source millions of years ago, and for some galaxies, billions of years ago. Pondering our insignificance, what our space represents compared to the apparent infinite size of the universe, is humbling.

Astronomers have been on the hunt for planets like ours, warm and friendly to carbon-based life with a circular orbit about their host star. Although planets have been found, so far none is like ours. The assumption in this hunt for Earthlike planets is that they might support life, given its biological source. If life could develop here on Earth, it may very well have developed elsewhere. It is difficult to argue against this assumption, but it is also just as difficult to discover that intelligent life on some other planet does in fact exist. Most of what astronomers view at night occurred so far in the distant past that whether those stars still exist is questionable.

The Law of Gravity as the Principle of Unity

What has been discovered in the grand scientific quest to understand the universe is the cosmic principle of unity: gravity. If not the biggest, gravity is one of the biggest scientific mysteries. Physicists can describe how gravity works, but they do not really know what it is or where it comes from. According to Einstein's theory of relativity, gravity is "curved space" pulling objects with mass together. This explanation of gravity tells us little about what the attraction actually is between objects with mass. Gravity is an intrinsic principle of mass, and mass is "configured energy," where $m = E/c^2$, which is a fundamental mystery. The reality of this mystery is that gravity is the functioning of unity. Angular momentum (motion) and the gravity of all celestial bodies balance each other in the universe.

If gravity is regarded as being situated in time and space, it is nothing more than an exoteric fact as a result of quantitative analysis. If gravity is viewed in terms of an idea applied to the essence of life, however, it becomes a unifying principle that ensures the stability as well as the continuity of the universe. This is the cosmic reality of gravity: *the functioning of a will to unite.*

With gravity as a unifying function, there is no true separation between any specific phenomenon (human beings included) and the rest of the universe. For example, Earth depends on the moon's gravity for its stability of rotation, as well as the sun for the energy and gravity to sustain biological life. There is also protection provided by Jupiter's gravity, which helps prevent errant objects from colliding with our home. The solar system itself, on which we are dependent, is part of a larger system called a galaxy. We now know that cosmic black holes are not just infinitely dense matter traps, but function as the source of galaxy formation through gravity, and that the naturally occurring elements that form our planet's physical composition were the result of a stellar explosion.

This balancing of motion and gravity, the fundamental characteristic and force of the universe, leads to a sequence of events whose outcome results in matter relating to itself in complex ways, thus resulting in the production of elements, from lighter ones to heavier ones, because of the stellar life cycle. Through an unknown force, this outcome (matter) became animated through the complex interaction of molecules, which contain the specifications of its expression (life), including its reproducibility (DNA) with the intention of continued existence.

The culmination of this sequence of events is the principle of existence, or Being, in which matter, in complex form, is conscious and self-aware. The final effect of the universe's characteristics is the human form, with its ability to perceive and measure. All aspects of the universe, whether they are behaviors or objects, are therefore measurements based on man's perception.

In his experiments, Heisenberg recognized the principle of unity:

> Actually the experiments have shown the complete mutability of matter. All the elementary particles can, at sufficiently high energies, be transmuted into other particles, or they can simply be created from kinetic energy and can be annihilated into energy, for instance into radiation. *Therefore, we have here actually the final proof for the unity of matter.* All the elementary particles are made of the same substance, which we may call energy or universal matter; they are just different forms in which matter can appear.[41] [Italics are author's emphasis.]

The characteristics of the universe emanate from an original cause, whatever that cause may be. What we can say about this unknown cause is that it is the origin and source of matter. Since matter is composed of polarized energy, the source is assumed to be nonpolarized energy.

The obvious conclusion is that the characteristics of the universe include, and thus will always lead to, a biological sequence of adaptability and change that ultimately is physiological Man. It is also an obvious conclusion that Man's conscious experience is a result of biological consciousness as it relates to the vital source of that consciousness. (This relationship is the definition of self-awareness.) Since consciousness is the result of the complex relationship matter experiences as it relates to itself, the source of biological consciousness must also be an intrinsic characteristic of the universe and contain the qualities that biological consciousness expresses.

The Law of Consciousness as the Principle of Self-Awareness

We also know that space is not a perfect vacuum as was once believed but rather an invisible bubbling ocean of virtual particles. Such is the physical makeup of the universe, yet the fact that we are here to perceive and explore this cosmos, as well as being the result of it, has yet to be explained.

Consciousness is absolutely a fact, just as much as the three spatial dimensions that make up physical reality. How can consciousness *not* be a function of the cosmos?

We humans function in the immediate, observable world as a result of brain activity. With its processing of external stimuli, the brain is the seat of consciousness, as neuroscience declares. There can be little question about that. Being self-aware, however, having the ability to reflect on one's Self, requires another variable that is not so easily described or understood. This reflection is an exchange of information between two "things" or natures. There is also the question of what specifically is being reflected, and to whom (or what) it is reflecting.

Nonetheless, the exchange of information—knowledge—which occurs continuously, does so as a result of the environment acting on itself. Our physical bodies, brains included, are a part of this. This is the one nature that is reflective to another nature. This second nature can only be described as the essence of who a person is. In the English language, three words suffice: *me, myself,* and *I.* A more formal term is *Self.* Self-awareness can be described as the brain's ability (one nature) to be cognizant of the Self (another nature). The unknown force that causes matter to become animate through the complex interaction of molecules must be conscious Self.

The Law of Subjectivity

Such a definition of consciousness and the Self implies a duality of existence where there is no evidence for such a duality. In a similar manner, the precepts of the quantum world conflict with our understanding of the Newtonian world.

What are we to make of this?

These two natures are nothing more than different perspectives on the same thing, just as science and religion are different perspectives or different ways of explaining the same thing. The result is that science in general claims there is no evidence for God's existence, while religion claims that the evidence is everywhere. Both are equally tied to an exoteric methodology to explain the unexplainable. Science fails with its rhetoric, since it cannot coherently explain the origin of consciousness; and religion fails in its dogma, since it cannot coherently explain the origin of the universe.

Yet both agree, in principle, through what I call the law of subjectivity, which means that the original cause (God) cannot be proved or disproved. Religion claims that God made us. If we really are made by God, and God is in all of nature, then aren't we all, as an aggregate (through our state of consciousness), God experiencing Godself?

Science claims that we are made of stardust. If we really are made from the elements of stars, as science suggests, aren't we, then, through our state of consciousness, simply the universe trying to understand itself? What is the difference between these two approaches? The differences are nothing more than terminology and perception.

This reasoning leads to the conclusion that the nature of the source (nonpolarized energy) is consciousness, which is manifest as nature, through a complex sequence of events, as biological consciousness. Since biological consciousness must be contained within the source (as everything that exists is required to be), a unity persists that to us appears dualistic, as exemplified in physics by the quantum and Newtonian worlds. We are just as much a function of the quantum world as we are of the Newtonian world.

Based on all the evidence, experiential included, there is no more fitting conclusion than that Man is just as much the cause as he is the effect of the cosmos. Man's nature is the source (the abstract Cosmos or Consciousness), and the physical cosmos (matter) is a vehicle for expression of that source, particularly as a creative and self-perceiving entity. With the principle of the anthropocosm, science and religion together explain the universe in a holistic, intimate, and meaningful way. But the explanation is specific to every culture and to each individual.

From Hydrogen to Humans

With the principle of the anthropocosm bridging the explanations that science and religion offer, the evolution of life becomes an integral part of the creative process, although the notion that random genetic mutations are the impetus for speciation must be discarded in favor of a more accommodating concept. Genetic changes do occur. The impetus for those changes, however, ultimately has to explain the human form and its innate ability of self-awareness. While DNA does explain differences in form (among species as well as individuals), only the indisputable fact of consciousness as a fundamental and intrinsic characteristic of the universe truly bridges that mysterious gap between simple and complex organisms.

How the first single-celled animal became mankind is a mysterious process, despite all the scientific rhetoric. Envisioning how life on Earth began and developed into the multitude of forms present today is impossible. It began with the creation of hydrogen atoms, and through the stellar process, hydrogen atoms were fused into helium atoms. These two elements make up nearly 98 percent of the atoms in the universe. Only though the destruction of stars did the heavier elements come to be.

Through the destruction of an unknown star, the birth of our solar system occurred. At first it was an enormous cloud of hot interstellar gas and dust, but it gradually cooled and condensed into a rotating disk of gas and debris. Over time, the gas cooled even more and, under the spell of gravity, became planets. Those planets nearest the sun benefited from an abundance of heavier elements, which cooled to form a solid surface.

In the early solar system there were three times the number of inner planets as there are today. Those with conflicting orbits eventually collided, forming the four inner planets we now recognize. There were also comets, blasting onto the inner planets from the Kuiper belt or the Oort cloud. Earth became a planet of water and rock.

Earth was struck approximately ten times during its early history. But the last impact, of a planetoid roughly the size of Mars, struck Earth a glancing blow. After the maelstrom engendered by that collision, our terrestrial home was endowed with an iron core and acquired a satellite that acted as stabilizer and governor. The result was that Earth rotates once every twenty-four hours instead of once every six hours. But at that time, the moon was so close to Earth that its gravity created tidal waves of monumental proportions. As the waves scoured the land and pulled bits of rock into the primordial ocean, a chemical soup was formed from which life mysteriously appeared. How it appeared remains a mystery.

Some speculate that Earth's first microbial life came from Mars or was deposited by comets. Whatever the case may be, how that microbial life eventually became mankind is an even greater mystery.

When life first appeared on Earth, there was as yet no oxygen in the atmosphere. Only after several more "total evaporation impacts" (of an asteroid large enough to completely evaporate Earth's surface on impact), followed by planet-wide deep freezes, did oxygen make its appearance.

Through volcanic activity, after the first deep freeze, carbon dioxide gradually built up in the atmosphere, creating a greenhouse effect and returning to the Earth's surface the warmth it once knew. As this warm-up occurred, *Cyanobacteria* (blue-green algae) thrived in the carbon dioxide–rich environment, releasing oxygen, its excrement, into the atmosphere.

After a second deep freeze, oxygen reached 20 percent of the atmosphere and became the fuel for the sudden appearance of complex multicellular macroscopic organisms in a scientific landmark event called the Cambrian explosion. After 530 million years and four mass extinctions, hominids appeared, and very quickly (in geologic time) became mankind as we know it today.

Isn't it easier to say that God created the heavens and Earth? It is, because envisioning how single-celled animals somehow banded together to form multicelled creatures is beyond the reach of the most imaginative science-fiction author. Moreover, from the perspective of the first self-aware human, the perception of suddenly being created is correct, and from that perspective, creationism is just as true as evolution. Perhaps it is even truer, since evolution itself is not creative. Rather, as Schwaller states, evolution "is the continuity of the creative function that causes evolution."[42] The expansion of consciousness is the esoteric and fundamental truth. Evolution is merely its physical manifestation, its exoteric appearance.

Because all aspects of the evolution of the cosmos lead to the development of a form capable of reflecting its own characteristics, where form is the universe itself as well as all conscious men and women, plants and animals, size becomes a perceptual stumbling block. When compared to the vast expanse of the universe, a single person is less significant than a single atom is to a human being. How can Mankind actually be the universe, in its totality, if the universe is so vast?

Our perceived insignificance because of size is a false perception based on the assumption that we are somehow apart from the universe. This mode of thinking goes back to the beginning of the age of science, when the first modern scientists came up against the power and authority of the church in describing reality. Earth, scientifically not being the center of the universe, made human existence unessential to the workings of that universe. It is not.

Size, as it turns out, is of utmost importance for the cosmic principle of self-awareness. The universe will always generate a form in the size of our immediate world. For a complex form to exist that is capable of self-awareness, size does matter. Too large, and the exchange of information, limited by the speed of light, takes too long; too small, and the form is not complex enough to be self-aware.[43]

The difference between Man's cosmic insignificance and centrality depends on the way in which thought is organized in establishing a science. There is the defined object, which leads to reductionism and rationalism, but there is also the activity, the function of natural laws, which leads to an esoteric understanding of nature. The latter leads to a true sense of identity that does not depend on Earth being the physical center of the universe, or, for that matter, on the material possessions or status one has within a given social structure.

Life is the effect of this identity in which the anthropocosm invites us to seek within ourselves. This identity is the revealer that enables us to observe. In essence, it is the mystery of the observer—the objective measurement—at the heart of traditional science. True science is perfect in its mystical and magical character. Where science is esoteric,[44] the man of science seeks the real truth.

Uroboros: Cycles of Consciousness

From our own terrestrial ecosystem to the solar system, galaxy, and beyond, it is apparent that all of nature functions in a cyclical manner. During the 1980s and 1990s, cosmologists wondered at what point in time the universe would finally run out of momentum and, through gravity, come crashing back together in a "big crunch." The latest investigation suggests that the big crunch will never happen, and the universe will continue expanding forever.

The big bang origin of the universe has been the model of choice for cosmologists for many decades now, but it has always been a scientific paradox. The laws of physics break down at the moment of the big bang. More paradoxically, how do we arrive at the universe that we experience from nothing? Perhaps the big bang is only a perspective to explain the current body of scientific data, and not necessarily an accurate cosmogonic model of the universe. As is all of nature, perhaps the universe is cyclical and oscillates between the never-ending destruction and creation of galaxies.

We don't really know how the universe began, or if it had a beginning. But what we do know and can be certain of is our conscious experience. Perhaps we should listen to the secret wisdom of our ancestors and view life as an eternal cycle of conscious experience, rather than a physical finality.

It has been the subject of this book to investigate the past and explain that the ancient Egyptians were a scientific society in their own right, just as much as our civilization is today, and how their knowledge was disseminated and often misunderstood, whether innocently or not, by subsequent civilizations. It has also been the subject of this book to discover how the ancient Egyptians maintained a thriving culture for such a long period in a world where empires came and went in a matter of centuries.

The secret to the Egyptians' grand civilization, I believe, was its practicality based on a functional scientific method. As a result, their approach to social structure, based on the principles of natural harmony, was moral and just. Their knowledge of Man served as the foundation for civil growth and wealth. It has been handed down to us as the secret wisdom of sacred science, and was immortalized in stone at Luxor's Temple of Amun-Mut-Khonsu, as described in Schwaller de Lubicz's *The Temple of Man*.

This temple was not only an ancient institution of learning but also a testament to the Egyptians' sophistication. Today it serves as evidence, proof positive, for their scientific methodology and the anthropocosmic philosophy that they expressed in everything they did. What has been naively interpreted as a primitive menagerie of gods was really a systematic way of describing the principles of nature, terrestrial as well as cosmic.

Ancient cultures that encountered Egypt were inspired by its knowledge and wisdom enough to create systems of thought that were religious (such as the Hebrew and Christian traditions) or intellectual (such as the classic Greek philosophical tradition). With the true meaning now revealed of the Logos and the Trinity, both scientific representations of nature's creative principle derived from the creative nature of man himself and knowing that the principle is Egyptian in origin, our history and our heritage now stand corrected. Just as the ancient Uroboros—the circular serpent biting its tail—symbolizes, through our modern scientific endeavors we have come full circle in understanding ourselves.

No one knows for sure in what culture or at what time the Uroboros was first fashioned as a symbol, but it is the most ancient deity of the prehistoric world.[45] From the fourth century BCE, Plato tells us in the *Timaeus* that this serpent was self-sufficient, since nothing outside of him existed, so nothing went into or came from him. Movement was right for his spherical structure, and so he was made to move in a circular manner. As a result of his own limitations, he revolves in a circle, and from this motion the universe was created.[46]

From Egypt's Ptolemaic period, the Uroboros's meaning is also clear. The artist who drew the *Chrysopoeia* (gold making) *of Kleopatra* wrote within the circular serpent: *The All Is One*. The serpent biting its tail is the ancient Egyptian symbol depicting self-creation and the source of life: "It slays, weds, and impregnates itself," writes Erich Neumann; "it is man and woman, beginning and conceiving, devouring and giving birth, active and passive, above and below, at once."[47] In the *Papyrus of Ani: Egyptian Book of the Dead,* Atum, meaning the All and symbolized by the serpent, springs from the primordial waters and declares:

> You shall be for millions on millions of years, a lifetime of millions of years. I will dispatch the Elders and destroy all that I have made; the Earth shall return to the Primordial Water, to the surging flood, as in its original state. But I will remain with Osiris, I will transform myself into something else, namely a serpent, without men knowing or the gods seeing.[48]

The Uroboros, the serpent, represents the creative principle of the cosmos, as well as the cosmos itself. It is the Man Cosmos—Schwaller's primordial scission from One to Two, symbolized by the serpent, and their creative relationship functioning in the number Three. The Uroboros, however, is not just an ancient mythical symbol. Nor is it the fabricated imagery of the primitive mind. Rather, it is Man's identification with the seamless, eternal state of oneness, whose essence is a deep memory of the origin that words cannot explain.

Despite the fact that modern science declares the question of our origin to be unanswerable, the human psyche, according to Neumann, "which can neither be taught or led astray by the self-criticism of the conscious mind always poses this question afresh as one that is essential to it."[49] As such, the Uroboros's esoterism is as valid today as it was at the dawn of Man, as physicist Joel Primack and his wife, Nancy Ellen Abrams, demonstrate in their book *The View from the Center of the Universe,* a modern cosmological treatise in which the Uroboros is portrayed as representing the unity of all that exists within the vast differences of the cosmic scale.

Yet Primack was not the first physicist to realize the esoterism built into the nature of being. Forty-eight years ago Erwin Schrödinger, the 1933 Nobel laureate in physics (along with Paul Dirac) and one of the creators of quantum physics, inspired by what he had discovered in his illustrious career, in the end turned to the Eastern philosophical traditions, specifically the Upanishads. In his final work, *Mind and Matter,* Schrödinger calls to our attention what he calls the "arithmetical problem." Our perception is scientifically indescribable "because it is itself that world picture. It is identical with the whole and therefore cannot be contained in it as a part of it."[50] This creates a problem, because there is a multitude of individuals experiencing consciousness but there is only one world. One answer to this paradox is that each of us experiences a unique world, which Schrödinger summarily dismisses:

> There is obviously only one alternative, namely the unification of minds or consciousness. There multiplicity is only apparent, in truth there is only one mind.[51]

Today, physicist Fred Alan Wolf agrees that "there really is only one Mind":[52]

> Distinctions [in the physical world] are not real. They are fleeting whispers of an all-pervading, subtle, non-expressive potential reality. The world is not made of separate things. Mind is not separate from matter. And you are not separate from any other being, animal, vegetable, living, dead, or seemingly inanimate matter. . . . In you, like a coiled serpent waiting to spring forth from your deepest shadows, lies every creative moment that exists, has ever been, and will ever be.[53]

There may be no clearer expression of the Man Cosmos from science's best and brightest minds.

The sophistication of thought is the end product of scientific activity, whether that science is functional or based on dialectics, whether it occurs today or occurred thousands of years ago. When science is esoteric, as opposed to technical, knowledge gained is meaningful, and through civilization it is expressed in a meaningful way. The creativity that is at the heart of the musician, the sculptor, painter, or the writer is the same natural function that propels the heart of the scientist who seeks an answer to the mystery that is presented to every one of us. As it is with modern civilization, so it is with the civilization that built the pyramids of Giza, the megaliths of Nabta Playa, the trilithon of Baalbek, which inspired the ancient Egyptians to carve the mysterious glyphs depicting aircraft above the entranceway to Seti's Temple at Abydos.

Fig. 13.1. The Abydos "aircraft" (courtesy of Christopher Dunn)

Paradigm Paralysis

Sometime in 1998, the Abydos aircraft picture surfaced somewhere on the Internet showing a helicopter and two airplanes, possibly jet aircraft, as they lack a propeller, or maybe something else, since wings are not obvious. These three carvings appeared on the same panel with a fourth glyph that has been likened to a tank. Such a picture smacks of fraud and was duly called a hoax. After further review, however, the picture was authenticated, but then explained as a palimpsest (a later overlay): the ancient Egyptians supposedly covered the original glyphs in plaster and carved new ones into a fresh surface. Several thousand years later, this plaster flaked off, leaving behind images that just happen to look like modern aircraft (see fig. 13.1). A coincidence? Certainly, the notion that the ancient Egyptians carved airplanes and helicopters is nonsense. But in a close-up of this panel, there appears to be very little erosion of the glyphs in question.

Nevertheless, it is likely that those who believe the Abydos aircraft is a palimpsest will continue to do so, and those who believe it represents the technical capabilities of the ancient Egyptians, or pre-pharaonic Egyptians, will also continue to do so. Whether scientific, religious, or something else, belief and paradigm are powerful, even paralyzing. So the logical conclusion is that these glyphs, as well as the pyramids of Giza, the megaliths of Nabta Playa, and the trilithon of Baalbek, will forever retain their builders' secrets. Even so, together these anomalies whisper of the heritage of a forgotten technology, quietly declaring that we, as a civilization, have been here before.

In the ancient Hindu tradition, the principle of Yuga explains how the consciousness that defines civilization exists as a continuous oscillation from an epoch of renaissance to one of turmoil and back again to a reawakening. The idea of a cyclical collective consciousness as it may apply to humankind is idiotic to the cadres of the peer-reviewed establishment. Yet such a notion is vindicated by the fact that our conscious experience is the only thing that everyone in the world knows is real and can agree on as being real. Ironically, the imperial brain trust of today's science-oriented technicians—today's scribes—is formally at a loss to explain what the phenomenon of consciousness actually is, let alone understand how it functions.

Despite this built-in aversion to change, today's new science is discovering that physical reality, through observation, is intricately tied to the conscious experience of us all. We appear to be as much the system itself as we are the final product of the system. As paradoxical as it may seem, a preponderance of evidence from this new science leads to this deduction, and to the deduction that the universe was never created but has always been, just as we have always been. The physical world we observe is our own creation through the process of consciousness expansion, constantly creating and re-creating through the collapse of the wave function. Such insights into the structure of reality allow for diverse beliefs to be spread across humanity, specific to individual interpretation, while symbolizing the unification of us all, Creationists and Darwinians alike.

Those who object to a view such as this are likely to cite the lack of concrete evidence as the measure of its rejection. Even so, they are forced to admit that the source for all that exists is but an imaginary one.

Where does this leave us?

Belief must be understood for what it is: the defining element of the conscious experience, even in those who adhere to the rigorousness of the scientific method. Detachment is simply unattainable, and data are meaningless aside from beliefs, which are imposing as much as they are enlightening. The interpretation of data always leads to the sacred, whatever the interpreter's beliefs may be. Call it a mystery or call it a miracle, if you like. We not only live on a miracle planet; we live in a miracle universe. We are the miracle.

Long ago, with an appreciation for the majesty of life and the sacredness at the heart of science, in their wisdom the ancient Egyptians understood the mystery and the miracle. As science continues to strive to understand the universe, it is as if we are returning to ourselves. The ancient Egyptians understood the principles of the cosmos that we are now just beginning to realize, through the purity of scientific investigation and its philosophical consequences: There is simply no other way to explain how matter became organized and animate, how order appeared as the cosmos. Consciousness is all there is, or was, or ever will be.

Notes

Introduction: The Nature of Knowledge

1. Fritjof Capra, *The Tao of Physics* (Boston: Shambhala, 2000), 328.
2. Ibid., 325–27.
3. Ibid., 9.
4. Ibid., 8.
5. John Anthony West, *Serpent in the Sky* (Wheaton, Ill.: Quest, 1993), 18.
6. Ian Tattersall, *Monkey in the Mirror* (New York: Harcourt, 2002), 78.
7. "The Ice Age (Pleistocene Epoch)," U.S. Environmental Protection Agency, www.epa.gov/gmpo/edresources/pleistocene.html, accessed 11/20/2003.
8. Richard A. Muller, "A Brief Introduction to Ice Age Theories," muller.lbl.gov/pages/IceAgeBook/IceAgeTheories.html, accessed 11/20/2003.
9. John Gribbin, "Powerful Forces Beneath the Ocean Waves May Wreak Havoc on Our Climate, Driven by Global Warming," www.firstscience.com/site/articles/gribbin.asp, accessed 11/20/2003.
10. *Cracking the Ice Age,* PBS television documentary, September 30, 1997, see www.pbs.org/wgbh/nova/transcripts/2320crac.html for a full transcript.
11. Donald Patton, *The Biblical Flood and the Ice Epoch* (Seattle: Pacific Meridian, 1966), 16–24.
12. Antony Sutcliffe, *On the Track of Ice Age Animals* (Cambridge, Mass.: Harvard University Press, 1986), 108.
13. "Mammoth Load of Ivory from Pleistocene," *National Geographic* magazine, January 1992, 146.
14. Charles Hapgood, *Path of the Pole* (Kempton, Ill.: Adventures Unlimited, 1999), 327.
15. For an in-depth dissertation on the physics of pole shift, see F. Barbiero, "On the Possibility of Very Rapid Shifts of the Poles," www.esterni.unibg.it/dmsia/dynamics/poles.html, accessed 11/30/2003.
16. Hapgood, *Path of the Pole,* 94–95.
17. Paul LaViolette, *Earth Under Fire* (Rochester, Vt.: Bear & Co., 2005), 88.
18. Ibid., 164.
19. Ibid.

20. Frank Hibben, "Evidence of Early Man in Alaska," *American Antiquity* 8 (1943): 254–59.
21. Immanuel Velikovsky, *Earth in Upheaval* (New York: Doubleday, 1955), 59.
22. George McCready Price, *The New Geology: A Textbook for Colleges, Normal Schools, and Training Schools; and for the General Reader* (Nampa, Idaho: Pacific Press Publishing Association, 1923), 579.
23. John Massey Stewart, "Frozen Mammoths from Siberia Bring the Ice Ages to Vivid Life," *Smithsonian* 8, no. 9 (1977): 67.
24. *Raising the Mammoth,* Discovery Channel video, March 12, 2000; also on DVD by Discovery Home Studios, July 23, 2002.

Chapter 1. Matter: Mystery of Quantum Reality

1. Evan Harris Walker, *The Physics of Consciousness* (New York: Perseus Publishing, 2000), 131.
2. Harold E. Puthoff, "The Energetic Vacuum: Implications for Energy Research," *Speculations in Science and Technology* 13 (1990): 247.
3. Bernard Haisch, Alfonso Rueda, and Harold Puthoff, "Inertia as a Zero-Point Lorentz Field," *Physical Review* A 49, no. 2 (1994): 678–94.
4. Lynne McTaggart, *The Field* (New York: Quill, 2003), 31–32.

Chapter 2. Consciousness: Mystery of the Observer

1. Kristel Halter, "Waking Up During Surgery: A Living Nightmare," Columbia News Service, www.jrn.columbia.edu/studentwork/cns/2004-03-01/563.asp, accessed 8/31/2005.
2. William Arntz, Betsy Chasse, and Matthew Hoffman, *What the Bleep Do We Know!?* (Captured Light and Lord of the Winds Films, 2004).
3. Heisenberg, *Physics and Philosophy,* 83.
4. Rudy Rucker, *Infinity and the Mind* (Princeton, N.J.: Princeton University Press, 2005), 35–36.
5. Ray S. Jackendoff, *Consciousness and the Computational Mind* (Cambridge, Mass.: MIT Press, 1987).
6. Ken Wilber, "An Integral Theory of Consciousness," *Journal of Consciousness Studies* 4, no. 1 (February 1997): 71–92. See also Alwyn Scott, *Stairway to the Mind* (New York: Springer-Verlag Copernicus, 1995).
7. Cate Montana, "A Quantum Mechanical View of Bi-location," www.whatthebleep.com/herald5/articles.shtml, accessed 9/14/2005.
8. Stuart Hameroff, "Overview: Could Life and Consciousness Be Related to the Fundamental Quantum Nature of the Universe?" www.quantumconsciousness.org/overview.html, accessed 3/1/2007.
9. Heisenberg, *Physics and Philosophy,* 58.
10. Ibid., "The Representation of Nature in Contemporary Physics," *Daedalus* 87 (Summer 1958): 95–108.

11. Stephen Hawking, *A Brief History of Time* (New York: Bantam, 1990), 56.
12. Paul Davies, *God and the New Physics* (New York: Simon & Schuster, 1983), 102.
13. John Gribbin, *In Search of Schrödinger's Cat* (New York: Bantam, 1984), 5.
14. Heisenberg, *Physics and Philosophy,* 55.
15. www.whatthebleep.com/frommakers.html, accessed 9/20/2005.
16. www.healthyplace.com/Communities/Thought_Disorders/schizo/nimh/chemical.asp, accessed 9/22/2005.
17. Hawking, *Brief History of Time,* 56.
18. Evan Harris Walker, "Quantum Theory of Consciousness," *Noetic Journal* 1 (1998): 100–7.
19. Ibid., "The Nature of Consciousness," *Mathematical Biosciences* 7 (1970): 131–78.
20. Walker, *Physics of Consciousness,* 182.
21. Ibid., 221.
22. Ibid., 223.
23. Ibid., 229.
24. Ibid.
25. Ibid., 231.
26. Ibid., 223.
27. Ibid., 37.
28. Ibid., 231.
29. Ibid., 233.
30. Ibid., 235.
31. Ibid.
32. Ibid., 291.
33. Ibid., 95.
34. Ibid., 234.
35. Ibid., 237.
36. Ibid., 307.
37. Ibid., 308.
38. Ibid., 309.
39. Ibid., 326.
40. Ibid., 329.
41. Ibid., 335.
42. Benjamin Libet, *Mind Time: The Temporal Factor in Consciousness* (Cambridge, Mass.: Harvard University Press, 2004), 184.
43. Ibid., 161.
44. Ibid., 163.
45. Ibid., 165.
46. Ibid.
47. Ibid., 166.

48. Ibid., 169.
49. Ibid., 180.

Chapter 3. Cosmos: Mystery of Life

1. Saul Perlmutter, "Supernovae, Dark Energy, and the Accelerating Universe," *Physics Today,* April 2003, 53–59.
2. Michael Mallary, *Our Improbable Universe* (New York: Thunder's Mouth Press, 2004), 3–11.
3. Ibid., 3.
4. Ibid., 4.
5. Ibid., 5.
6. Ibid.
7. Ibid., 5–6.
8. Ibid., 6.
9. Ibid., 6–7.
10. Ibid., 7–8.
11. Ibid., 8.
12. Ibid.
13. Ibid., 8–9.
14. Ibid., 9.
15. Ibid., 10.
16. Ibid., 10–11.
17. Bruce Lipton, *The Biology of Belief: Unleashing the Power of Consciousness, Matter, and Miracles* (Santa Rosa, Calif.: Mountain of Love/Elite Books, 2005), 66.
18. Bruce Lipton, "The New Biology," www.brucelipton.com/newbiology.php, accessed 2/10/2005.
19. Lipton, *Biology of Belief,* 37, 130–31.
20. Deborah Jordan Brooks, "Substantial Numbers of Americans Continue to Doubt Evolution as Explanation for Origin of Humans," Gallup News Service, March 5, 2001.
21. George H. Gallup Jr., "Americans' Spiritual Searches Turn Inward," Gallup Organization, February 11, 2003.
22. Stefan Lovgren, "Evolution and Religion Can Coexist, Scientists Say," National Geographic News, October 18, 2004, news.nationalgeographic.com/news/2004/10/1018_041018_science_religion.html#main, accessed 3/1/2007.
23. Mark Edwards, "100 Scientists, National Poll Challenge Darwinism, Discovery Institute," www.reviewevolution.com/press/pressRelease_100Scientists.php, accessed 3/1/2007.
24. "Evolution," PBS website, www.pbs.org/wgbh/evolution/library/glossary/glossary.html#gould_stephen_jay, accessed 3/1/2007, and www.blackwellpublishing.com/ridley/classictexts/eldredge.asp.

25. Hillary Mayell, "Documentary Redraws Humans' Family Tree," National Geographic News, January 21, 2003, news.nationalgeographic.com/news/2002/12/1212_021213_journeyofman.html, accessed 11/4/2005.
26. *Search for the Ultimate Survivor,* National Geographic Channel, Warner Home Video, April 12, 2005.
27. Lee Berger, speaking in the documentary *Search for the Ultimate Survivor.*
28. Ian Tattersall, *Monkey in the Mirror,* 38.
29. Ibid., 39.
30. Ibid., 46.
31. Joel Primack and Nancy Abrams, "'In a Beginning . . .' Quantum Cosmology and Kabbalah," physics.ucsc.edu/cosmo/primack_abrams/htmlformat/inabeginning.html, accessed 3/1/2007.
32. A. Kaoru Tsuda, Yoshiaki Kikkawa, Hiromichi Yonikawa, and Yuichi Tanabe, "Extensive Interbreeding Occurred Among Multiple Matriarchal Ancestors During the Domestication of the Dog: Evidence from Inter- and Intraspecies Polymorphisms in the D-Loop Region of Mitochondrial DNA Between Dogs and Wolves," *Genes and Genetic Systems* 72 (1997): 229–38.

Chapter 4. Cosmogony: The Origin Mystery

1. R. A. Schwaller de Lubicz, *The Temple of Man* (Rochester, Vt.: Inner Traditions, 1998), 80.
2. Ibid., 83.
3. Ibid., 90.
4. R. A. Schwaller de Lubicz, *A Study of Numbers: A Guide to Constant Creation of the Universe* (Rochester, Vt.: Inner Traditions, 1986). (Originally published in France, 1917.)
5. Ibid., 35–36.
6. Ibid., 37.
7. Ibid., 40.
8. Ibid., 42.
9. Ibid., 45.
10. Ibid., 46, 49.
11. Ibid., 50.
12. Ibid., 50.
13. Ibid., 5.
14. R. A. Schwaller de Lubicz, *Esoterism and Symbol* (Rochester, Vt.: Inner Traditions, 1985), 17.
15. Ibid., 34.
16. Schwaller de Lubicz, *Study of Numbers,* 57.
17. Ibid., 55.
18. Ibid., 60.
19. Ibid., 60.

20. Ibid., 61.
21. Ibid., 62.
22. Ibid., 62.
23. Ibid., 65.
24. Ibid., 66.
25. Ibid., 67.
26. Ibid., 67.
27. Ibid., 69–70.
28. Ibid., 68.
29. Ibid., 68.
30. Hawking, *Brief History of Time,* 50.
31. Ibid., 136.
32. Julian Barbour, *The End of Time: The Next Revolution in Physics* (Cambridge, Mass.: Oxford University Press, 1999), 229.
33. Ibid., 247.
34. Ibid., 325.
35. Ibid., 319.
36. mathworld.wolfram.com/SchwarzReflectionPrinciple.html, accessed 3/1/2007.
37. Rucker, *Infinity and the Mind,* 50–51.
38. Peter N. Spotts, "Building Blocks of Life—in Outer Space," 3/28/2002, www.csmonitor.com/2002/0328/p11s01-stss.html, accessed 11/20/2005.

Chapter 5. Changing Paradigms: Science and the Origin of Religion

1. Guy Nolch, "Mungo Man's DNA Shakes the *Homo* Family Tree," *Australasian Science* 22 (2) (March 2001): 29.
2. Steve Olson, *Mapping Human History* (Boston: Mariner, 2002), 127.
3. *Supervolcanoes,* BBC2, February 3, 2000. See also www.bbc.co.uk/science/horizon/1999/supervolcanoes_script.shtml, accessed 3/1/2005.
4. Ibid.
5. Stanley Ambrose, "Late Pleistocene Human Population Bottlenecks, Volcanic Winter, and Differentiation of Modern Humans," *Journal of Human Evolution* 34 (1998): 628.
6. Ben Marwick, "Pleistocene Exchange Networks as Evidence for the Evolution of Language," *Cambridge Archaeological Journal* 13, no. 1 (2003): 67–81.
7. Ibid., 67.
8. A. M. T. Moore, "The Neolithic of the Levant" (Cambridge: Oxford University, 1978), 87–99. Doctoral thesis, online at ancientneareast.tripod.com/07.html, accessed 11/4/2004.
9. R. M. Albert, O. Bar-Yosef, L. Meignen, and S. Weiner, "Quantitative Phytolith Study of Hearths from the Natufian and Middle Palaeolithic Levels of

Hayonim Cave (Galilee, Israel)," *Journal of Archaeological Science* 30 (2003): 461–80.

10. Ofer Bar-Yosef, "The Natufian Culture in the Levant, Threshold to the Origins of Agriculture," *Evolutionary Anthropology* 6 (5) (1998): 159–77.
11. Ibid.
12. Ibid., 162.
13. Ibid.
14. Ibid.
15. Ibid., 171.
16. Ibid., 164.
17. James Mellaart, *The Neolithic of the Near East* (New York: Scribner's, 1975). See ancientneareast.tripod.com/03.html, accessed 11/4/2004.
18. Bar-Yosef, 166.
19. Ibid., 171.
20. Eitan Tchernov, "Two New Dogs, and Other Natufian Dogs, from the Southern Levant," *Journal of Archaeological Science* 24 (1997): 65–95.
21. Ibid., 93.
22. Bar-Yosef, 165.
23. Ibid., 172.
24. Yuval Goren, A. Nigel Goring-Morris, and Irena Segal, "The Technology of Skull Modelling in the Pre-Pottery Neolithic B (PPNB): Regional Variability, the Relation of Technology and Iconography and Their Archaeological Implications," *Journal of Archaeological Science* 28 (2001): 671–90.
25. Ibid.
26. Louise Martin and Nerissa Russell, "Animal Bone Report," *Çatalhöyük 1997 Archive Report.* See www.catalhoyuk.com/archive_reports/1997/ar97_14.html, accessed 3/1/2007.
27. Christine Hastorf and Julie Near, "Archaeobotanical Archive Report," *Çatalhöyük 1997 Archive Report,* www.catalhoyuk.com/archive_reports/1997/ar97_13.html, accessed 3/1/2007.
28. Naomi Hamilton, "Figurines," *Çatalhöyük 1997 Archive Report.* See www.catalhoyuk.com/archive_reports/1997/ar97_16.html, accessed 3/1/2007.
29. Mary Settegast, *Plato Prehistorian: 10,000 to 5,000 BC—Myth, Religion, Archeology* (Hudson, N.Y.: Lindisfarne, 1990), 168, 170.
30. Ibid., 171.
31. R. A. Cooper, P. C. Molan, and K. G. Harding, "The Sensitivity to Honey of Gram-Positive Cocci of Clinical Significance Isolated from Wounds," *Journal of Applied Microbiology* 93 (2002): 857–63.
32. Settegast, 194.
33. Ibid., 190.
34. Ibid., 192–93.

35. Merlin Stone, *When God Was a Woman* (New York: Barnes & Noble, 1993), 10.
36. Ibid., 13.
37. Ibid., 12.

Chapter 6. The Historical Basis of the Logos: Greek Philosophy's Influence on Religious Thinking

1. Nicene Creed, www.sacred-texts.com/chr/nicene.htm.
2. Elaine Pagels, *The Gnostic Gospels* (New York: Vintage, 1989), xvii.
3. Ibid., xix.
4. *Jewish Encyclopedia,* "Ophites," www.jewishencyclopedia.com, 407–408.
5. John 3:14–15.
6. *Encyclopedic Theosophical Glossary,* Theosophical University Press, 1999, Collation of Theosophical Glossary. See "Ophites," www.theosophy-nw.org/theosnw/ctg/om-oz.1htm.
7. E. A. Wallis Budge, *The Egyptian Book of the Dead* (New York: Dover, 1967), 337.
8. Martin Bernal, *Black Athena* (New Brunswick, N.J.: Rutgers University Press, 1987), 110.
9. Genesis 1:26.
10. R. A. Schwaller de Lubicz, *Sacred Science: The Kingdom of Pharaonic Theocracy* (Rochester, Vt.: Inner Traditions, 1982).
11. Ibid., 23.
12. Ibid., 24.
13. Bernal, 162.
14. Dick Teresi, *Lost Discoveries: The Ancient Roots of Modern Science—from the Babylonians to the Maya* (New York: Simon & Schuster, 2003), 17.
15. Plato, *The Republic,* translated by Francis MacDonald Cornford (New York: Oxford University Press, 1970), 245.
16. Ibid., 242.
17. Ibid.
18. Aristotle, *The Basic Works of Aristotle, De Caelo* (Book I), edited by Richard McKeon (New York: Modern Library, 2001), 398.
19. Arnold Hermann, *To Think Like God: Pythagoras and Parmenides* (Las Vegas: Parmenides Publishing, 2004), 101–102.
20. Manly P. Hall, *The Secret Teachings of All Ages* (San Francisco: H. S. Crocker, 1928), 71.
21. Ibid.
22. Aristotle, *De Caelo,* 398.
23. "The Pythagorean Science of Numbers," *Theosophy* 27, no. 7 (May 1939): 301–6.
24. Acts 17:23.

Chapter 7. Egypt: The Ancient Source of Knowledge

1. E. A. Wallis Budge, *Egyptian Magic* (London: Kegan, Paul, Trench and Trübner & Company, 1901), xi.
2. Ibid., xiii.
3. Ibid., xiii–xiv.
4. The Shabaka Stela, line 53, maat.sofiatopia.org/shabaka.htm, accessed on 12/11/2005.
5. Ibid., lines 55–58.
6. Henri Frankfort, *Ancient Egyptian Religion* (New York: Harper, 1948), 29.
7. James Henry Breasted, "The Philosophy of a Memphite Priest," *Zeitschrift für ägyptische Sprache und Altertumskunde* 39 (1901): 39–54.
8. Tom Hare, *Remembering Osiris: Number, Gender, and the Word in Ancient Egyptian Representational Systems* (Palo Alto, Calif.: Stanford University Press, 1999), 179.
9. Bernal, 140.
10. Ibid.
11. Schwaller de Lubicz, *Sacred Science,* 187–88.
12. Isha Schwaller de Lubicz, *Her-Bak* 2 (New York: Inner Traditions, 1978), 156.
13. Schwaller de Lubicz, *Sacred Science,* 192.
14. Lucie Lamy, *Egyptian Mysteries: New Light on Ancient Knowledge* (New York: Thames & Hudson, 1981), 9.
15. Schwaller de Lubicz, *Sacred Science,* 188.
16. Ibid., 157.
17. Ibid., 83.
18. Ibid., 163.
19. Ibid., 164.
20. Ibid.
21. Ibid., 165.
22. Ibid.
23. Ibid.
24. Ibid., 17.
25. Ibid., 168.
26. Ibid., 236.
27. Ibid., 197.
28. Andre Vandenbroeck, *Al-Kemi: Hermetic, Occult, Political, and Private Aspects of R. A. Schwaller de Lubicz* (Rochester, Vt.: Inner Traditions, 1987), 56.
29. Cheikh Anta Diop, *The African Origin of Civilization: Myth or Reality* (Chicago: Lawrence Hill, 1974), 230.
30. Ibid.
31. Mary Lefkowitz, *Not Out of Africa: How Afrocentrism Became an Excuse to Teach Myth as History* (New York: Basic, 1997), 150.
32. Ibid., 140–41.

33. Ibid., 141.
34. Ibid.
35. Diop, 231.
36. Ibid.
37. Hermann, 62.
38. George G. M. James, *Stolen Legacy: Greek Philosophy Is Stolen Egyptian Philosophy* (Trenton, N.J.: Africa World Press, 1992), 145.
39. Roger Penrose, *The Emperor's New Mind: Concerning Computers, Minds, and the Laws of Physics* (New York: Oxford University Press, 1989), 95.

Chapter 8. Moses and the Mystery School: Egyptian Esoterism and the Birth of Religion

1. Genesis 37:28, 36.
2. Genesis 50:7, 9.
3. David Rohl, *Pharaohs and Kings: A Biblical Quest,* Discovery Channel Video, 1998.
4. Exodus 2:19.
5. Joshua 24:2.
6. Book of Jubilees 11:14.
7. *Who Was Moses?* BBC Warner Video, 2000.
8. Ibid.
9. Ibid.
10. "The Riddle of Mount Sinai: Archaeological Discoveries at Har Karkom," www.harkarkom.com, accessed 8/25/2005.
11. "The Religion of Upper Egypt," www.touregypt.net/emac3.htm.
12. Numbers 21:9.
13. Exodus 7:12.
14. Genesis 2:8–14.
15. David Rohl, *In Search of Eden,* Discovery Channel Video, 2002.
16. Ibid.
17. Ibid.
18. Ibid.
19. James Mellaart, *Earliest Civilizations of the Near East* (New York: McGraw-Hill, 1965), 63.
20. Peter M. M. G. Akkermans and Marie Le Miere, "Field Report: The 1988 Excavations at Tell Sabi Abyad, a Later Neolithic Village in Northern Syria," *American Journal of Archeology* 96, no. 1 (January 1992). See www.ajaonline.org/archive/96.1/akkermans_peter_mmg_.html, accessed 6/12/2003.
21. Ibid.
22. Mehrdad R. Izady, "Exploring Kurdish Origins," *Kurdish Life* no. 7 (Summer 1993). See www.xs4all.nl/~tank/kurdish/htdocs/his/orig.html, accessed 9/13/2003.

23. McGuire Gibson, "Patterns of Occupation at Nippur." See oi.uchicago.edu/research/projects/nip/pon.html, accessed 3/1/2007.
24. Clarence S. Fisher, *Excavations at Nippur* (Philadelphia: McCown and Haines, 1905), 20.
25. Mellaart, *Earliest Civilizations of the Near East,* 68.
26. Christian O'Brien and Joy O'Brien, *The Genius of the Few: The Story of Those Who Founded the Garden of Eden* (Wellingborough: Turnstone, 1985).
27. Andrew Collins, *From the Ashes of Angels: The Forbidden Legacy of a Fallen Race* (Rochester, Vt.: Bear & Company, 2001), 205.
28. Marija Gimbutas, "The Age of the Great Goddess: An Interview with Kell Kearns" (Boulder, Colo.: Sounds True Recordings, 1992).
29. Ibid.
30. Ibid.
31. West, 59.
32. Genesis 1:1–5.

Chapter 9. The Son of Man: Esoterism and the Message of Christ

1. Acts 26:22–23.
2. Jeremiah 31:33.
3. Matthew 5:17.
4. George Robert Stowe Mead, cited in E. J. Langford Garstin, *Theurgy or the Hermetic Practice: A Treatise on Spiritual Alchemy* (Berwick, Maine: Ibis Press, 204), 27.
5. Daniel 7:13.
6. Ibid., 7:14.
7. Ernst Herzfeld, *Zoroaster and His World* (Princeton, N.J.: Princeton University Press, 1947), 835–40.
8. Matthew 26:62–66.
9. Elizabeth Boyden Howes, *Intersection and Beyond* (San Francisco: Guild for Psychological Studies, 1986).
10. From a lecture by Dr. Walter Wink, "The Son of the Man: The Stone That the Builders Rejected," Ottawa Lay School of Theology, January 14, 2000. Wink quotes Elizabeth Boyden Howes, "Son of Man—Expression of the Self," 174. See www.olst.ca/wink.htm, accessed 12/16/2004.
11. Matthew 9:2–9.
12. Luke 7:9.
13. Ibid., 17:20–21.
14. Mark 6:1–6 and Matthew 13:53–58.
15. John 14:20.
16. Ibid., 14:6.
17. Ibid., 1:12–13.
18. Ibid., 14:20.

19. Ibid., 1:1–5.
20. Rudolf Steiner, *Christianity as Mystical Fact* (1902), See "Part IX: Plato as a Mystic," www.tphta.ws/RS_CHMYF.HTM.
21. Ibid.
22. Ibid.
23. Iamblichus, *Theurgia or The Egyptian Mysteries,* translated from the Greek by Alexander Wilder (London & New York: William Rider, The Metaphysical Publishing Co., 1911). See chapter 17, "The Personal Dæmon, One Guardian Dæmon Only to an Individual," www.esotericarchives.com/oracle/iambl_th.htm, accessed 5/4/2005.
24. Ibid.
25. Luke 20:40.
26. Ibid., 20:34–38.
27. Matthew 24:1.
28. Ibid., 24:3.
29. John 8:58.
30. Ibid., 4:8, 16.

Chapter 10. Secret Wisdom: Esoterism and Inspiration

1. "Constantine the Great," *New Advent Catholic Encyclopedia,* www.newadvent.org/cathen/04295c.htm, accessed 5/18/2005.
2. Garth Fowden, *The Egyptian Hermes: A Historical Approach to the Late Pagan Mind* (Princeton, N.J.: Princeton University Press, 1993), xxii.
3. Johannes Kepler, cited in R. A. Schwaller de Lubicz, *The Temple of Man* (Rochester, Vt.: Inner Traditions, 1998), 273–74.
4. *Da Vinci and the Mystery of the Shroud,* National Geographic video, airdate January 30, 2006.
5. "Solving Fermat: Andrew Wiles," Nova Online, www.pbs.org/wgbh/nova/proof/wiles.html, accessed 1/20/2005.
6. Eric W. Weisstein, "Pythagorean Triple," MathWorld: A Wolfram Web Resource, mathworld.wolfram.com/PythagoreanTriple.html, accessed 1/20/2005.
7. R. A. Schwaller de Lubicz, *Symbol and the Symbolic: Ancient Egypt, Science, and the Evolution of Consciousness* (New York: Inner Traditions: 1981), 77.
8. "The Chaplet," *New Advent Catholic Encyclopedia,* www.newadvent.org/fathers/0304.htm, accessed 5/18/2005.
9. Edward M. Blaiklock, *The Archaeology of the New Testament* (Nashville: Thomas Nelson, 1984), 62–63.
10. Schwaller de Lubicz, *Sacred Science,* 156.
11. Ibid., 214.
12. Ibid.
13. Ibid., 244.

14. Ibid., 245.
15. Schwaller de Lubicz, *Sacred Science,* 144.
16. Ibid.
17. Ibid.
18. Ibid., 143
19. Ibid., 144.
20. Adrian Gilbert, *Magi: The Quest for a Secret Tradition* (London: Bloomsbury, 1999), 226–29.
21. J. Quaegebeur, *Ritual and Sacrifice in the Ancient Near East* (Leuven: Peeters, 1993), 175–89. Also see Jacke Phillips, ed., *Ancient Egypt, the Aegean, and the Near East: Studies in Honour of Martha Rhoads Bell* 1 (San Antonio, Texas: Van Siclen, 1997), 169–178.

Chapter 11. Sacred Science: Ancient Wisdom of the Modern World

1. Brian P. Copenhaver, *Hermetica* (New York: Cambridge University Press, 1992), xiv.
2. Samuel A. B. Mercer, *The Pyramid Texts* (New York: Longmans, Green, 1952), 1.
3. Ibid.
4. Geraldine Pinch, *Egyptian Mythology: A Guide to the Gods, Goddesses, and Traditions of Ancient Egypt* (New York: Oxford University Press, 2002), 9–10.
5. Mercer, 2.
6. Miroslav Verner, *The Pyramids: The Mystery, Culture, and Science of Egypt's Great Monuments,* trans. by Steven Rendall (New York: Grove, 2001), 41.
7. Pinch, 10–11.
8. Timothy Freke and Peter Gandy, *Wisdom of the Pharaohs* (New York: Tarcher, 1999), 18.
9. Garth Fowden, 195.
10. Ibid., 198.
11. Ibid., 195.
12. Ibid., 180.
13. Ibid., 182.
14. Anthony S. Mercatante, *Who's Who in Egyptian Mythology,* edited and revised by Robert Steven Bianchi (New York: Barnes & Noble, 1998), 189.
15. Ibid., 190.
16. Pinch, 209.
17. Fowden, 22.
18. Mercatante, 190.
19. Fowden, 22.
20. Ibid., 23.
21. Mercatante, 190.

22. Fowden, 25.
23. Pinch, 72.
24. Ibid., 209.
25. Fowden, 26.
26. Ibid.
27. Pinch, 45.
28. Ibid., 43.
29. Ibid., 211.
30. Copenhaver, 4.
31. Ibid., 26.
32. Ibid., 20.
33. Ibid., 27.
34. Ibid., 32.
35. Ibid., 35.
36. Ibid., 36.
37. Ibid., 40.
38. Ibid., 41.
39. Pinch, 45.
40. Thomas Gerald Massey, *Ancient Egypt: The Light of the World* (London: T. Fisher Unwin Adelphi Terrace, 1907), 2.
41. Miroslav Verner, 237.
42. Massey, 6.
43. Ibid., 13.
44. James Hillman, *The Souls' Code: In Search of Character and Calling* (New York: Random House, 1996), 10.
45. Schwaller de Lubicz, *Sacred Science,* 10.
46. Ibid.
47. Ibid., 226.
48. Ibid., 22.
49. Ibid., 167.
50. Ibid., 238.
51. Ibid.
52. Ibid., 240.
53. Ibid., 246–47.
54. *Rameses: Wrath of God or Man?* (Silver Spring, Md.: Discovery Communications, 2003), film documentary.
55. "Heb-Sed festival," *Encyclopedia Britannica,* from Encyclopedia Britannica Premium Service, www.britannica.com/eb/article-9366811/Heb-Sed-festival, accessed 2/13/2006.
56. Margaret A. Murray, *The Osireion at Abydos* (London: Bernard Quaritch, 1904), 33.
57. Ibid.

58. Ibid., 28.
59. Ibid., 27.
60. Ibid., 32.
61. Ibid., 33.
62. *Rameses: Wrath of God or Man?*
63. Pinch, 213.
64. Ibid., 216–14.
65. Ibid., 127.
66. Ibid., 128.
67. Murray, 25.
68. Ibid.
69. Schwaller de Lubicz, *Sacred Science,* 240.
70. Ibid., 226.
71. Schwaller de Lubicz, *Esoterism and Symbol,* 62.
72. Ibid., 64.
73. Ibid.
74. Vandenbroeck, 178.
75. Dean Radin, *The Conscious Universe* (San Francisco: Harper, 1997), 259.
76. Ibid.; Radin quotes Alfred N. Whitehead, *Science and the Modern World* (New York: Cambridge University Press, 1933).
77. Lecomte du Noüy, *Human Destiny* (New York: Longmans, Green, 1947), 83–84.
78. Julian Barbour, 277.
79. Ibid., 247.
80. Ibid., 229.
81. Ibid., 52.
82. Ibid., 259.
83. Ibid., 267.
84. Ibid., 325.
85. Ibid., 312.
86. Ibid., 335.
87. Interview with Michio Kaku, *Coast to Coast AM* radio broadcast, January 22, 2006.

Chapter 12. Whispers of a Forgotten Technology: The Testimony of Atlantis

1. Michel M. Alouf, *History of Baalbek* (Beirut: American Press, 1949), 99.
2. Schwaller de Lubicz, *Sacred Science,* 96.
3. University of Wales (Aberystwyth) Tana Project, "Geology," www.aber.ac.uk/~qecwww/tana/geology.htm, accessed 11/24/2005.
4. Schwaller de Lubicz, *Sacred Science,* 96.
5. *Decoding the Past: The Other Nostradamus,* History Channel, aired February 18, 2006.

6. Schwaller de Lubicz, *Sacred Science,* 96 (footnote 29).
7. Michael Grant, *The Ancient Historians* (New York: Barnes & Noble, 1994), 3 (Grant quotes Herodotus II 77).
8. Schwaller de Lubicz, *Sacred Science,* 96.
9. Walter B. Emery, *Archaic Egypt* (New York: Penguin, 1961), 192.
10. Grant, 4.
11. Thomas G. Brophy, *The Origin Map: Discovery of a Prehistoric, Megalithic, Astrophysical Map and Sculpture of the Universe* (New York: Writers Club Press, 2002), 10.
12. Ibid., 54.
13. Verner, 449.
14. *Mysteries of the Pyramids* (Bethesda, Md.: Discovery Channel/BBC Video, 1999).
15. Verner, 449.
16. Michael A. Hoffman, *Egypt Before the Pharaohs* (New York: Dorsett, 1979), 270–71.
17. Ibid., 278.
18. Verner, 45.
19. *Where Did It Come From? Ancient Egypt: Iconic Structures,* Popular Arts Entertainment, The History Channel, aired 9/21/2006.
20. Verner, 70.
21. Ibid., 69.
22. Ibid.
23. Ibid., 70.
24. Ogden Goelet cited in Raymond Faulkner, trans., *The Egyptian Book of the Dead: The Book of Going Forth by Day* (San Francisco: Chronicle, 1994), 13.
25. Ibid.
26. Verner, 89.
27. Ibid., 91.
28. Ibid., 197.
29. Ibid.
30. Ibid., 199.
31. Ibid., 202.
32. Nuclear Energy Information Service, www.neis.org/literature/Brochures/npfacts.htm, accessed 2/19/2006.
33. Verner, 23.
34. Rudolf Gantenbrink, "Ascertaining and Evaluating Relevant Structural Points Using the Cheops Pyramid as an Example," www.cheops.org/startpage/publications/publications.htm, accessed 2/22/2006.
35. Diop, 234.
36. Christopher Dunn, *The Giza Power Plant* (Santa Fe: Bear & Company, 1998), 3.

37. M. E. Abdel-Salam, "Construction of Underground Works and Tunnels in Ancient Egypt," *Tunnelling and Underground Space Technology* 17, no. 3 (July 2002): 295–304.
38. Chance Gardner, "Episode 8: Cosmology," *Magical Egypt: A Symbolist Tour,* Cydonia Inc., 2002, documentary series.
39. Laird Scranton, *Science of the Dogon* (Rochester, Vt.: Inner Traditions, 2006), 81.
40. Ibid., 58.
41. Ibid., 93.
42. Ibid., 58.
43. Ibid., 59
44. Marcel Griaule and Germaine Dieterlen, *Conversations with Ogotemmeli: An Introduction to Dogon Religious Ideas* (New York: Oxford University Press, 1975), 17.
45. Ibid., 18.
46. Ibid., 19.
47. Ibid., 18.
48. Scranton, 44.
49. Ibid., 44.
50. Teresa Vergani, "Ethnomathematics and Symbolic Thought: The Culture of the Dogon," *Zentralblatt für Didaktik der Mathematik* (International Reviews on Mathematical Education), February 1999, 66.
51. Scranton, 76.
52. Ibid., 76.
53. Ibid., 77.
54. Ibid., 18.
55. Ibid., 80.
56. Ibid., 102.
57. Ibid., 81.
58. Ibid., 82.
59. Mercatante, 8.
60. Budge, *The Egyptian Book of the Dead,* 288.
61. Scranton, 83.
62. Mercatante, 85.
63. Ibid., 106.
64. Christian O'Brien and Barbara Joy O'Brien, *The Shining Ones* (Cirencester, England: Dianthus, 1997), 62.
65. Ibid., 87.
66. Ibid.
67. Ibid., 71–72.
68. Ibid., 74.
69. Ibid., 82
70. Ibid.

71. Ibid., 83.
72. Ibid., 86.
73. Ibid., 93
74. Ibid., 91.
75. Ibid., 94.
76. Ibid., 96–97.
77. Ibid., 98.
78. Ibid., 181.
79. Ibid., 184.
80. Ibid.
81. Ibid.
82. Ibid., 186.
83. Ibid., 149.
84. Ibid.
85. Ibid.
86. Ibid., 150.
87. Ibid., 151.
88. Ibid.
89. Ibid., 154.
90. Ibid.
91. Ibid.
92. Genesis 6:4.
93. "Mesopotamia: The Code of Hammurabi," www.wsu.edu/~dee/meso/code.htm, accessed 5/2/2006.
94. "World History and Wisdom: *Homo sapiens sapiens,* the Small Print," www.goldenageproject.org.uk/homo.html, accessed 5/2/2006.
95. Alouf, 107.
96. Andrew Lawler, "Iran Reopens Its Past," *Science* 302 (November 7, 2003): 971.
97. Ibid., "Jiroft Discovery Stuns Archeologists," *Science* 302 (November 7, 2003): 973.
98. Richard Covington, "What Was Jiroft?" www.saudiaramcoworld.com/issue/200405/what.was.jiroft..htm, accessed 5/15/2006.
99. Ibid.
100. "Experts Prepare Jiroft's 5,000-Year Map Source," Iranian Cultural Heritage News Agency, 10/25/04, www.payvand.com/news/04/oct/1192.html, accessed 3/8/2007.
101. Ibid.
102. Covington.
103. "Experts Prepare Jiroft's 5,000-Year Map Source."
104. David Lewis-Williams, *The Mind in the Cave* (London: Thames & Hudson, 2002), 97.
105. Rossella Lorenzi, "Lost City of Atlantis Found in Spain?" Discovery News, dsc.discovery.com/news/briefs/20040607/atlantis.html.

Chapter 13. The Path of Osiris: The Principles of Consciousness

1. Schwaller de Lubicz, *Sacred Science,* 217.
2. Ibid.
3. Ibid.
4. Pinch, 79.
5. Ibid.
6. Ibid., 80.
7. Schwaller de Lubicz, *Sacred Science,* 224.
8. Budge, *Egyptian Book of the Dead,* 106.
9. Madame Helena Petrovna Blavatsky, "Egyptian Immortality," *Theosophy* 15, no. 11 (September 1927): 499–507.
10. Budge, *Egyptian Book of the Dead,* 355.
11. Ibid., 356.
12. Ibid., 355–56.
13. Ibid., 356.
14. Ibid.
15. Ibid.
16. Ibid.
17. Jon E. Lewis, ed., *Ancient Egypt: The Mammoth Book of Eyewitness* (New York: Carroll & Graf, 2003), 47–48.
18. Ibid., 48.
19. Schwaller de Lubicz, cited in Vandenbroeck, 173.
20. Ibid., 178.
21. Ibid., 179.
22. Schwaller de Lubicz, *Sacred Science,* 73.
23. Ibid.
24. Ibid.
25. Ibid., 80.
26. Schwaller de Lubicz, cited in Vandenbroeck, 284.
27. www.naturalfacts.com.au/biochem.html, accessed 2/28/2006.
28. Schwaller de Lubicz, *The Temple of Man,* 35.
29. Schwaller de Lubicz, *Nature Word,* 84.
30. Schwaller de Lubicz, *Esoterism and Symbol,* 61.
31. Michael Shermer, "Mr. Skeptic Goes to Esalen: Science and Spirituality on the California Coast," *Scientific American* (December 2005): 38.
32. Ibid.
33. Ibid.
34. Amit Goswami, "Consciousness and Biological Order: Toward a Quantum Theory of Life and Its Evolution," *Integrative Physiological & Behavioral Science* 32, no. 1 (January–March 1997): 86.
35. Amit Goswami, *The Physicist's View of Nature, Part 2: The Quantum Revolution* (New York: Springer, 2002), 273.

36. Johnjoe McFadden, "Quantum Evolution," www.surrey.ac.uk/qe/O4.htm, accessed 5/12/2006.
37. Goswami, *Physicist's View of Nature,* 95.
38. McFadden.
39. Goswami, *Physicist's View of Nature,* 97.
40. Verner, 41.
41. Heisenberg, *Physics and Philosophy,* 160.
42. Schwaller de Lubicz, *The Temple of Man,* 32.
43. Joel R. Primack and Nancy Ellen Abrams, *The View from the Center of the Universe* (New York: Riverhead, 2006), 285–86.
44. Heisenberg, *Physics and Philosophy,* 141.
45. Erich Neumann, *The Origin and History of Consciousness* (Princeton, N.J.: Princeton University Press, 1993), 10.
46. Plato, *Timaeus and Critias* (New York: Penguin, 1977), 45.
47. Neumann, 10.
48. Budge, *Egyptian Book of the Dead,* plate 29.
49. Neumann, 7.
50. Erwin Schrödinger, *What Is Life with Mind and Matter and Autobiographical Sketches* (New York: Cambridge University Press, 1992), 128.
51. Ibid., 129.
52. Fred Alan Wolf, *Dr. Quantum's Little Book of Big Ideas* (Needham, Mass.: Moment Point Press, 2005), 25.
53. Ibid., 18.

Selected Bibliography

Abdel-Salam, M. E. "Construction of Underground Works and Tunnels in Ancient Egypt." *Tunnelling and Underground Space Technology* 17, no. 3 (July 2002): 295–304.

Albert, R.M., O. Bar-Yosef, L. Meignen, and S. Weiner. "Quantitative Phytolith Study of Hearths from the Natufian and Middle Palaeolithic Levels of Hayonim Cave (Galilee, Israel)." *Journal of Archaeological Science* 30 (2003): 461–80.

Alouf, Michel M. *History of Baalbek.* Beirut: American Press, 1949.

Ambrose, Stanley. "Late Pleistocene Human Population Bottlenecks, Volcanic Winter, and Differentiation of Modern Humans." *Journal of Human Evolution* 34 (1998): 628.

Aristotle. *The Basic Works of Aristotle.* New York: Modern Library, 2001.

Bar-Yosef, Ofer. "The Natufian Culture in the Levant, Threshold to the Origins of Agriculture." *Evolutionary Anthropology* 6 (5) (1998): 159–77.

Barbour, Julian. *The End of Time: The Next Revolution in Physics.* Cambridge: Oxford University Press, 1999.

Bernal, Martin. *Black Athena.* New Brunswick, N.J.: Rutgers University Press, 1987.

Brophy, Thomas G. *The Origin Map: Discovery of a Prehistoric, Megalithic, Astrophysical Map and Sculpture of the Universe.* New York: Writers Club Press, 2002.

Budge, E. A. Wallis. *The Egyptian Book of the Dead.* New York: Dover, 1967.

———. *Egyptian Magic.* London: Kegan, Paul, Trench and Trübner & Company, 1901.

Capra, Fritjof. *The Tao of Physics.* Boston: Shambhala, 2000.

Collins, Andrew. *From the Ashes of Angels: The Forbidden Legacy of a Fallen Race.* Rochester, Vt.: Bear & Company, 2001.

Copenhaver, Brian P. *Hermetica.* New York: Cambridge University Press, 1992.

Davies, Paul. *God and the New Physics.* New York: Simon & Schuster, 1983.

Diop, Cheikh Anta. *The African Origin of Civilization: Myth or Reality.* Chicago: Lawrence Hill, 1974.

du Noüy, Lecomte. *Human Destiny.* New York: Longmans, Green, 1947.
Dunn, Christopher. *The Giza Power Plant.* Santa Fe: Bear & Company, 1998.
Emery, Walter B. *Archaic Egypt.* New York: Penguin, 1961.
Faulkner, Raymond, trans. *The Egyptian Book of the Dead: The Book of Going Forth by Day.* San Francisco: Chronicle Books LLC, 1994.
Fowden, Garth. *The Egyptian Hermes: A Historical Approach to the Late Pagan Mind.* Princeton, N.J.: Princeton University Press, 1993.
Frankfort, Henri. *Ancient Egyptian Religion.* New York: Harper, 1948.
Freke, Timothy, and Peter Gandy. *Wisdom of the Pharaohs.* New York: Tarcher, 1999.
Garstin, E. J. Langford. *Theurgy or the Hermetic Practice: A Treatise on Spiritual Alchemy.* Berwick, Maine: Ibis, 2004.
Gilbert, Adrian. *Magi: The Quest for a Secret Tradition.* London: Bloomsbury, 1999.
Goswami, Amit. "Consciousness and Biological Order: Toward a Quantum Theory of Life and Its Evolution." *Integrative Physiological & Behavioral Science* 32, no. 1 (January–March 1997): 86.
———. *The Physicist's View of Nature, Part 2: The Quantum Revolution.* New York: Springer, 2002.
Grant, Michael. *The Ancient Historians.* New York: Barnes & Noble, 1994.
Griaule, Marcel, and Germaine Dieterlen. *Conversations with Ogotemmeli: An Introduction to Dogon Religious Ideas.* New York: Oxford University Press, 1975.
Gribbin, John. *In Search of Schrödinger's Cat.* New York: Bantam, 1984.
Haisch, Bernard, Alfonso Rueda, and Hal Puthoff. "Inertia as a zero-point Lorentz field." *Physical Review* A 49, no. 2 (1994): 678–94.
Hall, Manly P. *The Secret Teachings of All Ages.* San Francisco: H. S. Crocker, 1928.
Hameroff, Stuart. "Overview: Could Life and Consciousness Be Related to the Fundamental Quantum Nature of the Universe?" www.quantumconsciousness.org/overview.html, accessed 4/21/07.
Hapgood, Charles. *Path of the Pole.* Kempton, Ill.: Adventures Unlimited, 1999.
Hare, Tom. *Remembering Osiris: Number, Gender, and the Word in Ancient Egyptian Representational Systems.* Palo Alto, Calif.: Stanford University Press, 1999.
Hawking, Stephen. *A Brief History of Time.* New York: Bantam, 1990.
Heisenberg, Werner. *Physics and Philosophy.* Amherst, N.Y.: Prometheus, 1999.
———. "The Representation of Nature in Contemporary Physics." *Daedalus* 87 (Summer 1958).
Hermann, Arnold. *To Think Like God: Pythagoras and Parmenides.* Las Vegas: Parmenides Publishing, 2004.
Hillman, James. *The Souls' Code: In Search of Character and Calling.* New York: Random House, 1996.

Hoffman, Michael A. *Egypt Before the Pharaohs.* New York: Dorsett, 1979.

Iamblichus. *Theurgia or The Egyptian Mysteries.* Translated from the Greek by Alexander Wilder. London: William Rider, Metaphysical Publishing, 1911.

LaViolette, Paul. *Earth Under Fire.* Rochester, Vt.: Bear & Co., 2005.

Lewis, Jon E., ed. *Ancient Egypt: The Mammoth Book of Eyewitness.* New York: Carroll & Graf, 2003.

Lewis-Williams, David. *The Mind in the Cave.* London: Thames & Hudson, 2002.

Libet, Benjamin. *Mind Time: The Temporal Factor in Consciousness.* Cambridge, Mass.: Harvard University Press, 2004.

Lipton, Bruce. *The Biology of Belief: Unleashing the Power of Consciousness, Matter, and Miracles.* Santa Rosa, Calif.: Mountain of Love/Elite Books, 2005.

Mallary, Michael. *Our Improbable Universe.* New York: Thunder's Mouth Press, 2004.

Marwick, Ben. "Pleistocene Exchange Networks as Evidence for the Evolution of Language." *Cambridge Archaeological Journal* 13, no. 1 (2003): 67–81.

Massey, Thomas Gerald. *Ancient Egypt: The Light of the World.* London: T. Fisher Unwin Adelphi Terrace, 1907.

McTaggart, Lynne. *The Field.* New York: Quill, 2003.

Mellaart, James. *Earliest Civilizations of the Near East.* New York: McGraw-Hill, 1965.

———. *The Neolithic of the Near East.* New York: Scribner's, 1975.

Mercatante, Anthony S. *Who's Who in Egyptian Mythology.* New York: Barnes & Noble, 1998.

Mercer, Samuel A. B. *The Pyramid Texts.* New York: Longmans, Green, 1952.

Murray, Margaret A. *The Osireion at Abydos.* London: Bernard Quaritch, 1904.

Neumann, Erich. *The Origin and History of Consciousness.* Princeton, N.J.: Princeton University Press, 1993.

Nolch, Guy. "Mungo Man's DNA Shakes the *Homo* Family Tree." *Australasian Science,* 22 (2) (March 2001).

O'Brien, Christian, and Joy O'Brien. *The Genius of the Few: The Story of Those Who Founded the Garden of Eden.* Wellingborough: Turnstone, 1985.

Olson, Steve. *Mapping Human History.* Boston: Mariner, 2002.

Pagels, Elaine. *The Gnostic Gospels.* New York: Vintage, 1989.

Penrose, Roger. *The Emperor's New Mind: Concerning Computers, Minds, and the Laws of Physics.* New York: Oxford University Press, 1989.

Perlmutter, Saul. "Supernovae, Dark Energy, and the Accelerating Universe." *Physics Today* (April 2003): 53–59.

Pinch, Geraldine. *Egyptian Mythology: A Guide to the Gods, Goddesses, and Traditions of Ancient Egypt.* New York: Oxford University Press, 2002.

Plato. *The Republic of Plato.* Trans. by Francis MacDonald Cornford. New York: Oxford University Press, 1970.

———. *Timaeus and Critias.* New York: Penguin, 1977.

Primack, Joel R., and Nancy Ellen Abrams. *The View from the Center of the Universe.* New York: Riverhead, 2006.

Puthoff, Hal E. "The Energetic Vacuum: Implications for Energy Research." *Speculations in Science and Technology* 13 (1990): 247.

Radin, Dean. *The Conscious Universe.* San Francisco: Harper, 1997.

Rucker, Rudy. *Infinity and the Mind.* Princeton, N.J.: Princeton University Press, 1995.

Schrödinger, Erwin. *What Is Life with Mind and Matter and Autobiographical Sketches.* New York: Cambridge University Press, 1992.

Schwaller de Lubicz, R. A. *A Study of Numbers: A Guide to Constant Creation of the Universe.* Rochester, Vt.: Inner Traditions, 1986.

———. *Esoterism and Symbol.* Rochester, Vt.: Inner Traditions, 1985.

———. *Nature Word.* Rochester, Vt.: Inner Traditions, 1985.

———. *Sacred Science: The King of Pharaonic Theocracy.* Rochester, Vt.: Inner Traditions, 1982.

———. *The Temple of Man.* Rochester, Vt.: Inner Traditions, 1998.

Scranton, Laird. *Science of the Dogon.* Rochester, Vt.: Inner Traditions, 2006.

Settegast, Mary. *Plato Prehistorian: 10,000 to 5,000 BC—Myth, Religion, Archeology.* Hudson, N.Y.: Lindisfarne, 1990.

Shermer, Michael. "Mr. Skeptic Goes to Esalen: Science and Spirituality on the California Coast." *Scientific American* (December 2005): 38.

Stone, Merlin. *When God Was a Woman.* New York: Barnes & Noble, 1993.

Sutcliffe, Antony. *On the Track of Ice Age Animals.* Cambridge, Mass.: Harvard University Press, 1986.

Tattersall, Ian. *Monkey in the Mirror.* New York: Harcourt, 2002.

Tchernov, Eitan. "Two New Dogs, and Other Natufian Dogs, from the Southern Levant." *Journal of Archaeological Science* 24 (1997): 65–95.

Teresi, Dick. *Lost Discoveries: The Ancient Roots of Modern Science—from the Babylonians to the Maya.* New York: Simon & Schuster, 2003.

Vandenbroeck, Andre. *Al-Kemi: Hermetic, Occult, Political, and Private Aspects of R. A. Schwaller de Lubicz.* Rochester, Vt.: Inner Traditions, 1987.

Vergani, Teresa. "Ethnomathematics and Symbolic Thought: The Culture of the Dogon." *Zentralblatt für Didaktik der Mathematik* (International Reviews on Mathematical Education), February 1999.

Verner, Miroslav. *The Pyramids: The Mystery, Culture, and Science of Egypt's Great Monuments.* Trans. by Steven Rendall. New York: Grove, 2001.

Walker, Evan Harris. "The Nature of Consciousness." *Mathematical Biosciences* 7 (1970): 131–78.

———. *The Physics of Consciousness.* New York: Perseus, 2000.

———. "Quantum Theory of Consciousness." *Noetic Journal* 1 (1998): 100–107.

West, John Anthony. *Serpent in the Sky.* Wheaton, Ill.: Quest, 1993.
Wilber, Ken. "An Integral Theory of Consciousness." *Journal of Consciousness Studies* 4, no. 1 (February 1997).
Wolf, Fred Alan. *Dr. Quantum's Little Book of Big Ideas.* Needham: Moment Point, 2005.

Index